304

Topics in Current Chemistry

Topics in Current Chemistry
Recently Published and Forthcoming Volumes

Microfluidics: Technologies and Applications
Volume Editor: Bingcheng Lin
Vol. 304, 2011

Photocatalysis
Volume Editor: Carlo Alberto Bignozzi
Vol. 303, 2011

Computational Mechanisms of Au and Pt Catalyzed Reactions
Volume Editors: Elena Soriano, José Marco-Contelles
Vol. 302, 2011

Reactivity Tuning in Oligosaccharide Assembly
Volume Editors: Bert Fraser-Reid, J. Cristóbal López
Vol. 301, 2011

Luminescence Applied in Sensor Science
Volume Editors: Luca Prodi, Marco Montalti, Nelsi Zaccheroni
Vol. 300, 2011

Chemistry of Opioids
Volume Editor: Hiroshi Nagase
Vol. 299, 2011

Electronic and Magnetic Properties of Chiral Molecules and Supramolecular Architectures
Volume Editors: Ron Naaman, David N. Beratan, David H. Waldeck
Vol. 298, 2011

Natural Products via Enzymatic Reactions
Volume Editor: Jörn Piel
Vol. 297, 2010

Nucleic Acid Transfection
Volume Editors: Wolfgang Bielke, Christoph Erbacher
Vol. 296, 2010

Carbohydrates in Sustainable Development II
Volume Editors: Amélia P. Rauter, Pierre Vogel, Yves Queneau
Vol. 295, 2010

Carbohydrates in Sustainable Development I
Volume Editors: Amélia P. Rauter, Pierre Vogel, Yves Queneau
Vol. 294, 2010

Functional Metal-Organic Frameworks: Gas Storage, Separation and Catalysis
Volume Editor: Martin Schröder
Vol. 293, 2010

C-H Activation
Volume Editors: Jin-Quan Yu, Zhangjie Shi
Vol. 292, 2010

Asymmetric Organocatalysis
Volume Editor: Benjamin List
Vol. 291, 2010

Ionic Liquids
Volume Editor: Barbara Kirchner
Vol. 290, 2010

Orbitals in Chemistry
Volume Editor: Satoshi Inagaki
Vol. 289, 2009

Glycoscience and Microbial Adhesion
Volume Editors: Thisbe K. Lindhorst, Stefan Oscarson
Vol. 288, 2009

Templates in Chemistry III
Volume Editors: Broekmann, P., Dötz, K.-H., Schalley, C.A.
Vol. 287, 2009

Tubulin-Binding Agents: Synthetic, Structural and Mechanistic Insights
Volume Editor: Carlomagno, T.
Vol. 286, 2009

Microfluidics

Technologies and Applications

Volume Editor: Bingcheng Lin

With Contributions by

S. Basuray · L. Capretto · H.-C. Chang · H.-W. Chen · L.-J. Cheng · W. Cheng · T.D. Chung · H. Gai · X. Gong · M. Hill · K. Kawai · H.C. Kim · Y. Li · B. Lin · X. Liu · S.K. Njoroge · J. Noh · D. Noort · J.K. Osiri · J. Qin · S. Senapati · H. Shadpour · W. Shi · S. Shoji · Z. Slouka · S.A. Soper · L. Wang · H. Wen · W. Wen · M.A. Witek · H. Xie · E.S. Yeung · S. Zeng · C. Zhang · X. Zhang

Springer

Editor
Dr. Bingcheng Lin
Dalian Institute of Chemical Physics
Chinese Academy of Sciences
457 Zhongshan Road
Dalian 116023
China
bclin@dicp.ac.cn

ISSN 0340-1022 e-ISSN 1436-5049
ISBN 978-3-642-23049-3 e-ISBN 978-3-642-23050-9
DOI 10.1007/978-3-642-23050-9
Springer Heidelberg Dordrecht London New York

Library of Congress Control Number: 2011934439

Printed on acid-free paper

Springer is part of Springer Science+Business Media (www.springer.com)

Volume Editor

Dr. Bingcheng Lin

Dalian Institute of Chemical Physics
Chinese Academy of Sciences
457 Zhongshan Road
Dalian 116023
China
bclin@dicp.ac.cn

Editorial Board

Topics in Current Chemistry Also Available Electronically

Topics in Current Chemistry is included in Springer's eBook package *Chemistry and Materials Science*. If a library does not opt for the whole package the book series may be bought on a subscription basis. Also, all back volumes are available electronically.

For all customers with a print standing order we offer free access to the electronic volumes of the series published in the current year.

If you do not have access, you can still view the table of contents of each volume and the abstract of each article by going to the SpringerLink homepage, clicking on "Chemistry and Materials Science," under Subject Collection, then "Book Series," under Content Type and finally by selecting *Topics in Current Chemistry*.

You will find information about the

- Editorial Board
- Aims and Scope
- Instructions for Authors
- Sample Contribution

at springer.com using the search function by typing in *Topics in Current Chemistry*.

Color figures are published in full color in the electronic version on SpringerLink.

Aims and Scope

The series *Topics in Current Chemistry* presents critical reviews of the present and future trends in modern chemical research. The scope includes all areas of chemical science, including the interfaces with related disciplines such as biology, medicine, and materials science.

The objective of each thematic volume is to give the non-specialist reader, whether at the university or in industry, a comprehensive overview of an area where new insights of interest to a larger scientific audience are emerging.

Thus each review within the volume critically surveys one aspect of that topic and places it within the context of the volume as a whole. The most significant developments of the last 5–10 years are presented, using selected examples to illustrate the principles discussed. A description of the laboratory procedures involved is often useful to the reader. The coverage is not exhaustive in data, but rather conceptual, concentrating on the methodological thinking that will allow the non-specialist reader to understand the information presented.

Discussion of possible future research directions in the area is welcome.

Review articles for the individual volumes are invited by the volume editors.

In references *Topics in Current Chemistry* is abbreviated *Top Curr Chem* and is cited as a journal.

Impact Factor 2010: 2.067; Section "Chemistry, Multidisciplinary": Rank 44 of 144

Preface

The field of microfluidics, or "lab on a chip", is now regarded as one of the key sciences and technologies of miniaturization. It is advancing at a rapid pace and has been developed to demonstrate unprecedented abilities for practical applications in biology, chemistry and engineering in the past decade.

This volume entitled "Microfluidics: Technologies and Applications" presents the current status of selected areas of this broad discipline. It features 11 chapters in total written by authors from 10 leading groups all over the world. Its content covers a spectrum of topics pertaining to fundamentals, basic technologies and applications.

The first part is made up of seven chapters and deals with fundamentals and basic technologies: Shuichi Shoji and his colleague from Waseda University, Japan, give a general overview of the methods and devices of the flow control in microfluidics; Weijia Wen and his colleagues from Hong Kong University of Science and Technology describe the electrorheological fluid technology; scientists in University of Southampton, UK, Xunli Zhang and his colleagues focus on mixtures in micro channels, which is one of the most important phenomena in microfluidics. For the detection approach, Hongwei Gai and Edward S Yeung from Hunan University, China and Ames National Lab, US respectively, discuss the optical detection systems in more detail; Taek Dong Chung and his colleagues in Seoul University, Korea, give a comprehensive outline of biosensors; and Hsueh-Chia Chang and his colleagues from University of Notre Dame, US, describe a nanomembrane-based nucleic acid sensing platform. Then Xin Liu from University of Cambridge, UK, together with the editor's group present the recent works of droplet, which is an another important mode in microfluidics besides the channel based one.

This volume is finalized with four reviews on applications: Steven A Soper and his colleagues from Louisiana State University, US, kindly contribute two chapters on applications of microfluidic technology on DNA and proteins respectively; in the chapter written by Danny van Noor and his colleagues in National University of Singapore write a chapter on the cell in microfluidics; and the last chapter is contributed by the editor's group, Dalian Institute of Chemical Physics, CAS,

China, where the authors summarize the multicellular organism (C.elegans) study on the chips.

As the editor, I hope that the collection of above articles reflects the current status of this important field of microfluidics in a timely fashion, and this book becomes a helpful resource for readers, especially students, engineers, and even advanced non-specialist scientists who work outside of our field by providing a deep and thorough understanding of the mechanisms, technology and future promise of microfluidics for research and applications. I believe that the scope and the variety of topics covered in this volume will attract readers from different communities such as chemistry, physics, biology, medicine and engineering.

As the editor of this special review book, I would like to thank all of authors for their contributions with high quality articles and all of reviewers for their constructive comments, and to thank my colleague, Dr. Hua Xie, for her valuable help. I also appreciate the team at Springer for all the practical help and continuous support of this special issue.

Dalian, China Bingcheng Lin

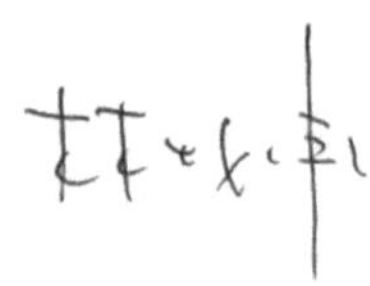

Contents

Flow Control Methods and Devices in Micrometer Scale Channels 1
Shuichi Shoji and Kentaro Kawai

Micromixing Within Microfluidic Devices 27
Lorenzo Capretto, Wei Cheng, Martyn Hill, and Xunli Zhang

Basic Technologies for Droplet Microfluidics 69
Shaojiang Zeng, Xin Liu, Hua Xie, and Bingcheng Lin

Electrorheological Fluid and Its Applications in Microfluidics 91
Limu Wang, Xiuqing Gong, and Weijia Wen

Biosensors in Microfluidic Chips 117
Jongmin Noh, Hee Chan Kim, and Taek Dong Chung

A Nanomembrane-Based Nucleic Acid Sensing Platform for Portable Diagnostics 153
Satyajyoti Senapati, Sagnik Basuray, Zdenek Slouka, Li-Jing Cheng, and Hsueh-Chia Chang

Optical Detection Systems on Microfluidic Chips 171
Hongwei Gai, Yongjun Li, and Edward S. Yeung

Integrated Microfluidic Systems for DNA Analysis 203
Samuel K. Njoroge, Hui-Wen Chen, Małgorzata A. Witek, and Steven A. Soper

Integrated Multifunctional Microfluidics for Automated Proteome Analyses 261
John K. Osiri, Hamed Shadpour, Małgorzata A. Witek, and Steven A. Soper

Cells in Microfluidics 295
Chi Zhang and Danny van Noort

Microfluidic Platform for the Study of *Caenorhabditis elegans* 323
Weiwei Shi, Hui Wen, Bingcheng Lin, and Jianhua Qin

Index 339

Top Curr Chem (2011) 304: 1–25
DOI: 10.1007/128_2011_146

Published online: 28 April 2011

Flow Control Methods and Devices in Micrometer Scale Channels

Shuichi Shoji and Kentaro Kawai

Abstract Recent advances in the fabrication of microflow devices using micro-electromechanical systems (MEMS) technology are described. Passive and active liquid flow control and particle-handling methods in micrometer-scale channels are reviewed. These methods are useful in micro total analysis systems (μTAS) and laboratory-on-a-chip systems. Multiple flow control systems (i.e., arrayed micro-valves) for advanced high-throughput microflow systems are introduced. Examples of microflow devices and systems for chemical and biochemical applications are also described.

Keywords Laminar flow · Micro-electromechanical systems · Micro total analysis systems · Microchannel · Microfluidics · Microvalve

Contents

1 Introduction 2
2 Flow Control in Microchannels: Continuous Flow 3
2.1 Passive Flow Control 3
2.2 Active Flow Control 8
3 Multiple Flow Control Systems: Arrayed Microvalves 17
4 Microflow Devices and Systems for Chemical and Biochemical Applications 19
4.1 Sample-Metering and Injection Devices 19
4.2 Multifunctional Sample Injection Arrayed Microwells 19
5 Summary and Outlook 20
References 21

S. Shoji (✉)
Department of Electronic and Photonic Systems, Faculty of Science and Engineering, Waseda University, 3-4-1, Ohkubo, Shinjuku-ku, Tokyo 169–8555, Japan
e-mail: shojis@waseda.jp

K. Kawai
Department of Precision Science and Technology, Graduate School of Engineering, Osaka University, 2–1, Yamada-Oka, Suita, Osaka 565–0871, Japan

Abbreviations

μTAS	Micro total analysis systems
CFD	Computational fluid dynamics
CMOS	Complementary metal-oxide semiconductor
DEMUX	Demultiplexer
DEP	Dielectrophoresis
EOF	Electro-osmotic flow
MEMS	Micro-electromechanical systems
ODEP	Optically induced dielectrophoretic
PDMS	Poly(dimethylsiloxane)
Re	Reynolds number
TGP	Thermoreversible gelation polymer

1 Introduction

Microvalves and micropumps fabricated using silicon and glass three-dimensional (3D) technologies based on photolithography are at the frontier of work on micro-electromechanical systems (MEMS) [1]. In the early stages, many types of "classic" microflow control devices, (i.e., microvalves and micropumps) using various types of small actuators were developed and evaluated [2]. Piezoelectric, electrostatic, and thermopneumatic actuators are used in these devices. Recently, new MEMS applications, namely micro total analysis systems (μTAS) or laboratory-on-a-chip (LOC) systems have been widely accepted in chemistry and biochemistry. The advantages of μTAS are their small size, the need for only a small volume of samples and reagents, and rapid response as a result of the chemical reaction and detection being completed in micrometer-scale channels. The size effect is the most remarkable feature of μTAS. In practical applications, classic microflow control devices are not as widely accepted as μTAS. Since miniaturization of these devices is limited by the actuator size, it is difficult to integrate them in systems. On the other hand, polymer MEMS technologies open up new device fabrication capabilities for compact microflow devices. Pneumatic microvalves and micropumps have been developed using poly(dimethylsiloxane) (PDMS), because of its easy fabrication by molding and its elasticity. Microflow devices for chemical and biochemical applications have been developed using PDMS structures. Large-scale integrated microvalve arrays have also been fabricated for complicated parallel chemical reaction systems and high-throughput chemical and biochemical analysis systems.

Many flow control methods based on liquid flow behaviors in micrometer-scale channels have been reported and demonstrated [3]. Hydrodynamic flow control methods using two-dimensional (2D) or 3D microstructures have been developed and used for continuous-flow devices and systems. Passive flow control of spiral flow, sheath flow, flow focusing, and particle separation are achieved in microchannels without movable mechanical structures. In this case, flows in the microchannel are driven by an off-chip pump, for example, a syringe pump. Active flow controls using on-chip pneumatic microvalves are realized using PDMS microstructures. Dielectrically, optically, and electromagnetically controlled active flow devices have also been developed. Particle-handling is also achieved using the above flow control methods. These methods are applicable in biological applications, such as cell sorting or handling biomolecules.

Recent progress in microflow devices and systems is described in this chapter. Examples of passive and active flow control methods applicable in practical μTAS are described in Sect. 2. Multiple flow control systems, i.e., arrayed microvalves, for advanced high-throughput microflow systems are then introduced in Sect. 3. Examples of microflow devices and systems for chemical and biochemical applications are described in Sect. 4.

2 Flow Control in Microchannels: Continuous Flow

Liquid flow is incompressible, so, in micrometer-scale channels, the flow has a small Reynolds number (Re), usually less than 1, and the flow in simple microchannels is laminar, thus chaotic or turbulent flows are not observed [1]. Many types of microfluidic device have been developed on the basis of this flow behavior. Functional flow control methods based on laminar flow profiles have been proposed and applied in microflow devices and systems. Passive and active flow control methods and their applications are introduced in this section.

2.1 Passive Flow Control

2.1.1 Spiral Flow: Twisting Flow

3D spiral flow along a microchannel is generated using slanted grooves, i.e., obliquely oriented grooves, on the channel wall (Fig. 1a) [4]. Much efficient spiral flow is achieved with short channel lengths by fabricating slanted grooves on the three walls (Fig. 1b) [5]. This flow behavior is useful for enhancing transverse components of flow in stretching and holding over a cross-section of the channel, which is necessary for effective mixing of materials in simple channels. The advantages of these devices are their simple structures and low pressure drops. Functional chaotic mixers have been fabricated using microchannels with staggered herring-bone (chevron) grooves and additional embedded barriers [4, 6]. These devices are fabricated by a soft lithographic method using PDMS.

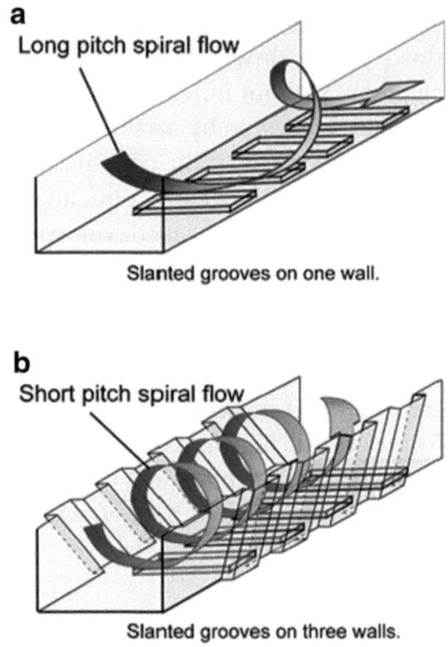

Fig. 1 3D spiral flow microdevices. (**a**) Slanted grooves on the bottom wall [4]. (**b**) Slanted grooves on the bottom and side walls [5]

2.1.2 Sheath Flow and Flow Focusing

Hydrodynamic focusing is a useful function in the handling and sorting of reagents, particles, and biomolecules in microchannels. 2D flow focusing, i.e., 2D sheath flow, is obtained by sandwiching the target flow with two side carrier flows. 2D flow focusing is used in microflow cytometry for cell sorting [7]. 2D flow focusing using a flow lens structure is also used in multichamber cell-cultivation systems [8]. 3D flow focusing, i.e., 3D sheath flow, has been achieved in microchannels that can localize the position of reagents or particles at the cross-sectional central region of the microchannel, and can minimize interaction with the channel walls. Ideal 3D flow focusing of a cylindrical sheath flow has been obtained with a pulled glass capillary and PDMS (Fig. 2a) [9]. Core–shell hydrogel microwires were fabricated using cascade connections of this device [10]. However, these devices are not suitable for mass fabrication. Using planar fabrication processes, 3D sheath flow devices have been fabricated with silicon, glass, and polymer substrates. 3D sheath flow achieved by the two-step introduction of carrier flows to constrain the sample flow in a bonded silicon glass structure has been reported (Fig. 2b) [11]. A 3D sheath flow device with four vertical and horizontal carrier inlets was obtained by a membrane sandwich method using PDMS (Fig. 2c) [12]. These devices have good performance within a narrow Rerange. However 3D flow focusing over a wide Rerange, based on microfluidic drafting, was obtained with a single-layer planar

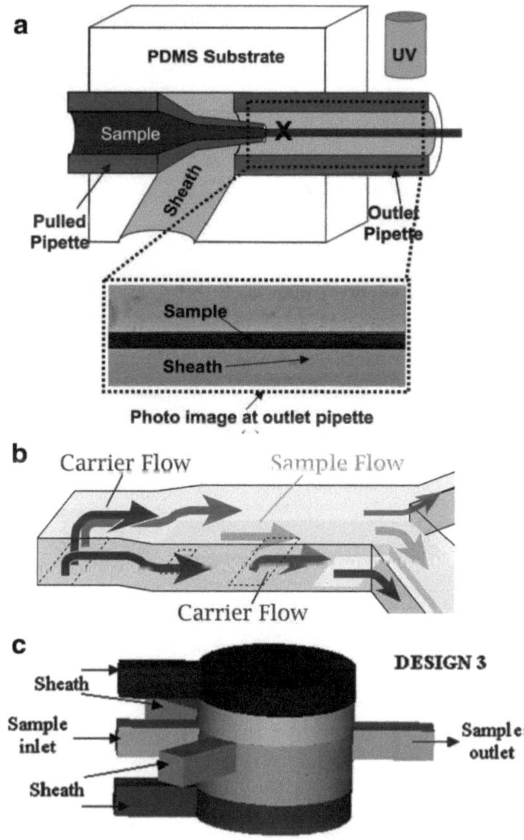

Fig. 2 3D sheath flow devices. (**a**) Cylindrical sheath flow [9]. (**b**) Two-step introduction of carrier flow [11]. (**c**) Four carrier flows of vertical and horizontal carrier inlets [12]. Reprinted with kind permissions from [9] Copyright 2004 Royal Society of Chemistry and [12] Copyright 2004 IEEE

flow device [13]. 3D flow focusing over a wide Re range has also been developed using laminated PDMS structures [14]. High flow rate 3D focusing structures using multistep carrier flow injection from symmetric side channels have been proposed [15, 16]. Multistep injection methods for carriers sometimes need complicated device structures and precise flow rate control of the carriers. In order to simplify the structure and reduce the carrier flow inlets, core sheath sample transfer was developed [17]. The design and principle of this device are illustrated in Fig. 3. The slanted grooves on both side-walls, and the herring bone (chevron) grooves on the top wall, form a PDMS microchannel. These structures form a centrifugal flow, symmetrical around the flow direction, and the sample flow is shifted to the center of the flow channel. This device needs only one carrier inlet so it can be produced by a simple planar fabrication process. Multiple core sheath flow was obtained by connecting this device in series and in parallel. 3D sheath flow was also achieved by a combination of flow focusing of a 2D sheath flow and flow shift using top and bottom herring bone (chevron) grooves [18].

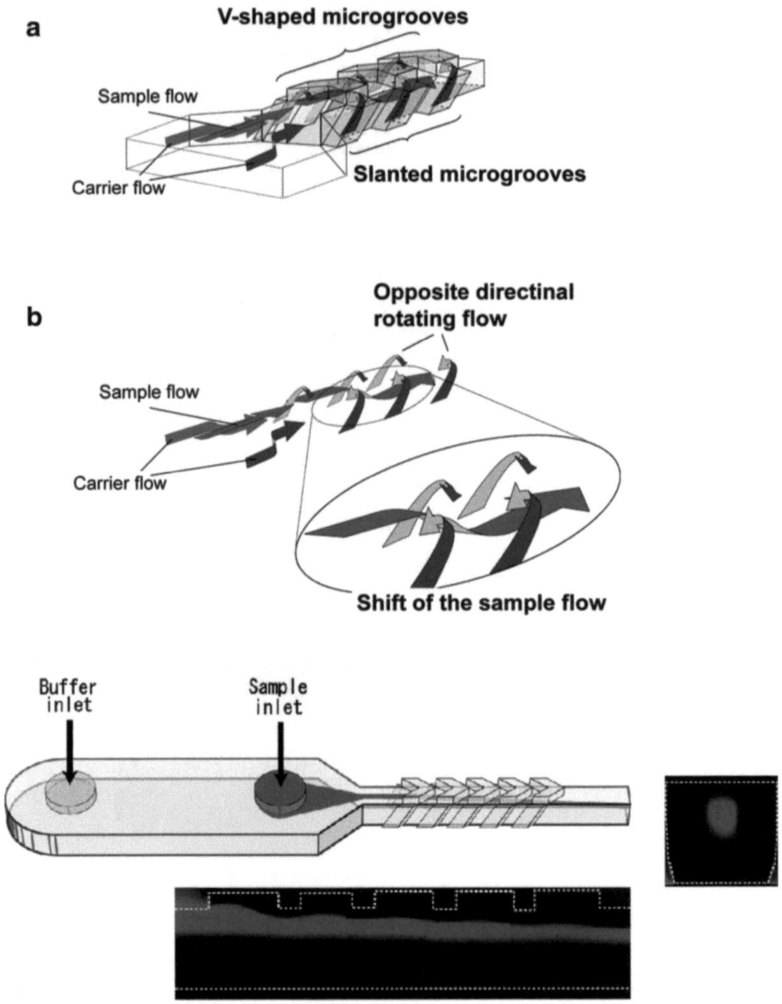

Fig. 3 3D sheath flow with core sheath sample transfer by side grooves [17]

2.1.3 Fluid Dynamics Separation

Particle handling in continuous size separation and filtration, using hydrodynamic flow behavior in T-shaped microchannels has been developed. The principle of pinched flow fractionation and the design of a size-separation device using asymmetric pinched flow fractionation is illustrated in Fig. 4 [19, 20]. Particles are aligned on the upper wall of the pinched segment and separated according to their size by the spreading flow profile at the inlet of the wide channel, the so called pinched flow effect, as shown in Fig. 4a. The size-separation device has two inlets, one for sample flow and one for carrier flow, and five outlets, as shown in Fig. 4b.

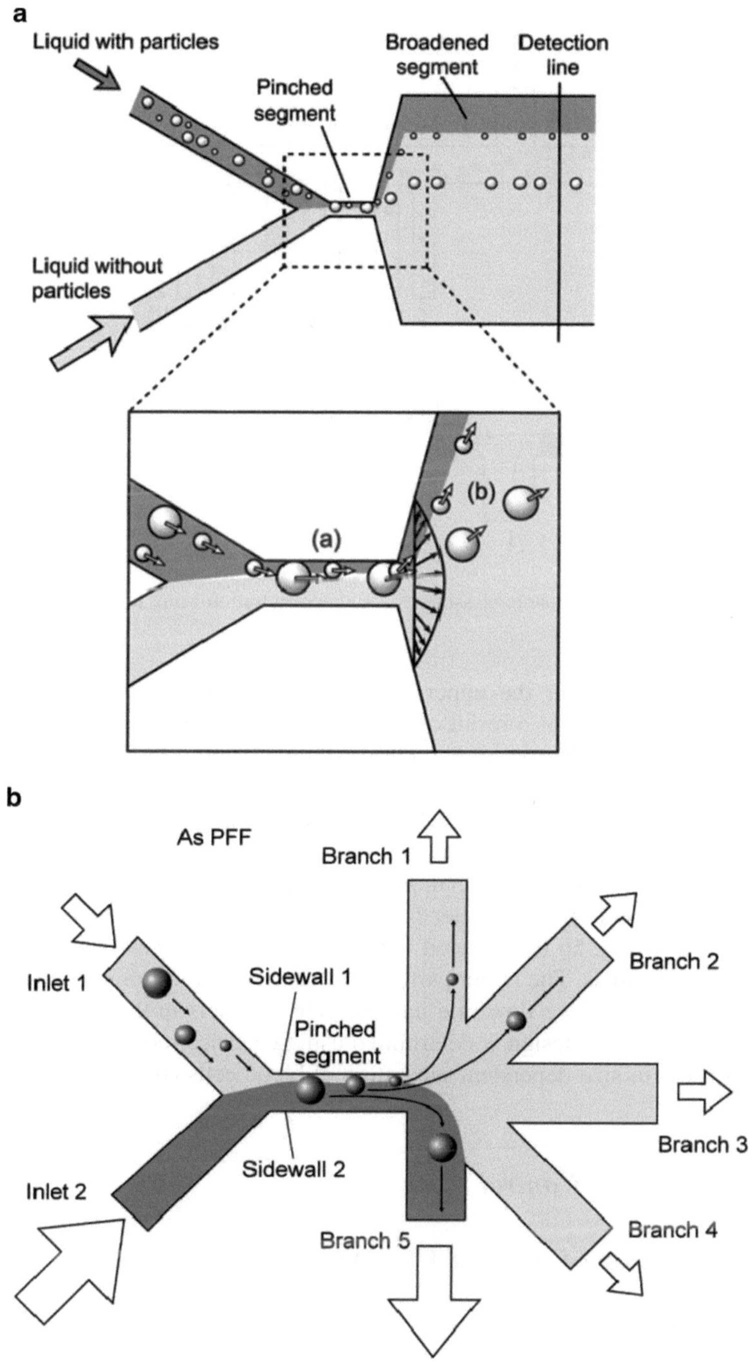

Fig. 4 Pinched flow fractionation devices. (**a**) Principle of pinched flow [19]. (**b**) Size separation device using asymmetric pinched flow fractionation [20]. Reprinted with kind permission from [20] Copyright 2005 Royal Society of Chemistry

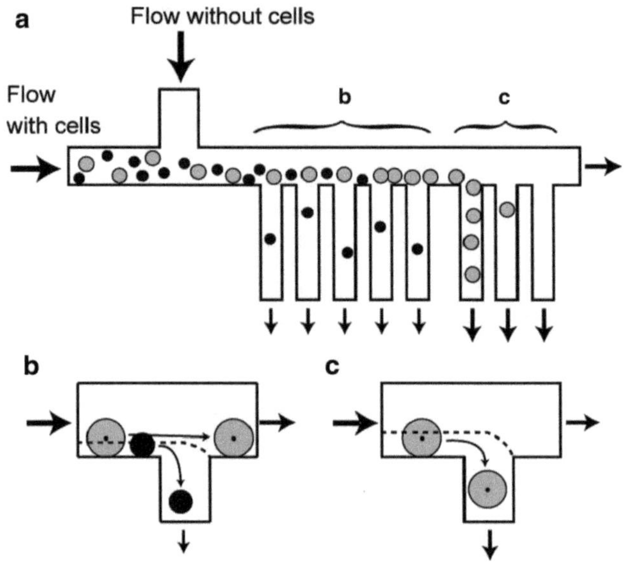

Fig. 5 Principle of hydrodynamic classification and concentration using multiple T-shaped side channels [23]

The particles flow along the upper side-wall and flow out at different outlets according to their sizes, as a result of the spreading flow profile. By using outlet 5 as a drain channel, a five-outlet particle separator is obtained. Size separation of particles by slanted grooves, using steric hindrance, has also been reported [21]. This method has been applied to cell-cycle synchronization [22]. Hydrodynamic classification and concentration using multiple T-shaped side channels, as shown in Fig. 5, have been proposed [23]. Sample flow, including particles, is pushed to the lower side-wall, and all the smaller particles flow out to the side channels in the region shown in Fig. 5b by repeated removal of small amounts of liquid through multiple side channels. The larger particles are then removed from the main stream by increasing the relative flow rate into the side channels in the region shown in Fig. 5c. The optimum design is determined using electrical circuit design, and the system is used for size-dependent separation of blood cells and liver cells [23, 24].

2.2 *Active Flow Control*

2.2.1 Pneumatically Controlled Flow Device

Active Sheath Flow

Flow switching in 2D sheath flow devices was obtained by controlling the carrier flow balance, as shown in Fig. 6a. A flow device with one inlet and two outlets has

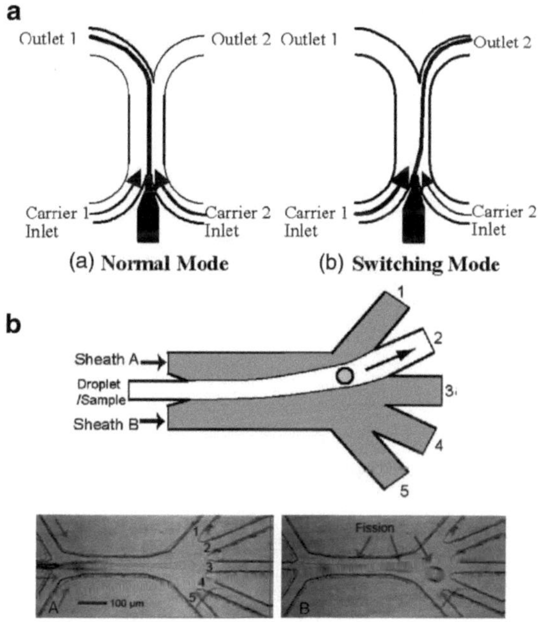

Fig. 6 Active sample flow switching with sheath flow balance change. (**a**) Principle of flow switch [25]. (**b**) Sorting of monodisperse oil–water emulsions [27]. Reprinted with kind permission from [27] Copyright 2007 Chemical & Biological Microsystems Society

been developed for cell sorting at an early stage [25]. One inlet and multi-outlet devices have also been developed for sample injection and particle separation [26]. The application of sheath flow control to the sorting of monodisperse oil–water emulsions has been reported, as shown in Fig. 6b [27]. The switching speed of these devices is limited by the response time of the carrier flow rate control; slow switching is a big problem when off-chip valve control or external pumping control is used. High-speed control of sheath flow switching was achieved using pneumatically driven on-chip microvalves with switching speeds of about 100 ms [28]. Sample injectors with two outlets or four outlets were fabricated as shown in Fig. 7. Nanoliter volume sampling is achieved with these devices. Cell-sorting devices with multiple outlets from four up to 16 chambers were fabricated as shown in Fig. 8. On-chip microvalves enable fast and accurate sorting [29]. In order to control carrier flow, a piezoelectric actuator is used for the switching operation in high-throughput cell sorting [30].

Horizontal Valve Actuation

For continuous flow rate and direction control in microchannels, passive and active microvalves formed along a channel provide microfluidic systems of simple structure. A flexible PDMS structure is useful for horizontally driven microvalves. A 2D venous-like valve consisting of two flexible cantilever bars, which act as valve flaps, was developed for this purpose, as shown in Fig. 9a [31]. Pneumatically

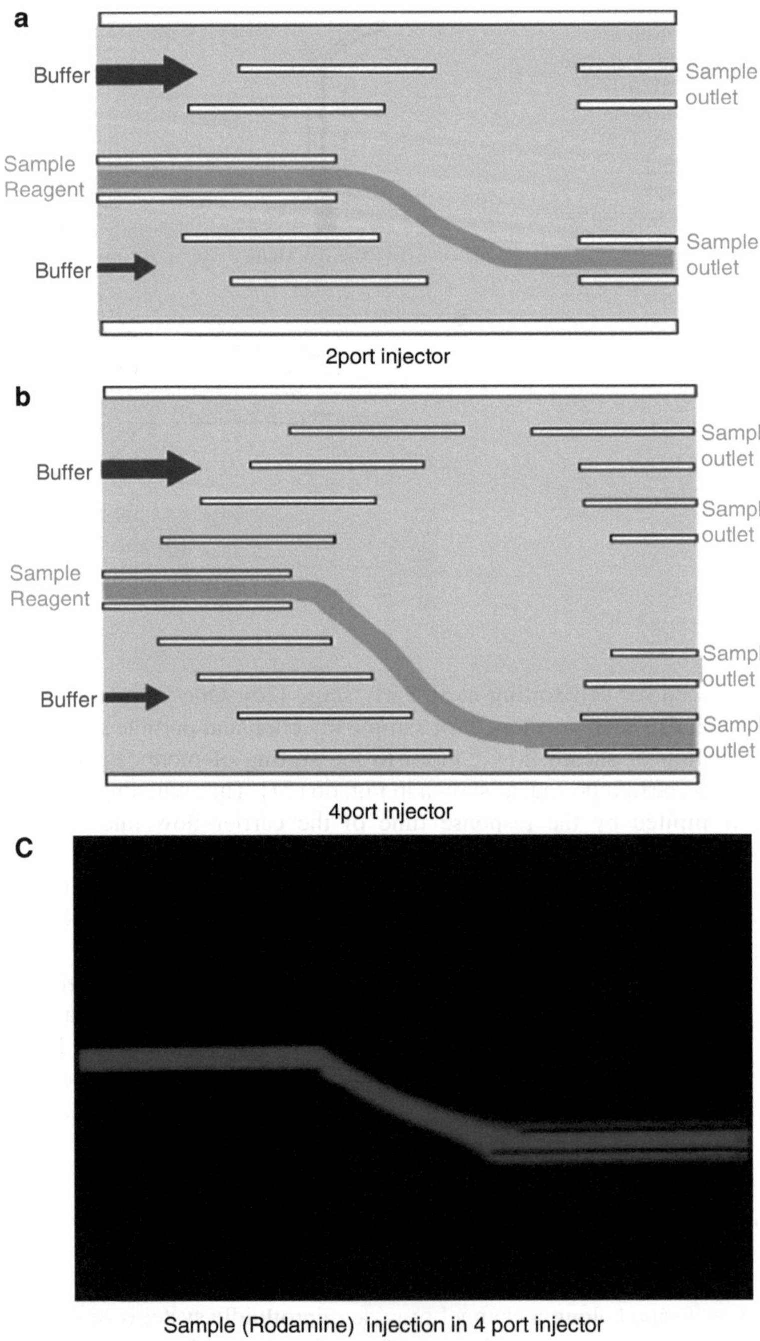

Fig. 7 Sheath flow sample injector with on-chip microvalves [28]. (**a**) Two-outlet sample injector. (**b**) Four-outlet sample injector. (**c**) Sample (rhodamine) injection in the four-outlet sample injector

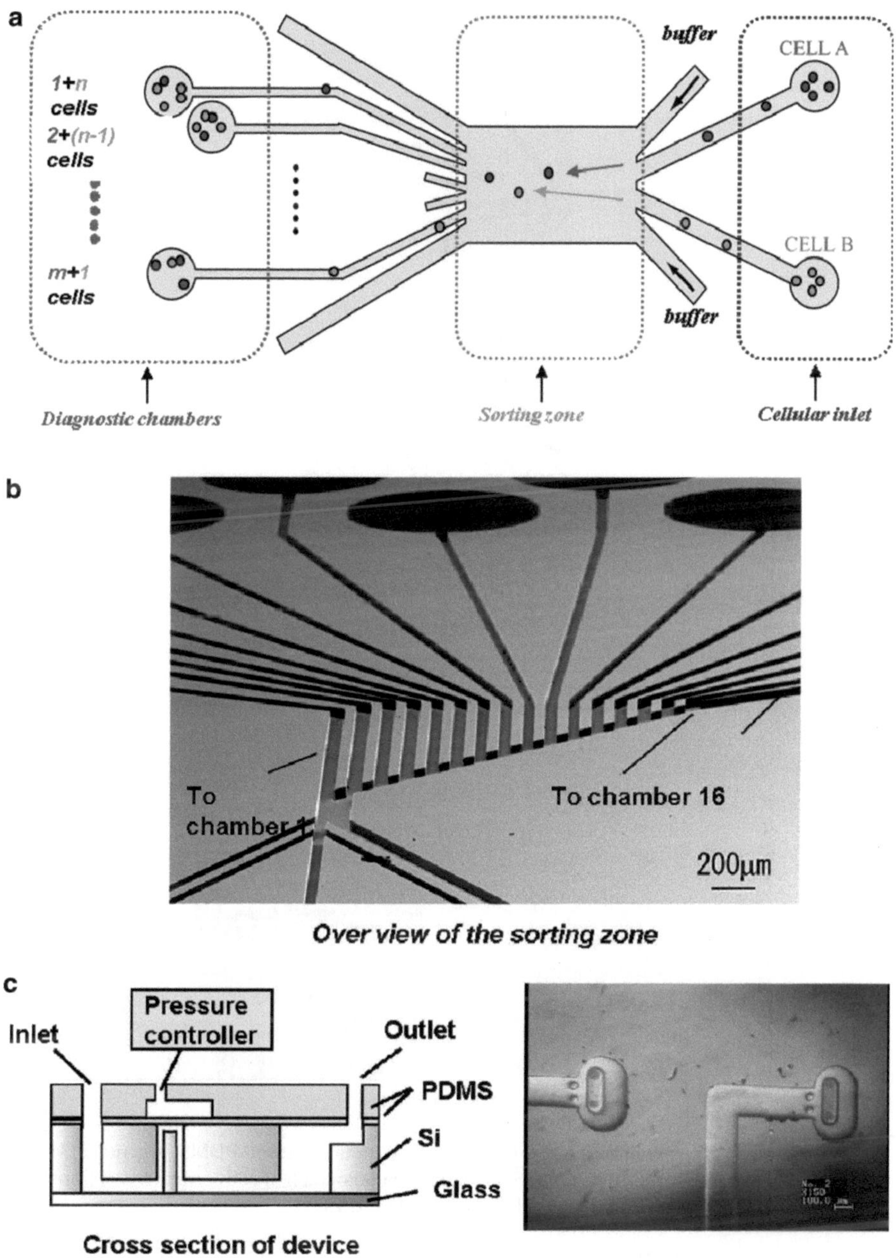

Fig. 8 Multicell sorting device using sheath flow control [29]. (**a**) Principle of multicell sorting device. (**b**) SEM image of 16-chamber multicell sorting device. (**c**) Pneumatically driven microvalve

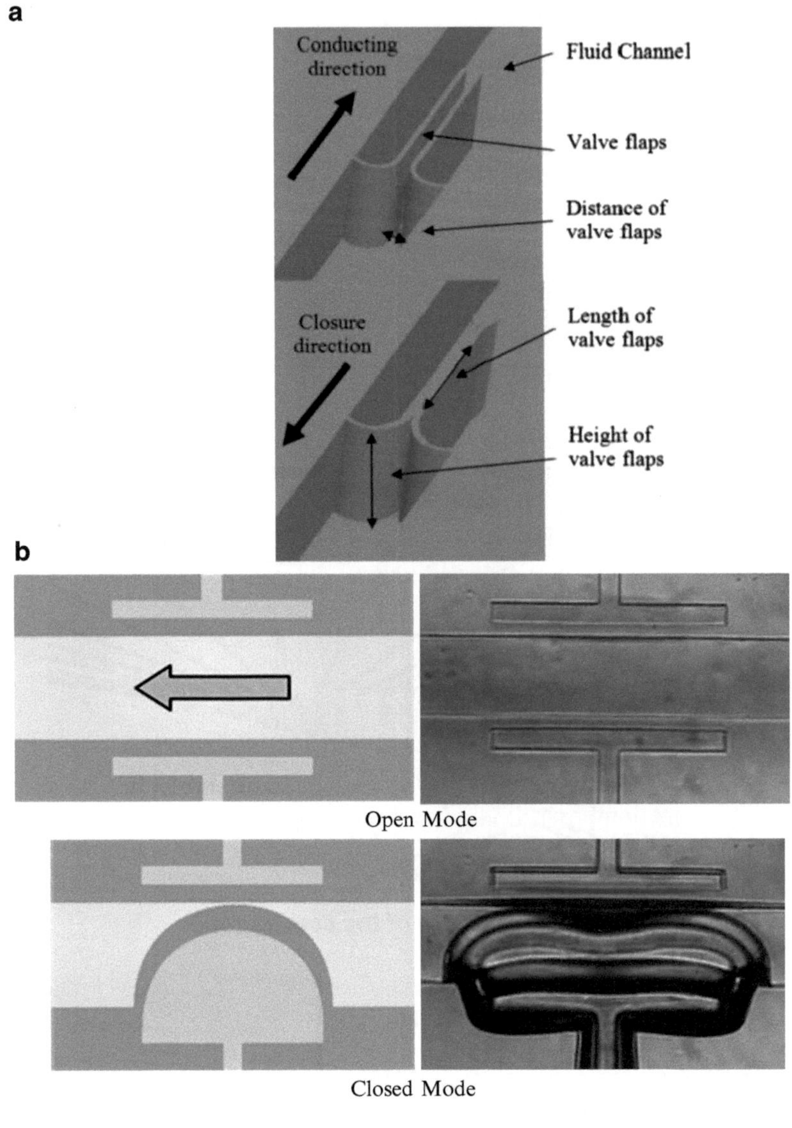

Fig. 9 Continuous flow control microvalves using horizontally moved PDMS structures [31, 32]. (**a**) Schematic of passive microvalve. (**b**) Schematic and photograph of active microvalve (*top view*). Reprinted with kind permission from [31] Copyright 2008 IEEE

controlled moving-wall microvalves have been developed for active flow control; the principle is illustrated in Fig. 9b. This type of microvalve was used for double emulsion formation and bead trapping and release in microfluidic devices [32, 33]. Multiple droplet sorting is obtained using flexible high-aspect-ratio PDMS walls, as shown in Fig. 10 [34]. The advantage of these devices is simple fabrication by one-step PDMS molding and bonding to a glass substrate. A similar pneumatic flow

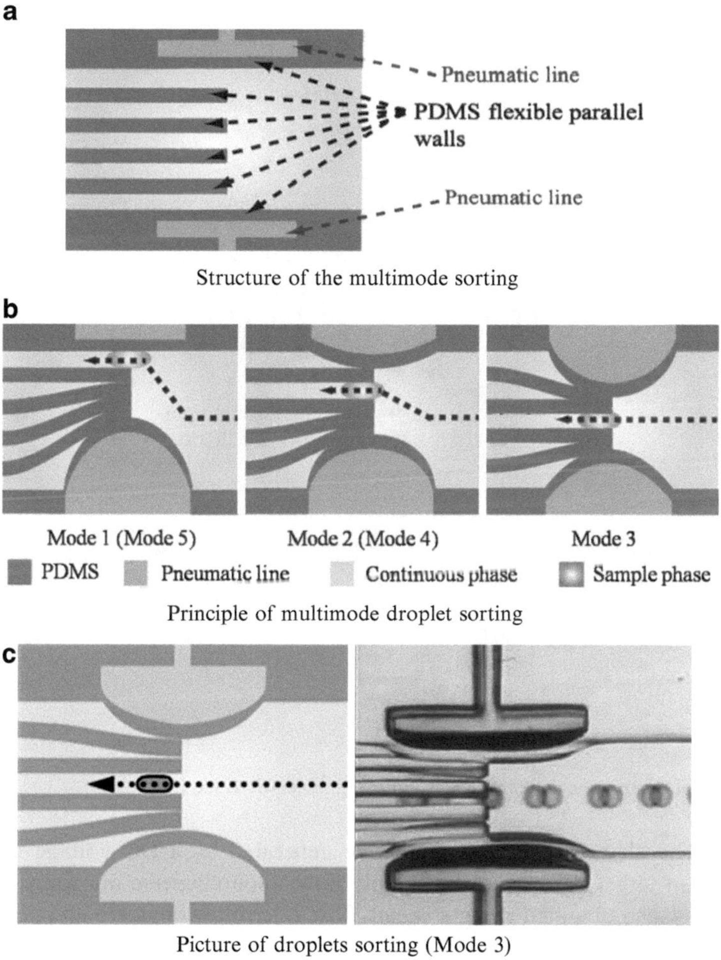

Fig. 10 Multiple-droplet sorting device using flexible high-aspect ratio PDMS walls [34]. (**a**) Structure of the multimode droplet sorting device. (**b**) Principle of multimode droplet sorting. (**c**) Photograph of droplet sorting (Mode 3)

control device, a pneumatic field effect transistor, was fabricated from a bonded silicon and glass structure [35].

2.2.2 Dielectrically Controlled Flow Devices

Dielectrically controlled sorting devices have the advantage of fast switching times. High-throughput particle collection is expected. A non-destructive cell-sorting system was fabricated from a bonded PDMS and glass structure [36]. The sorting part consists of X-shaped microchannels for sample and buffer flows, as shown in

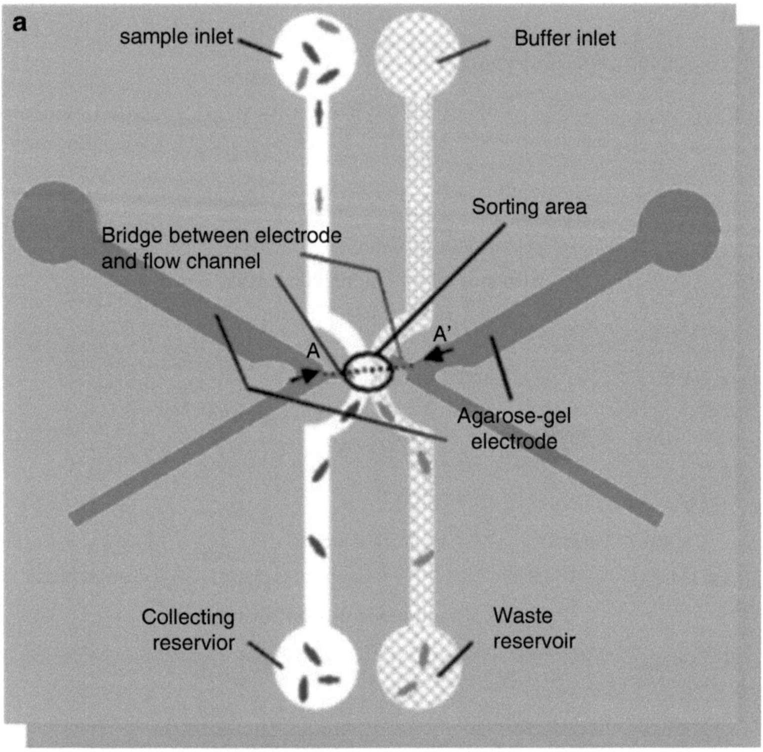

Fig. 11 Non-destructive cell sorting system using dielectric force [36]

Fig. 11. The electrostatic repulsion force generated by a DC voltage is used to remove non-target cells from the sample flow. Monodisperse oil–water emulsion sorting was also obtained using a sheath flow microflow device with one inlet and three outlets [37]. Neutral droplets and two types of charged droplet are generated and sorted to three outlets with plus and minus applied DC voltages larger than 100 V. Dielectrophoresis (DEP) flow switching using vertical electrodes in the side-wall of microchannels has been reported [38]. The non-uniform electric field distribution generated by the side-wall electrodes gives a DEP force along the lateral direction of the channel and the flow direction of the particles is changed by the DEP force. The voltage needed for DEP is of the order of tens of volts. Multiple sorting using five outlets was obtained by changing the voltage amplitude at a frequency of 10 MHz. Multistep particle separators consisting of serpentine microchannels and a planar spot electrode array have also been reported [39, 40].

2.2.3 Optically Controlled Flow Devices

Optically driven flow switches using sheath flow have been developed. Optical switch control gives simple flow device structures with no movable mechanical

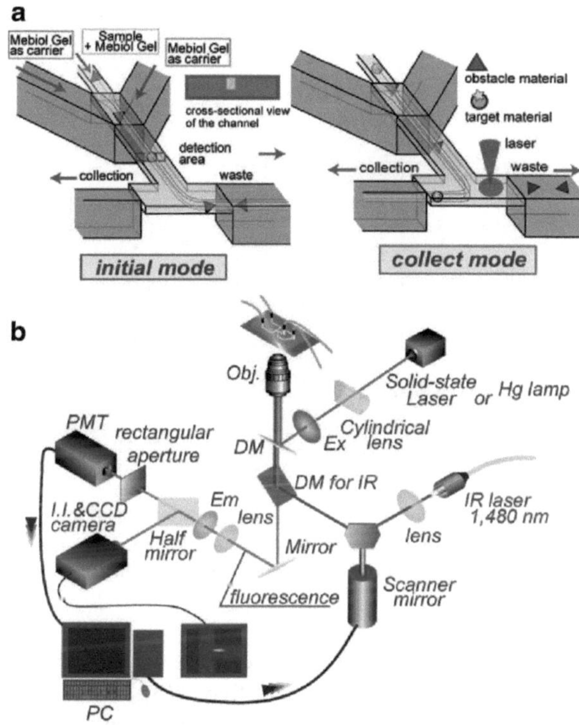

Fig. 12 Flow switching sheath flow device based on sol–gel transition of hydrogel (TGP) [42]. (**a**) Principle. (**b**) Optical setup

structures and no electrical contacts. Optical trapping or optical tweezers, obtained from radiation pressure forces of a focused optical beam were used for cell sorting [41]. Fast switching, 2–4 ms, is obtained. A flow switching device based on sol–gel transition of a hydrogel (thermoreversible gelation polymer, TGP) by local heating with focused infrared (IR) radiation has been proposed. The principle and optical setup of this device are shown in Fig. 12 [42]. The sample is mixed with TGP in water and the same TGP solution is also used as the carrier flow. Diluted TGP remains in the sol state at room temperature and changes to a gel on heating up above the critical temperature. An IR laser of wavelength 1,480 nm, which is near the local maximum wavelength of water absorbance, changes the flow resistance at the outlet channel by thermal gelation. Since the sol–gel transition of a TGP solution consisting of Mebiol gel (LCST 32°C, Mebiol Inc., Japan) takes about 1 ms, fast flow switching of the order of milliseconds is possible [43]. This sorting system has the advantage of controlling the flow in smaller microchannels than those needed in other systems. The method was therefore first developed for sorting small bioparticles, including mitochondria. In order to achieve high-throughput sorting, eight parallel channel sheath flow devices were fabricated, and sorting of 72 molecules/s with 90% recovery was obtained [44]. Sorting of multiple biomolecules in parallel using a sheath flow device with one inlet and five outlets has also been achieved [45]. A high-speed cell sorter was produced by improving this

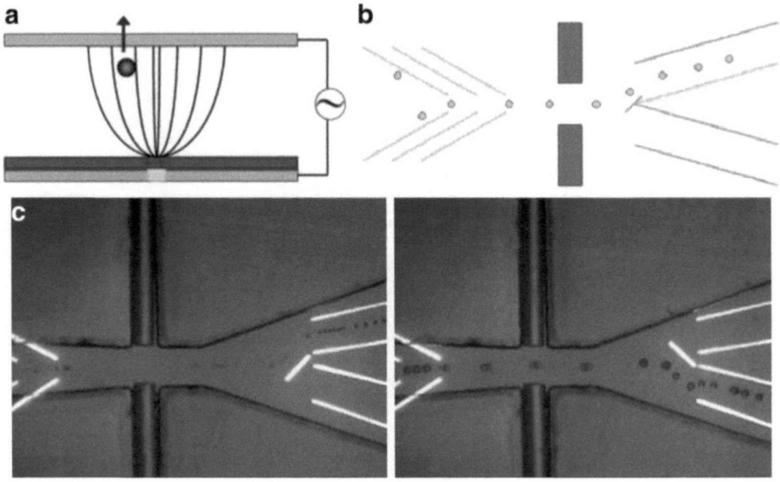

Fig. 13 Particle-sorting devices using negative ODEP force [47]. (**a**) Principle of negative ODEP. (**b**) Principle of two-outlet particle-sorting device. (**c**) Particle sorting with ODEP guide patterns. Reprinted with kind permission from [47] Copyright 2008 Chemical & Biochemical Microsystems Society

structure [46]. About 17,000 cells are sorted with a 75% recovery ratio and 90% purity at a throughput of about five cells/s.

Other optically driven particle-sorting devices were developed using a negative optically induced dielectrophoretic (ODEP) force. The principle and structure are shown in Fig. 13 [47]. A negative ODEP force is induced by an AC voltage when an illuminated amorphous silicon layer forms a non-uniform electric field, as shown in Fig. 13a. By projecting light patterns on an amorphous silicon layer, an AC voltage drop is generated across the illuminated area, as shown in Fig. 13b. The particles flow along the illuminated patterns as a result of the induced non-uniform electric field and the DEP force. Continuous particle sorting and separation were performed by this method as shown in Fig. 13c [47, 48].

2.2.4 Others

Flow control devices using magnetically driven microstructure have been developed. The principle of particle sorting is shown in Fig. 14 [49]. A magnetic microtool made of a neodymium powder composite is driven by a pair of magnetic needles; the needles focus the magnetic flux generated by electromagnetic coils at the tip. A switching frequency of 180 Hz is obtained with this device. This type of magnetic microtool has also been used for size control in oil–water emulsion droplet formation [50].

Particle separation methods based on the effects on suspended particles of exposure to an ultrasonic standing-wave field have been reported [51]. Carrier medium exchange in a laminar flow microchannel has also been achieved using

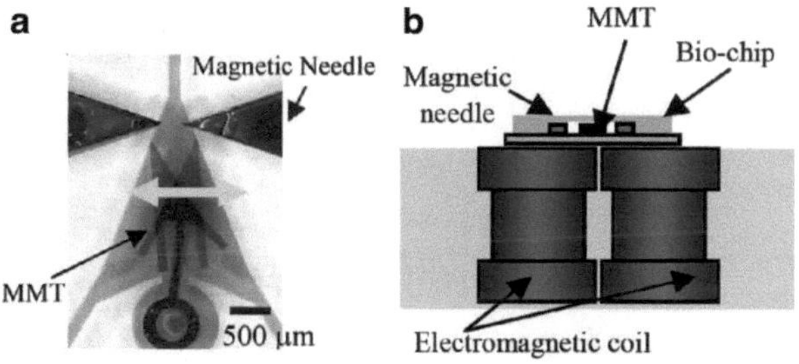

Fig. 14 Flow switch using magnetically driven microtool [49]. (**a**) Operation of magnetically driven microtool. (**b**) Electromagnetically driven setup. Reprinted with kind permission from [49] Copyright 2009 IEEE

ultrasonic acoustic forces [52]. Red blood cells are switched from blood to clean blood plasma by this method.

3 Multiple Flow Control Systems: Arrayed Microvalves

Microvalves are one of the most important fluidic components for fluid control in microfluidic systems. Although various methods and structures using on-chip valves have been reported achieving on-chip fluid control, pneumatic actuation is the most widely employed because of its simple structure and easy fabrication. However, because the operation of pneumatic valves requires several world-to-chip connectors and off-chip solenoid valves, integration techniques for "pneumatic" logic circuits have been developed. A combination of microvalves arranged in parallel can control many microchannels using a small number of connectors [53]. Individually controlled microvalves are only controllable one at a time, but n^2 microvalves can be controlled by $2n$ control lines. Normally close microvalves have been proposed in place of pressure-driven microvalves [54]. $2(n-1)$ latching-type microvalves are controlled by n control lines. By using gas-permeable PDMS membranes, the latching valves remain open or closed for several minutes. Figure 15 shows a Boolean logic circuit using the difference between the channel resistance and the pneumatic control line [55]. An 8-bit microprocessor has been demonstrated for fluid control and mixing. A Baille display is used as the pressure driver instead of air actuation [56]. The driving force using a Braille pin is less than that of an air-driven pressure driver, but instruments for pneumatic pressure are not required. Microvalve control with three thresholds has been proposed in place of open(0)/closed(1) microvalves [57]. Quaternary control has been obtained using four different widths of microchannels and control lines. The addressing valve control system has a complementary metal-oxide semiconductor (CMOS) memory-like design for the control line [58]. On-chip valve operation is carried out by three-step control of the drive lines; pressure is applied to the on-chip valves and the gate lines, which maintains the state of the on-chip valve. $m \times n$ on-chip valves can be individually controlled by $m + n$ control lines. The multiplexed valve control

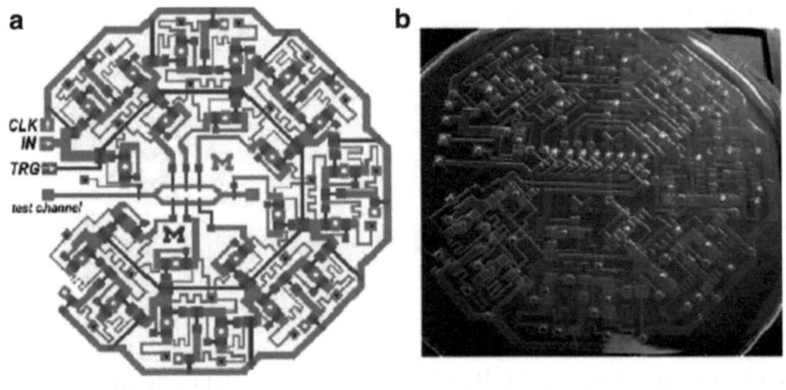

Fig. 15 Eight-bit digital pneumatic microprocessor based on pneumatic Boolean logic circuit [55]. Reprinted with kind permission from [55] Copyright 2009 Royal Society of Chemistry

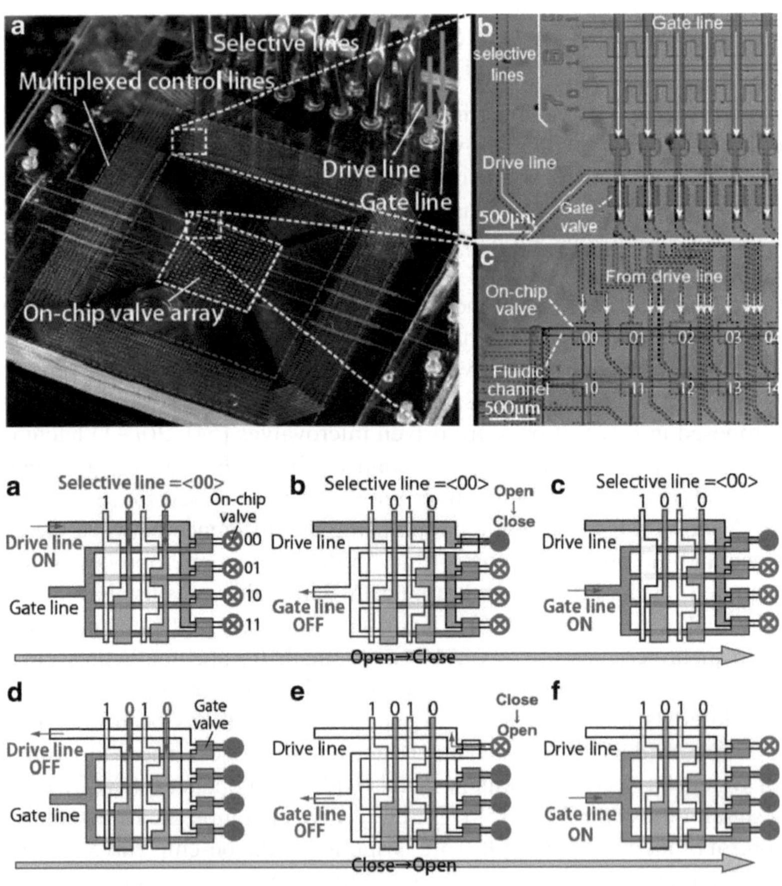

Fig. 16 Multiplexed valve control system [59]. (**a**) Overview of 2^8 (256) microvalve array. (**b**) Operating principle of multiplexed valve-control system

system has the same demultiplexer (DEMUX) arrangement as that shown in Fig. 16a [59]. This valve control system consists of pressure input lines for a drive line, a gate line, and selective lines. The drive line is used for switching the state of the on-chip valves. The gate line applies the pneumatic pressure of the gate valves that select the on-chip valve to be activated. The multiplexed selective lines choose one gate valve of the addressed on-chip valve. This method can control 2^n valve arrays with $2n + 2$ control lines, where n is the bit number, as shown in Fig. 16b.

4 Microflow Devices and Systems for Chemical and Biochemical Applications

4.1 *Sample-Metering and Injection Devices*

For chemical and biochemical experiments using microfluidic devices, precise control of the fluid volume or fluid flow is required. The technique of sample metering and injection uses the fluid properties of droplet formation in heterophase or laminar flow. The use of rectangular channels with hydrophobic valves provides good performance and reproducibility [60]. A number of nanoliter-sized droplets can be accurately dispensed and mixed with the aid of specific channels under pneumatic pressure. Use of a fluoro-polymer instead of PDMS reduces adsorption to the channel surface [61]. Valve actuation can control droplet generation [62]. Generation of droplets of volumes from femtoliters to several picoliters is performed by a small metering channel and valve-controlled channel flow [63]. An inert liquid is used to prevent dilution of small-volume droplets. A 3D sheath flow scanner selects the stimulation spot of a reagent by the surrounding buffer flow [64]. An on-chip electro-osmotic flow (EOF) pump provides real-time control of spot selection.

4.2 *Multifunctional Sample Injection Arrayed Microwells*

A 6/24/96-well plate is widely used in conventional biochemical screening. Arrayed systems in microfluidic devices are promising for application in high-functionality analysis and high-throughput screening. The method of fluidic access to microwell arrays is passive or active. The passive type has limited microwell selectivity, but large-scale integration and batch processing are possible. The active type is limited with respect to miniaturization of large-scale integration, but comparative experiments can be carried out in the same conditioned device. Microfluidic long-term cell-culture systems enable individual reagent injection to 96 microwells [65]. Although only single microwells can be controlled by one drive, a minute-level delay is not a big problem in long-term cell culture. Parallel

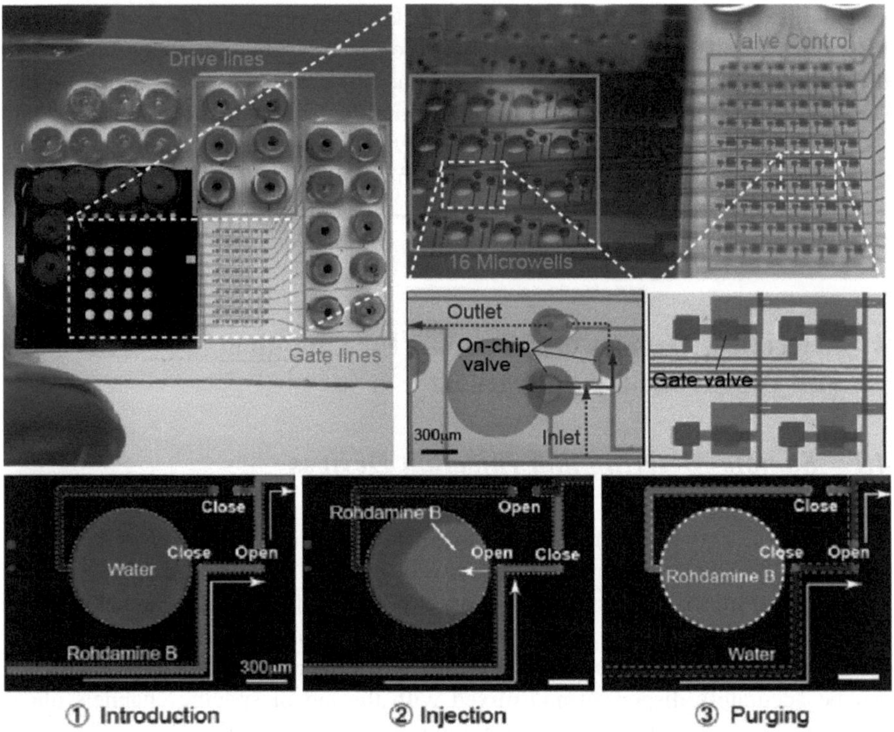

Fig. 17 Sixteen independently controlled microwell array for parallel reagent injection [68]

screening devices for DNA and protein expression have the merits of no cross-contamination or diffusion; this is because each reaction microwell is separate [66]. Microwell arrays in which the cell suspension is introduced passively enable chip sizes to be reduced [67]. To reduce the variation in cell numbers between each cell-trapping structure, a C-shaped ring is inserted in the microwell. Figure 17 shows the microwell array for parallel reagent control [68]. The fluid flow to each microwell is individually controlled by three microvalves for inlet, outlet, and purging. Each microwell is independent of the state of the other microwells, and parallel reagent injection to selected microwells can be carried out at the same time.

5 Summary and Outlook

Recent advances in the fabrication of microflow devices using MEMS technology are described from the technological point of view. The flow control methods and multiple flow control systems reported here are applicable in efficient chemical microreactors as well as in chemical analysis systems. For high-performance flow device design, computational dynamics (CFD) simulation is indispensable. The choice of structural materials and surface modifications when considering actual applications are very important. Recent trends in µTAS and LOC are high-throughput DNA and protein analysis, biological cell function analysis, etc. Close collaboration between MEMS researchers and chemical/biochemical researchers is still needed in order to open up new µTAS application fields.

Acknowledgments The authors would like to thank Dr. Jun Mizuno, Dr. Tetsushi Sekiguchi at Waseda University, Prof. Takashi Funatsu at the University of Tokyo, Dr. Takahiro Arakawa at Tokyo Dental & Medical University, Yoshitaka Shirasaki at Riken, and Dr. Masaki Kanai at Shimadzu Co.

References

1. Madou MJ (2002) Fundamentals of microfabrication. CRC, New York
2. Shoji S (1998) Technology in chemistry and life science. Top Curr Chem 194:163–188
3. Tabeling P (2005) Introduction to microfluidics. Oxford University Press, Oxford
4. Stroock AD, Dertinger SKW, Ajdari A, Mezic I, Stone HA, Whiteside GM (2002) Chaotic mixer for microchannels. Science 295:647–651
5. Sato H, Ito S, Tajima K, Orimoto N, Shoji S (2005) PDMS microchannels with slanted grooves embedded in three walls to realize efficient spiral flow. Sens Actuators A 119:365–371
6. Kim DS, Lee SW, Kwon TH, Lee SS (2002) Barrier embedded chaotic micromixer. In: Baba Y, Shoji S, Berg A (eds) Proceedings 6th international conference on micro total analysis systems (µTAS'02), Nara, Japan, 3–7 November 2002. Kluwer, The Netherlands, pp 757–759
7. Larsen UD, Blankenstein G, Branebjerg (1997) Microchip coulter particle counter. In: Proceedings international conference on solid-state sensors and actuators, Transducers'97, Chicago, IL, 16–19 June 1997, pp 1319–1322
8. Kawai K, Kanai M, Munaka T, Abe H, Murakami A, Shoji S (2008) Parallel and passive distribution to arrayed microwells using self-regulating pinched flow. Sens Mater 20(6):281–288
9. Jeong W, Kim J, Kim S, Lee S, Mensing G, Beebe DJ (2004) Hydrodynamic microfabrication via "on the fly" photopolymerization of microscale fibers and tubes. Lab Chip 4:576–580
10. Onoe H, Gojo R, Tssuda Y, Kiriya D, Takeuchi S (2010) Core-shell gel wires for the construction of large area heterogeneous structures with biomaterials. In: Proceedings IEEE 23rd international conference on micro electro mechanical systems, Wanchai, Hong Kong, 24–28 January 2010, pp 248–251
11. Tashiro K, Sekiguchi T, Shoji S, Funatsu T, Masumoto W, Sato H (2000) Design and simulation of particles and biomolecules handling microflow cells with three-dimensional sheath flow. In: Proceedings 4th international conference on micro total analysis systems (µTAS), Enschede, The Netherlands, 14–18 May 2000. Kluwer, The Netherlands, pp 209–212
12. Sundararajan N, Pio MS, Lee LP, Berlin A (2004) Three-dimensional hydrodynamic focusing in plolydimethylsiloxane (PDMS) microchannels. IEEE JMEMS 13(4):559–567
13. Mao X, Waldeisen JR, Huang TJ (2007) "Microfluidic drifting"-implementing three-dimensional hydrodynamic focusing with a single-layer planer microfluidic device. Lab Chip 7:1260–1262
14. Zhuang GS, Jensen TG, Kutter JP (2008) Thee-dimensional hydrodynamic focusing over a wide Reynolds number range using a two-layer microfluidic design. In: Proceedings 12th international conference on miniaturized systems for chemistry and life sciences (µTAS'08), San Diego, CA, 12–16 October 2008, pp 1357–1359

15. Shirasaki Y, Goto M, Sugino H, Arakawa T, Yoon D, Mizuno J, Shoji S, Funatsu T (2010) A microfluidic mammalian cell sorter with thermal gelation polymer solution. In: Proceedings 14th international conference on miniaturized systems for chemistry and life sciences (μTAS'10), Groningen, The Netherlands, 3–7 October 2010, pp 1571–1573
16. Gambin Y, Simonnet C, VanDelinder V, Deniz A, Groisman A (2010) Ultrafast microfluidic mixer with three-dimensional flow focusing for studies of biochemical kinetics. Lab Chip 10:598–609
17. Sato H, Sasamoto Y, Sekiguchi T, Homma T, Shoji S (2007) Multiple core-sheath liquid transfer using matrix arrangement of 3D sheath flows. In: Proceedings 11th international conference on miniaturized systems for chemistry and life sciences (μTAS'07), Paris, France, 7–11 October pp 1571–1573
18. Howell PB Jr, Golden JP, Hilliard LR, Erickson JS, Mott DR, Ligler FS (2008) Two simple and rugged designs for creating microfluidic sheath flow. Lab Chip 8:1097–1103
19. Seki M, Yamada M (2004) Pinched flow fractionation: continuous size separation of particles utilizing a laminar flow profile in a pinched microchannel. Anal Chem 76:5465–5471
20. Takagi J, Yamada M, Yasuda M, Seki M (2005) Continuous particle separation in a microchannel having asymmetrically arranged multiple branches. Lab Chip 5:778–784
21. Choi S, Song S, Choi C, Park J-K (2009) Hydrophoretic sorting of micrometer and submicrometer particles using anisotropic microfluidic obstacles. Anal Chem 81:50–55
22. Choi S, Song S, Choi C, Park J-K (2009) Microfluidic self-sorting of mammalian cells to achieve cell cycle synchrony by hydrophoresis. Anal Chem 81:1964–1968
23. Yamada M, Kano K, Tsuda Y, Kobayashi J, Yamato M, Seki M (2007) Microfluidic devices for size-dependant separation of live cells. Biomed Microdevices 9:637–645
24. Yamada M, Seki M (2005) Hydrodynamic filtration for on-chip particle concentration and classification utilizing microfluidics. Lab Chip 5:1233–1239
25. Blankenstein G, Scampavia L, Branebjerg J, Larsen UD, Ruzica J, (1996) Flow switch for analyte injection and cell/particle sorting. In: Proceedings 2nd international conference on miniaturized total analysis systems (μTAS'96), Basel, Switzerland, 19–22 November 1996, pp 82–84
26. Wolff A, Larsen UD, Blankenstein G, Philip J, Telleman P (1998) Rare event cell sorting in a microfluidic system for application in prenatal diagnosis. In: Harrison DJ, Berg A (eds). Proceedings micro total analysis systems workshop (μTAS'98), Banff, Canada, 13–16 October 1998. Kluwer, The Netherlands, pp 77–80
27. Fisher JS, Kuo TS, Poulos J, Lee AP (2007) Design parameters for 1XN microdroplet switch. In: Proceedings 11th international conference on miniaturized systems for chemistry and life sciences (μTAS'07), Paris, France, 7–11 October 2007, pp 1531–1533
28. Kanai K, Ikeda S, Tanaka J, Go GS, Nakanishi H, Shoji S (2004) The multiple sample injector using improved sheath flow to prevent sample dilution. Sens Actuators A 111:32–36
29. Ohtsuka S, Kanai M, Hayashi M, Nakanishi H, Shoji S (2004) Development of individual cell sorting system for intercellular reaction analysis. In: Proceedings 8th international conference on miniaturized systems for chemistry and life sciences (μTAS'04), Malmö, Sweden, 26–30 September 2004, vol 1, pp 30–32
30. Chen CH, Cho SH, Erten A, Lo YH (2008) High-throughput cell sorter with piezoelectric actuation. In: Proceedings 12th international conference on miniaturized systems for chemistry and life sciences (μTAS'08), San Diego, CA, 12–16 October 2008, pp 155–157
31. Klammer I, Buchenauer A, Dura G, Mokwa W, Schnakenberg U, (2008) A novel valve for microfluidic PDMS-based systems. In: Proceedings IEEE 21st international conference on micro electro mechanical systems, Tucson, AZ, 13–17 January 2008, pp 626–629
32. Lin Y-H, Lee C-H, Lee G-B (2008) A new droplet formation chip utilizing controllable moving-wall structure for double emulsion applications. In: Proceedings IEEE 21st international conference on micro electro mechanical systems, Tucson, AZ, 13–17 January 2008, pp 22–25

33. Iwai K, Takeuchi S (2009) A dynamic microarray with pneumatic valves for selective trapping and releasing of microbeads. In: Proceedings IEEE 22nd international conference on micro electro mechanical systems, Sorrento, Italy, 25–29 January 2009, pp 371–373
34. Wakui D, Takahashi S, Sekiguchi T, Shoji S (2010) Multi channel droplet sorting device with horizontal pneumatic actuation using single layer PDMS flexible parallel walls. In: Proceedings IEEE 23rd international conference on micro electro mechanical systems, Wanchai, Hong Kong, 24–28 January 2010, pp 144–147
35. Takao H, Tanaka N, Sugiura M, Sawada K, Ishida M (2009) Non-linear fluidic integrated circuits realized by pneumatic-field effect transducers with controllable output resistance. In: Proceedings IEEE 22nd international conference on micro electro mechanical systems, Sorrento, Italy, 25–29 January 2009, pp 503–506
36. Takahashi K, Hattori A, Suzuki I, Ichiki T, Yasuda K (2004) Non-destructive on-chip cell sorting system with real-time microscopic image processing. J Nanobiotechnology 2:5–12
37. Ahn B, Panchapakesan R, Lee K, Oh KW (2008) Fast, robust and simultaneous sorting with droplet generation by synchronized high switching frequency of electrostatic actuation. In: Proceedings 12th international conference on miniaturized systems for chemistry and life sciences (μTAS'08), San Diego, CA, 12–16 October 2008, pp 119–121
38. Wang L, Flanagan LA, Jeon NL, Monuki E, Lee AP (2007) Dielectrophoresis switching with vertical sidewall electrodes for microfluidic flow cytometry. Lab Chip 7:1114–1120
39. Zhang L, Bossche A (2009) A novel device for particle batch separation based on dielectrophoresis. In: Proceedings international conference on solid-state sensors, actuators and microsystems, Transducers'09, Denver, CO, 21–25 June 2009, pp 2151–2154
40. Chang S, Cho Y-H, (2007) A continuous multi-size particle separator using negative dielectrophoretic virtual pillars induced by a planar spot electrode array. In: Proceedings IEEE 20th international conference on micro electro mechanical systems, Hyogo, Japan, 21–25 January 2007, pp 19–22
41. Wang MM, Tu E, Raymond DE, Yang JM, Zhang H, Hagen N, Dees B, Mercer EM, Foster AH, Kariv I, Marchand J, Bulter WF (2005) Microfluidic sorting of mammalian cells by optical force switching. Nat Biotechnol 23(1):83–87
42. Shirasaki Y, Tanaka J, Makazu H, Tashiro K, Shoji S, Tsukita S, Funatsu T (2006) On-chip cell sorting system using laser-induced heating of a thermoreversible gelation polymer to control flow. Anal Chem 78(3):695–701
43. Arakawa T, Shirasaki Y, Aoki T, Funatsu T, Shoji S (2007) Three-dimensional sheath flow sorting microsystem using thermosensitive hydrogel. Sens Actuators A 135:99–105
44. Sugino H, Ozaki K, Shirasaki Y, Arakawa T, Shoji S, Funatsu T (2009) On-chip microfluidic sorting with fluorescence spectrum detection and multiway separation. Lab Chip 9:1254–1260
45. Ozaki K, Sugino H, Arakawa T, Shirasaki Y, Funatsu T, Shoji S, (2009) High performance multiple E. coli cell sorting system using thermosensitive hydrogel and fluorescence spectrum detection. In: Proceedings 13th international conference on miniaturized systems for chemistry and life sciences (μTAS'09), Jeju, Korea, 1–5 November 2009, pp 1856–1858
46. Ozaki K, Sugino H, Shirasaki Y, Aoki T, Arakawa T, Funatsu T, Shoji S (2010) Microfluidic cell sorter with flow switching triggered by a sol-gel Transition of a therm-resersible gelation polymer. Sens Actuators B 150:449–455
47. Lin W-Y, Lee G-B (2009) A new micro flow cytometer using optically-induced dielectrophoretic forces for continuous microparticle counting and sorting. In: Proceedings 12th international conference on miniaturized systems for chemistry and life sciences (μTAS'08) San Diego, USA, Korea, 1–5 November 2009, pp 47–50
48. Lee G-B, Lin Y-H, Lin W-Y, Wang W, Guo T-F, (2009) Optically-induced dielectrophoresis using polymer materials for biomedical applications. In: Proceedings international conference on solid-state sensors, actuators and microsystems, Transducers'09, Denver, CO, 21–25 June 2009, pp 2135–2138
49. Arai F, Sakuma S, Yamabishi Y, Onda K (2009) Powerful actuation of magnetized microtoll by focused magnetic field on a disposable microfluidic chip. In: Proceedings IEEE 22nd

international conference on micro electro mechanical systems, Sorrento, Italy, 25–29 January 2009, pp 51–54
50. Yamanishi Y, Feng L, Arai F (2010) On-demand and size-controlled production of emulsion droplet in microfluidic devices. In: Proceedings IEEE 23rd international conference on micro electro mechanical systems, Wanchai, Hong Kong, 24–28 January 2010, pp 1087–1090
51. Nilsson A, Petersson F, Jonsson H, Laurell T (2004) Acoustic control of suspended particles in microfluidic chips. Lab Chip 4:131–135
52. Petersson F, Nilsson A, Jonsson H, Laurell T (2005) Carrier medium exchange through ultrasonic particle switching in microfluidic channels. Anal Chem 77:1216–1221
53. Zhong JF, Chen Y, Marcus JS, Scherer A, Quake SR, Taylor CR, Weiner LP (2007) A microfluidic processor for gene expression profiling of single human embryonic stem cells. Lab Chip 8:68–74
54. Grover WH, Jensen EC, Lvester RHC, Mathies RA (2006) Programmable pneumatic logical circuits for microfluidic device control. In: Proceedings 10th international conference on miniaturized systems for chemistry and life sciences (μTAS'06), Tokyo, Japan, 5–9 November 2006, pp 506–508
55. Rhee M, Burns MA (2009) Microfluidic pneumatic logic circuits and digital pneumatic microprocessors for integrated microfluidic systems. Lab Chip 9:3131–3143
56. Gu W, Chen H, Tung YC, Meiners JC, Takayama S (2007) Multiplexed hydraulic valve actuation using ionic liquid filled soft channels and Braille displays. Appl Phys Lett 90 (3):033505
57. Lee DW, Cho Y-H (2009) A quaternary microfluidic multiplexer using dynamic control of pressure valves having different thresholds. In: Proceedings international conference on solid-state sensors, actuators and microsystems, Transducers'09, Denver, CO, 21–25 June 2009, pp 433–436
58. Kawai K, Kanai M, Shoji S (2007) Efficient addressable fluid control system using pneumatic valve array. In: Proceedings 11th international conference on miniaturized systems for chemistry and life sciences (μTAS'07), Paris, France, 7–11 October 2007, pp 32–34
59. Kawai K, Shibata Y, Shoji S (2009) 100 Picoliter droplet handling using 256 (2^8) microvalve array with 18 multiplexed control lines. In: proceedings international conference on solid-state sensors, actuators and microsystems, Transducers'09, Denver, CO, 21–25 June, 2009, pp 802–805
60. Yamada M, Seki M (2004) Nanoliter-sized liquid dispenser array for multiple biochemical analysis in microfluidic devices. Anal Chem 76(4):895–899
61. Sato T, Kawai K, Kanai M, Shoji S (2009) Development of all fluoroplastic microfluidic device applied as a nanoliter sample injector. Jpn J Appl Phys 48:No.06FJ03
62. Lin B-C, Su Y-C (2008) On-demand droplet metering and fusion utilizing membrane actuation. In: Proceedings 12th international conference on miniaturized systems for chemistry and life sciences (μTAS'08), San Diego, CA, 12–16 October 2008, pp 86–88
63. Fujii M, Kawai K, Shoji S (2009) Multi uniform picoliter volume droplets generation and sorting device for digital picoliter dispenser. In: Proceedings 13th international conference on miniaturized systems for chemistry and life sciences (μTAS'09), Jeju, Korea, 1–5 November 2009, pp 1356–1358
64. Suzuki Y, Kanai M, Kawai K, Nishimoto T, Shoji S (2007) Spatially focused reagent injection system for cell analysis using 3-D sheath flow scanner. In: Proceedings 14th international conference on solid-state sensors, actuators and microsystems, Transducers'07, Lyon, France, 10–14 June 2007, pp 25–28
65. Gomez-Sjoberg R, Leyrat AA, Pirone DM, Chen CS, Quake SR (2007) Versatile, fully automated, microfluidic cell culture system. Anal Chem 79:8557–8563. doi:10.1021/ac071311w
66. Fordyce PM, Gerber D, Tran D, Zheng J, Li H (2010) De novo identification and biophysical characterization of transcription-factor binding sites with microfluidic affinity analysis. Nat Biotechnol 28:970–975

67. Lee PJ, Hung PJ, Rao VM, Lee LP (2006) Nanoliter scale microbioreactor array for quantitative cell biology. Biotechnol Bioeng 94(1):5–14
68. Shibata Y, Kawai K, Kanai M, Shoji S (2009) Precise volume controlled multi reagents injective microwell array for efficient cell function analysis. In: Proceedings 13th international conference on miniaturized systems for chemistry and life sciences (μTAS'09), Jeju, Korea, 1–5 November 2009, pp 1488–1490

[illegible]

Top Curr Chem (2011) 304: 27–68
DOI: 10.1007/128_2011_150

Published online: 28 April 2011

Micromixing Within Microfluidic Devices

Lorenzo Capretto, Wei Cheng, Martyn Hill, and Xunli Zhang

Abstract Micromixing is a crucial process within microfluidic systems such as micro total analysis systems (μTAS). A state-of-art review on microstructured mixing devices and their mixing phenomena is given. The review first presents an overview of the characteristics of fluidic behavior at the microscale and their implications in microfluidic mixing processes. According to the two basic principles exploited to induce mixing at the microscale, micromixers are generally classified as being passive or active. Passive mixers solely rely on pumping energy, whereas active mixers rely on an external energy source to achieve mixing. Typical types of passive micromixers are discussed, including T- or Y-shaped, parallel lamination, sequential, focusing enhanced mixers, and droplet micromixers. Examples of active mixers using external forces such as pressure field, electrokinetic, dielectrophoretic, electrowetting, magneto-hydrodynamic, and ultrasound to assist mixing are presented. Finally, the advantages and disadvantages of mixing in a microfluidic environment are discussed.

Keywords Active micromixers · Microfluidics · Micromixing · Mixing principles · Passive micromixers

Contents

1 Introduction and Outline 29
2 The Microfluidic Environment and Mixing Principles 30
  2.1 Reynolds Number and Diffusion 30
  2.2 Mixing in Microfluidic Devices 32
3 Micromixers 33
  3.1 Passive Micromixers 33
  3.2 Active Micromixers 51

L. Capretto, W. Cheng, M. Hill, and X. Zhang (✉)
School of Engineering Sciences, University of Southampton, Southampton SO17 1BJ, UK
e-mail: XL.Zhang@soton.ac.uk

4 Why Microfluidic Mixers? 57
5 Conclusions 59
References 60

Symbols

A	Cross-sectional area (m^2)
Ca	Capillary number
D	Diffusion coefficient ($m^2\ s^{-1}$)
D_h	Hydraulic diameter (m)
f	Frequency of the disturbance action
h	Height of the channels (m)
j	Diffusion flux (mol $m^{-2}\ s^{-1}$)
k	Boltzmann's constant ($k = 1.381{\cdot}10^{-23}$J K^{-1})
n	Number of parallel fluid substreams
Pe	Peclét number
P_{wet}	Wetted perimeter (m)
Q_1	Volumetric flow rates for the lateral channels ($m^3\ s^{-1}$)
Q_2	Volumetric flow rates of the central inlet channel ($m^3\ s^{-1}$)
Q_3	Volumetric flow rates for the lateral channels ($m^3\ s^{-1}$)
Q_f	Volumetric flow rates of the focused stream ($m^3\ s^{-1}$)
R	Radius of the particles (or molecules) (m)
Re	Reynolds number
St	Strouhal number
t	Time (s)
T	Absolute temperature
u	Velocity of fluid (m s^{-1})
v_2	Average flow velocity of the flow within central inlet channel (m s^{-1})
v_f	Average flow velocity of the flow within focused stream (m s^{-1})
v_o	Average flow velocities of the flow within the mixing channel (m s^{-1})
w_2	Width of central inlet channel (m)
w_f	Width of the focused stream (m)
w_o	Width of the mixing channel (m)
x	Position of the species (m)

Greek Symbols

γ	Interfacial tension (N m^{-1})
φ	Species concentration (Kg m^{-3})
ρ	Fluid density (kg m^{-3})
μ	Fluid dynamic viscosity (Pa s)
v	Fluid kinematic viscosity ($m^2\ s^{-1}$)

Abbreviations

μTAS	Micro total analysis systems
ASM	Asymmetric serpentine micromixer
CDM	Circulation–disturbance micromixer
CGM	Connected-groove micromixer
CMM	Crossing manifold micromixer
EKI	Elecrokinetic instability
EWDO	Electrowetting on dielectrics
LOC	Lab on a chip
MHD	Magneto hydrodynamic
PCR	Polymerase chain reaction
PSM	Planar serpentine micromixer
SAR	Split-and-recombine micromixers, sequential lamination micromixers
SGM	Slanted-groove micromixer
SHM	Staggered-herringbone micromixers
SOC	Staggered overlapping crisscross micromixer

1 Introduction and Outline

Over the past two decades, lab-on-a-chip (LOC) technologies have driven considerable progress in the development of microsystems, particularly for chemical, biological, and medical applications. LOC technology has been applied in a wide range of processes such as nanoparticle crystallization [1, 2], extraction [3–5], polymerization [6–9], organic synthesis [10–12], enzyme assay [13, 14], protein folding [15], biological screening [16, 17],analytical assay [18–20], cell analysis [21, 22], bioprocess optimization [23, 24], clinical diagnostics [25, 26], and drug delivery studies [27].

The miniaturized systems, designed for the above cited applications, are generally implemented with a microscale mixer to provide an intimate contact between the reagent molecules for interactions/chemical reactions. Furthermore, beside their integration in more complex micro total analysis systems (μTAS) [28], microscale mixers could also work as stand-alone devices for applications where a superior control and a scaling-down of the mixing process are required.

The exponential increase of research in miniaturization and in microfluidic applications highlights the importance of understanding the theory and the mechanisms that govern mixing at the microscale level. This chapter will review the most recent research and developments in mixing processes within microfluidic devices. In order to better understand the rationale behind the design of the microfluidic mixers reported in the literature, Sect. 2 will discuss the unique physical characteristics and theory of the microfluidic environment and their implications in the context of mixing. Then, an up-to-date critical review of the different types and designs of micromixers will be provided in Sect. 3.

Owing to the increasing interest in "digital" or droplet-based microfluidics, the microfluidic generation of microdroplets, the associated active and passive mixing process, and the manipulation of microsized droplets in microfluidic devices will also be covered.

Finally, a section summarizing the general advantages of microfluidic mixers/reactors is presented. Although of high interest and importance, an in-depth review of microfluidic mixers in a diversity of microsystems for specific applications is not addressed since it falls out of the scope of this chapter. The reader is therefore directed to a number of excellent recent review articles on the specific subjects [9, 19, 29–36].

2 The Microfluidic Environment and Mixing Principles

In this section, we will summarize the basic theory of fluid flow and the implications of using microfluidic devices for mixing purpose. Generally, the same laws that describe the flow at a macroscale govern fluid flow in the microenvironment. However, miniaturization confers additional characteristics that can be leveraged to perform processes not possible at a macroscale. Microfluidic devices, indeed, are not merely a miniature version of their macroscale counterparts because many physical characteristics, such as surface area–to-volume ratio, surface tension and diffusion, do not simply scale linearly from large to small devices. Another important feature is the omnipresence of laminar flow conditions because in the microfluidic channel viscous forces dominate. These factors become significant at a microscale level, and their effects should be taken into account during the design and implementation of LOC devices.

In other words, it must be noted that, rather than design microfluidic mixer as just a scaled-down copy of a macroscale mixing device, they should be designed in ways that leverage the physical characteristic of the mixing in a confined space.

2.1 Reynolds Number and Diffusion

Fluid flow is generally categorized into two flow regimes: laminar and turbulent. Laminar flow is characterized by smooth and constant fluid motion, whereas turbulent flow is characterized by vortices and flow fluctuations. Physically, the two regimes differ in terms of the relative importance of viscous and inertial forces. The relative importance of these two types of forces for a given flow condition, or to what extent the fluid is laminar, is measured by the Reynolds number (*Re*):

$$Re = \frac{\rho u D_{\mathrm{h}}}{\mu} = \frac{u D_{\mathrm{h}}}{\nu}, \tag{1}$$

where ρ and μ are the fluid density and dynamic viscosity, respectively; ν is the kinematic viscosity; u is the velocity of fluid and D_{h} is the hydraulic diameter of the channel. The hydraulic diameter of the channel is a characteristic number that depends on the cross-sectional geometry of the channel, and is given by:

$$D_{\mathrm{h}} = \frac{4A}{P_{\mathrm{wet}}}, \tag{2}$$

where A and P_{wet} are the cross-sectional area and the wetted perimeter of the channel, respectively.

At low *Re*, the viscous effects dominate inertial effects and a completely laminar flow occurs. In the laminar flow system, fluid streams flow parallel to each other and the velocity at any location within the fluid stream is invariant with time when boundary conditions are constant. This implies that convective mass transfer occurs only in the direction of the fluid flow, and mixing can be achieved only by molecular diffusion [37]. By contrast, at high *Re* the opposite is true. The flow is dominated by inertial forces and characterized by a turbulent flow. In a turbulent flow, the fluid exhibits motion that is random in both space and time, and there are convective mass transports in all directions [38].

Between the definite regimes of laminar and turbulent flow there is a transitional *Re* range. The exact values of this number range are a function of many parameters, such as channel shape, surface roughness, and aspect ratio. The transition *Re* is generally expected to be in the range of 1,500 and 2,500 for most situations [39]. For microfluidic systems, *Re* are typically smaller than 100 and the flow is considered essentially laminar. This characteristic has a direct consequence on mixing within microfluidic devices.

In an environment where the fluid flow is restrictedly laminar, mixing is largely dominated by passive molecular diffusion and advection. Diffusion is defined as the process of spreading molecules from a region of higher concentration to one of lower concentration by Brownian motion, which results in a gradual mixing of material. Diffusion is described mathematically using Fick's law:

$$j = -D\frac{\mathrm{d}\varphi}{\mathrm{d}x}, \tag{3}$$

where φ is the species concentration, x is the position of the species, and D is the diffusion coefficient. For simple spherical particles, D can be derived by the Einstein–Stokes equation:

$$D = \frac{kT}{6\pi\mu R}, \tag{4}$$

where k is Boltzmann's constant, T is the absolute temperature, R is the radius of the particles (or molecules) and μ is the viscosity of the medium. The diffusion coefficient for a small molecule in water at room temperature has the typical value of $10^{-9}\ \mathrm{m}^2\ \mathrm{s}^{-1}$ [40].

Diffusion is a nonlinear process in which the time t required for a species to diffuse scales quadratically with the distance x covered. A simple case of diffusion can be modeled in one dimension by the equation:

$$x^2 = 2Dt, \tag{5}$$

where t is the average time for particles to diffuse over the distance x. Regarding the microfluidic channel, x represents the stream width of the fluid to be mixed along the microfluidic channel [41]. On a microfluidic length scale, the diffusion distance can be extremely small, particularly if the fluid streams are hydrodynamically focused. Because x varies with the square power, a decrease in distance dramatically reduces the time required for complete mixing. Therefore, diffusion becomes a viable method to move particles and mix fluid in microfluidic devices.

2.2 *Mixing in Microfluidic Devices*

At a macroscale level, mixing is conventionally achieved by a turbulent flow, which makes possible the segregation of the fluid in small domains, thereby leading to an increase in the contact surface and decrease in the mixing path. As discussed in the previous section, the *Re* is small in microfluidic systems, implying that hydrodynamic instability does not develop; therefore, the flows cannot be turbulent. Owing to this limitation, mixing in microfluidic devices is generally achieved by taking advantage of the relevant small length, which dramatically increases the effect of diffusion and advection.

Micromixers are generally designed with channel geometries that decrease the mixing path and increase the contact surface area. According to the two different basic principles exploited to induce mixing at the microscale, micromixers are generally classified as being passive or active.

Active micromixers use external energy input as well as fluid pumping energy to introduced time-dependent perturbations that stir and perturb the fluid for accelerating the mixing process [42]. The type of external force employed by active micromixers can be further categorized as pressure field-driven [43], acoustic (ultrasonic)-driven [44], temperature-induced [45] or magneto-hydrodynamic [46]. Generally, active micromixers have higher mixer efficiency [47]. However, the requirement to integrate peripheral devices such as the actuators for the external power source into the microdevice, and the complex and expensive fabrication process, limit the implementation of such devices in practical applications. In addition, in active mixing mechanisms such as ultrasonic waves, high temperature gradients can damage biological fluids. Therefore, active mixers are not a popular choice when applying microfluidics to chemical and biological applications [48].

Passive mixing devices rely entirely on fluid pumping energy and use special channel designs to restructure the flow in a way that reduces the diffusion length and maximizes the contact surface area. Passive mixers were the first microfluidic device reported, often entail less expense and more convenient fabrication than active micromixers, and can be easily integrated into more complex LOC devices. The reduction in mixing time is generally achieved by splitting the fluid stream using serial or parallel lamination [49, 50], hydrodynamically focusing mixing streams [51], introducing bubbles of gas (slug) or liquid (droplet) into the flow [52, 53], or enhancing chaotic advection using ribs and grooves designed on the channel walls [54, 55].

Micromixers are also commonly characterized accordingly to three nondimensional fluid parameters: *Re* (as discussed above), Peclét number *Pe*, and Strouhal number *St*. Peclét number is defined as:

$$Pe = \frac{uL}{D}, \tag{6}$$

which is a measure of the relative importance of advection and diffusion in providing the mass transport associated with the mixing. Advection is dominant at high *Pe*.

The Strouhal number is defined as:

$$St = \frac{fD_{\mathrm{h}}}{u}, \tag{7}$$

where f is the frequency of the disturbance action, is generally associated with active micromixers, and represents the ratio between the residence time of a species and the time period of disturbance [48, 56, 57].

3 Micromixers

3.1 Passive Micromixers

Passive micromixers rely on the mass transport phenomena provided by molecular diffusion and chaotic advection. These devices are designed with a channel geometry that increases the surface area between the different fluids and decreases the diffusion path. By contrast, the enhancement of chaotic advection can be realized by modifying the design to allow the manipulation of the laminar flow inside the channels. The modified flow pattern is characterized by a shorter diffusion path that improves the mixing velocity. In this section, an overview of the different types of passive micromixers is provided. Mixed phase passive micromixers can be categorized as:

1. T- and Y-shaped micromixers
2. Parallel lamination micromixers
3. Sequential lamination micromixers
4. Focusing enhanced mixers
5. Chaotic advection micromixers
6. Droplet micromixers

3.1.1 T- or Y-Shaped Micromixers

The easiest and most basic design for a micromixer is represented by either T- or Y-shaped channel micromixers [58–61]. A schematic of the general design of this type of mixer is shown in Fig. 1.

The mixing process in this type of micromixer is obtained by guiding the two liquids to be mixed in contact through a flow-through channel. It must be noted that, for the basic design of T- and Y-type micromixers, mixing solely depends on diffusion of the species at the interface between the two liquids, hence the mixing is rather slow and a long mixing channel is required. In order to enhance the mixing efficiency, different authors proposed slight modifications to the geometrical setup by adding obstacles or roughening the channel walls [54, 59, 62]. Further reduction of the mixing time could be achieved by using a high flow rate, hence high *Re*, where a chaotic flow is expected [63, 64], (Fig. 2). Veenstra et al. further reduced the mixing path in a T-shaped micromixer by a simple narrowing of the mixing channel and therefore shortening of the diffusion length [65].

3.1.2 Parallel Lamination

The concept of T- and Y-shaped micromixers can be improved by using more complicated designs that split the inlet main streams into *n* sub-streams and then rejoin them to form a laminate stream (Fig. 3) [66, 70–72]. This type of micromixer enhances the mixing process by decreasing the diffusion length and increasing the contact surface area between the two fluids.

According to Erbacher et al., the subdivision of each stream into *n* laminae leads to mixing that is faster by a factor of n^2, as reported in the following expression derived from (5) [66]:

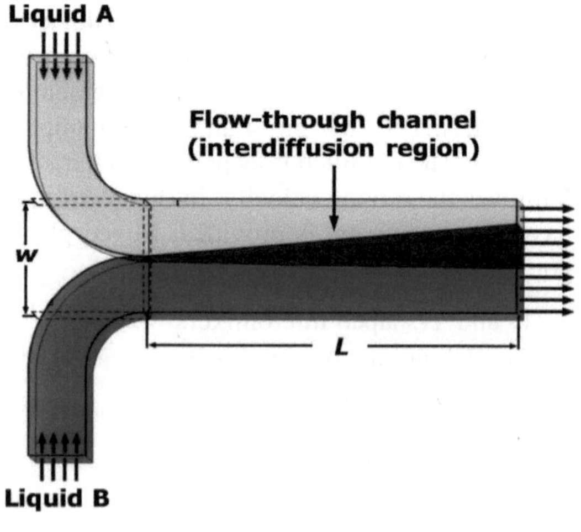

Fig. 1 T-shaped micromixer with two input fluids, each containing one diffusing species. *L* and *w* represent the length and width of the mixing channel, respectively (Adapted with permission from [58]. Copyright 1999 American Chemical Society)

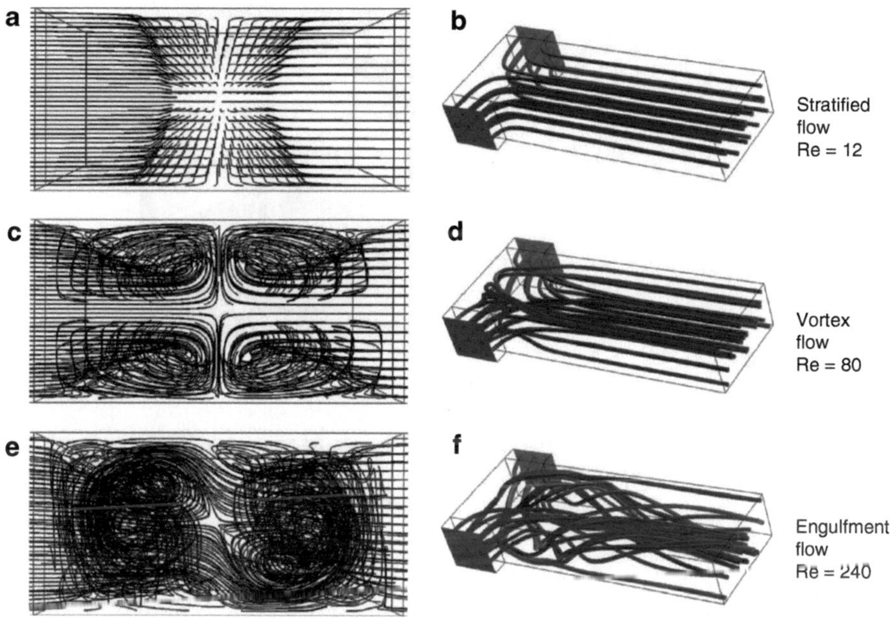

Fig. 2 Path lines (**a**, **c**, **e**) and streamlines (**b**, **d**, **f**) for different *Re* numbers of 12 (**a**, **b**), 80 (**c**, **d**) and 240 (**e**, **f**). The swirling of the fluid flow obtained at higher *Re* number results in better dispersion of the fluid within the channel volume and hence an improvement in the mixing quality (Reprinted from [61]. Copyright (2008) with permission from Elsevier)

$$t = \frac{x^2}{2n^2D}, \tag{8}$$

where n is the number of parallel fluid substreams.

Lamination of the fluids to be mixed can be achieved using two different feeds arrangements known as (1) bifurcation-type feeds and (2) parallel interdigital-type feeds. Bifurcation-type feeds [66, 70–72] are characterized by an alternate arrangement of feeds (Fig. 3a) that are later joined by passing through an inverse bifurcation channel pattern followed by a folded mixing channel in which the mixing takes place. By exploiting this configuration, Bessoth et al. [70] demonstrated that mixing was completed is less than 100 ms, while 95% of mixing was achieved in 40 ms.

Parallel-flow interdigital-type feeds is the most-used feeding concept. This type of micromixer comprises a feeding structure characterized by a co-[67, 73–76] or counterflow [77] interdigital array of microchannels. Similar to the previous types of feed concept, the microchannel array leads to an alternate lamellae arrangement of the liquid to be mixed. However, unlike bifurcation-type feeds, the way to obtain this pattern is based on a pressure-loss triggered flow equiparation (Fig. 3b). Generally, after the lamellae rearrangement, the multilaminated flow is focused through a geometrical constrain (mixing channel narrowing) [67, 71, 73, 75, 76] in

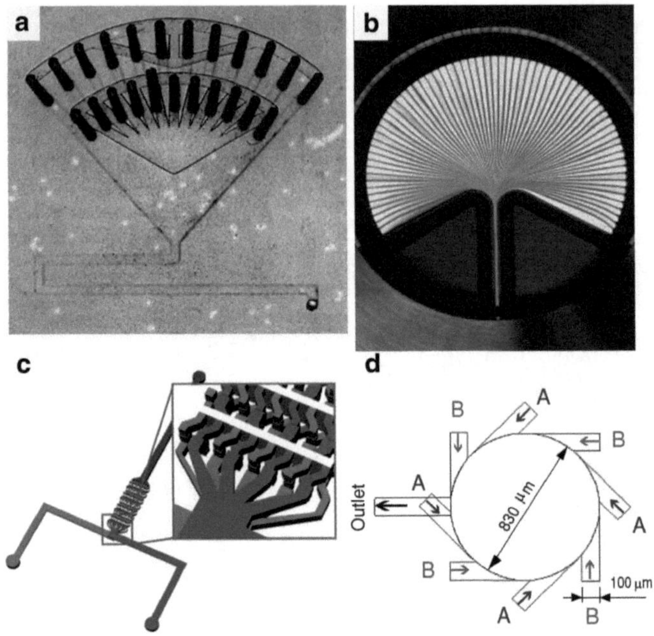

Fig. 3 Parallel lamination micromixer types: (**a**) Bifurcation-type feeds (Adapted from [66] with kind permission from Springer Science). (**b**) Interdigital-type feeds, super focus mixer (Reproduced from [67] with permission. Copyright Wiley-VCH). (**c**) Chessboard micromixer (Adapted from [68] with permission. Copyright IOP Publishing). (**d**) Circular micromixer (Adapted from [69] with permission. Copyright IOP Publishing)

order to decrease the diffusion length and enhance the mixing, using a concept similar to that presented by Veenstra et al. [65] for a T-type micromixer.

Drese et al. [73], developed a special interdigital-type feed micromixer, named the super focus mixer (Fig. 3b), in which the various lamellae have a different angle with respect to the channel direction and which is capable of obtaining 95% mixing in 4 ms.

Interdigital-type feed micromixers were recently applied as a reactor for a nitroxide-mediated radical polymerization, demonstrating a control over the molecular weight distributions as a result of an improved control of the co-polymerization reaction [78, 79].

Cha et al. [68] presented a novel micromixer design relying on a concept not far from that of the multilamination mixer, named a chessboard mixer. The mixer was able to complete the mixing in only 1.400 mm and the author claimed that the flow rate can be increased easily by using different arrays without affecting the performance (Fig. 3c).

A further interesting concept for the creation of multilaminated streams is that applied in circular micromixers [69, 80, 81]. Circular micromixers rely on the formation of a vortex due to the self-rotation of the fluid stream injected in a quasitangential orientation to the circular mixing chamber (Fig. 3d). Excellent mixing performance of this type of micromixer was reported at either high (Re = 150) [80] or low [81] Re number (Re = 4).

Lastly, StarLaminators are devices based on a multilamination concept and are capable of high liquid throughput up to 1,000 Lh^{-1} [56, 82, 83]. The mixing is provided by a stack of plates with star-like openings that leads to turbulent flow, which causes mixing by formation of eddies.

3.1.3 Sequential Lamination

Similar to parallel lamination micromixers, sequential lamination micromixers [also called split-and-recombine (SAR) micromixers] rely on an exponential increase in the contact surface area and decrease in the length path to achieve a shorter mixing time. The difference between the two types of micromixers is the method used to achieve lamination of the fluid to be mixed. As suggested by the name, the lamination in sequential lamination micromixers is obtained by sequential processes of splitting and rejoining the fluids (Fig. 4a) [84, 86–89].

Different geometries for SAR micromixers have been proposed, such as ramp-like [86] and curved-like [90] architectures. However, in order to achieve exponential sequential lamination, three steps are typically required: flow splitting, flow recombination, and flow rearrangement (Fig. 4a).

It must be noted that SAR mixers generally work at small Re. However, some secondary recirculation flows can be generated, as demonstrated by particle tracking simulation [90].

Recently, Fang et al. [85] proposed a SAR micromixer incorporating chaotic advection features named (SAR μ-reactor design) to mix fluids in a wide range of Re and viscosity (Fig. 4b–c). They compared the results with those obtained from a slanted-groove micromixer (SGM) (see Sect 3.1.5) [55], demonstrating better mixing efficiency of the SAR μ-reactor compared to SGM as result of the synergistic effect of the two mixing concepts.

Bertsch et al. [91] presented two micromixers, similar in concept to the SAR micromixer, with internal structures resembling conventional large-scale static mixers (Fig. 5). The first micromixer was characterized by introducing an internal structure with intersecting channels, which worked in a similar way to a SAR micromixer by splitting and recombining fluid streams. The second micromixer comprised a series of helical elements (Fig. 5a). Computational fluid dynamics (CFD) simulated results showed the higher mixing efficiency of the first type of micromixer over a helical based mixer.

Recently Lim et al. [92] presented a three-dimensionally micro-mixer, named crossing manifold micromixer (Fig. 5b). The micromixer was able to perform almost complete mixing of 90% channel length of 250 μm.

The main disadvantage of SAR mixers is the complex fabrication process required to make a 3D structure. However, an effect on the liquid stream similar to that exploited by SAR can be achieved by a planar, packed bed configuration that enhances trans-channel coupling. Melin et al. [93] fabricated and tested multiple

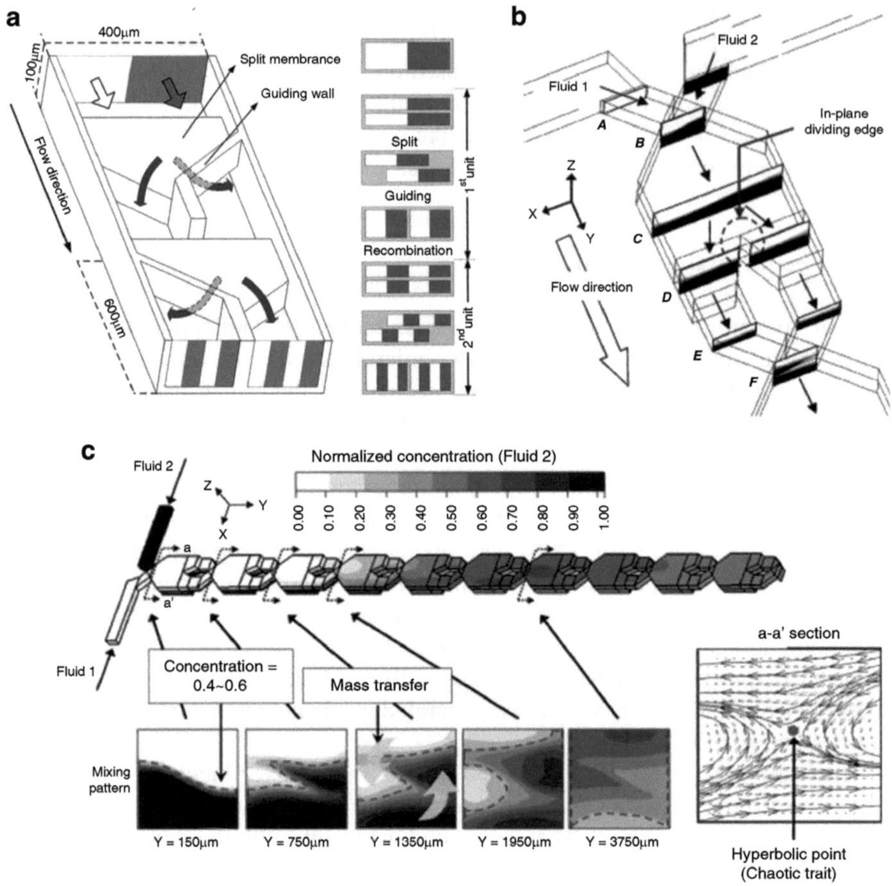

Fig. 4 Sequential lamination micromixer (I): (**a**) Mixing unit of the SAR micromixer and corresponding cross-section view of the laminated flow (Adapted from [84] with permission. Copyright IOP Publishing). (**b**) Mixing unit and (**c**) computed cross-section view of fluid arrangements along the SAR μ-reactor ($Re = 1$). Chaotic advection generated by the fluids overlapping causes the fluid interface to rotate, increasing the mixing efficiency (Reprinted from [85] with permission. Copyright 2009 Elsevier)

intersecting microchannels (known as a packed bed micromixer) that create a constantly changing flow pattern as the liquid samples pass through the mixing chamber, and achieved homogenous mixing of the two fluids in just 0.4 s. This concept was also applied to electrokinetically driven flow, as reported by He et al. [94] (Fig. 6a).

Another way of obtaining a SAR-like effect within a planar microfluidic chip was introduced by Sudarsan et al. [95] (Fig. 6b). It works with a multistep action: Initially, this geometry leveraged the generation of Dean vortices that arise in the vertical plane of curved channels to induce a 90° rotations in the fluid. At this point, the fluid is

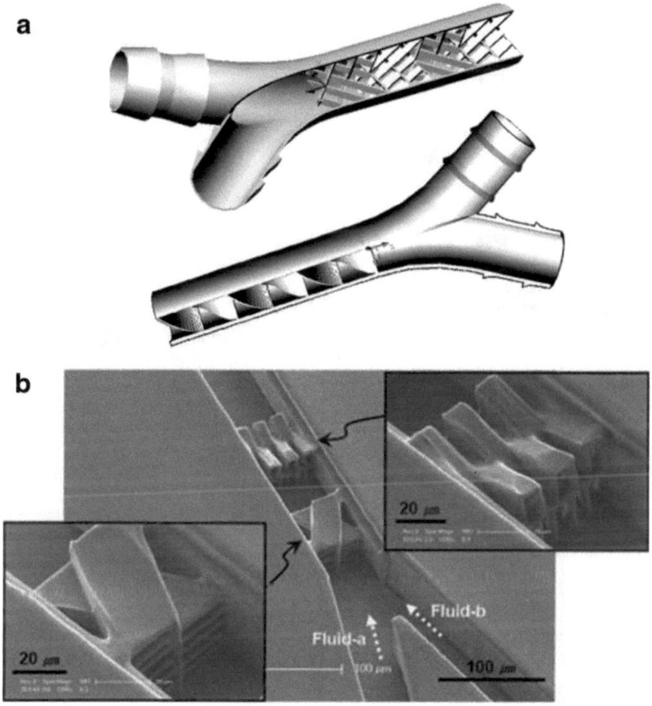

Fig. 5 Sequential lamination micromixer (II): (**a**) Static micromixer, with intersecting channel (*top*) and helical elements (*bottom*) (Reproduced from [91] by permission of The Royal Society of Chemistry). (**b**) Internal structure of crossing manifold micromixer (CMM) (Reproduced from [92] by permission of The Royal Society of Chemistry)

divided into many substreams that undergo the same 90° rotations in the fluid. At the end, the transformed substreams rejoin to create a multilamellae arrangement.

3.1.4 Flow Focusing

Another solution for reducing the mixing path is hydrodynamic focusing [51]. The basic design for hydrodynamic focusing is a long microchannel with three inlets (Fig. 7).

In hydrodynamic focusing, a central sample solution (supplied from the middle inlet) flows within the sheath of outer fluids (supplied from the side inlets), which constrain laterally the inner sample flow to achieve a smaller stream and thinner lamination width. The extent of the width decrease of the focused stream depends on the volumetric flow rate ratio between the sample flow and sheath flows. The greater the flow rate difference, the greater the degree of width reduction. As indicated by (5), mixing time is inversely proportional to the square of the diffusion path length (in this case represented by the focused stream width), therefore,

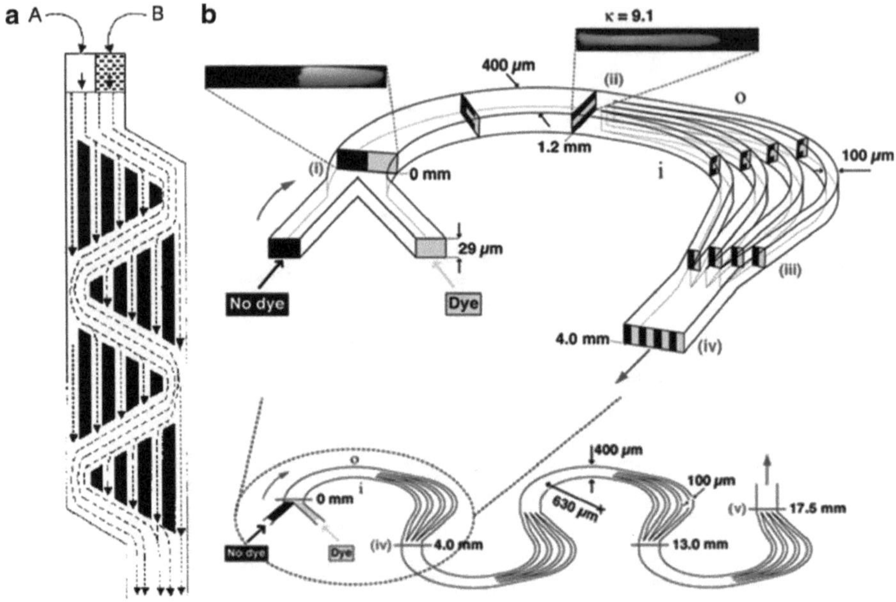

Fig. 6 Planar SAR-like micromixers: (**a**) Mixing unit of a packed bed micromixer (Adapted from [94] with permission. Copyright 2001 American Chemical Society). (**b**) Planar SAR micromixer that relies on Dean flows to generate alternate lamellae of fluid in a SAR-like fashion (Adapted from [95] with permission. Copyright 2006 National Academy of Sciences, USA)

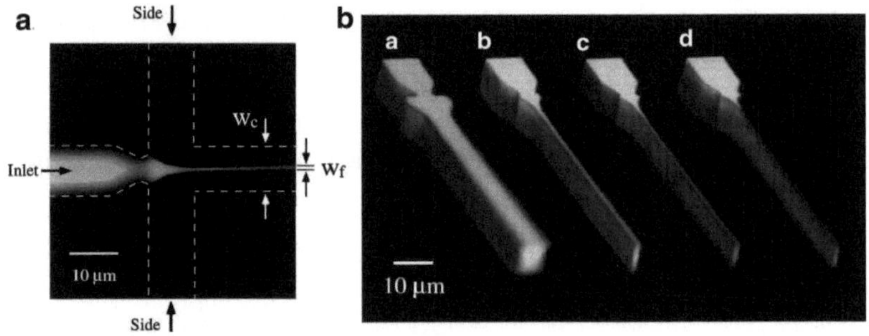

Fig. 7 (**a**) Focusing enhanced mixer. (**b**) Effect of ratio α of the side pressure to the inlet pressure on the width of the focused stream: (*a*) 0.5, (*b*) 1.0, (*c*) 1.1, and (*d*) 1.2 (Reprinted from [51] with permission. Copyright 1998 American Physical Society)

decreasing the stream width results in faster mixing. Notably, the position of the focused stream is also a function of the relative flow rate ratio of the three inlets. As a result, by changing the relative flow rate of the two side streams it is possible to direct the focused stream into a specific outlet [96].

The relative flow rate of the three streams is generally controlled using multiple external pressure sources or pumps (e.g., syringe pumps) (Fig. 7b). However, Stiles et al. proposed and tested the use of a vacuum-pumped microfluidic device using either a single suction pump or a capillary pumping effect to control the relative flow rates by varying the flow resistance of the input channels [97]. The analysis and prediction of the focused stream width employs a simple model based on mass conservation principles [98, 99]. The 2D focused stream width is computed under these simplified assumptions:

1. Flow in the microchannel is steady and laminar
2. Fluids are Newtonian
3. Fluids have the same density in the four channels (three inlet channels and one outlet channel)
4. Fluids flow in a rectangular microchannel
5. The four channels have the same height

According to the mass conservation principle, the volume of sample liquid that passes through the inlet channel (Q_2) must match the volume of the focused stream:

$$Q_2 = v_2 w_2 h = v_f w_f h = Q_f. \tag{9}$$

This leads to (10):

$$w_f = \frac{Q_2}{v_f h}, \tag{10}$$

In (9) and (10), w_f and w_2 represent the width of the focused stream and central inlet channel, respectively. Q_2 and Q_f are the volumetric flow rates of the central inlet channel and focused stream, respectively. h is the height of the channels, and v_2 and v_f are the average velocity of the flow in the central inlet channel and of the focused stream, respectively. The amount of fluid passing through the outlet channel (channel O) must be equal to the total amount of the fluid supplied from the three inlets:

$$Q_o = w_0 v_0 h = Q_1 + Q_2 + Q_3, \tag{11}$$

$$w_o = \frac{Q_1 + Q_2 + Q_3}{v_o h}, \tag{12}$$

where Q_1 and Q_3 are the volumetric flow rates for the two lateral channels, and v_o and w_o are the average velocities of the flow and width of the mixing channel, respectively. Combining (9) and (12), and assuming v_o and v_f have the same values, it is possible to obtain the relationship between the width of the focused stream and volumetric flow rate of the inlets:

$$\frac{w_f}{w_o} = \frac{Q_2}{Q_1 + Q_2 + Q_3}. \tag{13}$$

This equation provides a simple guideline for predicting the focused stream width. However, it does not reflect the effect of other factors such as device structure, channel surface, and fluidic property, which could affect the focusing process. In this respect, Lee et al. [99] proposed a similar model in which the effect of the density of the different fluids is taken into account. Moreover, Wu and Nguyen [100] presented a more complex method that considered the effect of the different viscosities of the sample stream and sheath streams.

In recent years, more complex channel geometry structures that rely on the hydrodynamic focusing to achieve mixing have been fabricated and examined. Wu and Nguyen [101] reported a mixer with two sample streams (solvent and solute streams). The two streams are brought into contact and then focused by two lateral sheath streams. The dramatic decrease in the diffusion path length improves mixing significantly. Park et al. [102] described a novel five-inlet port mixer in which the additional two diagonal sheath flows served as a barrier between solutions flowing from the center and the two side channels during the focusing process. In that configuration, the additional sheath reduced premixing before the formation of the focused jet without compromising rapid mixing by diffusion. Nguyen and Huang [103] reported a microfluidic mixer that relied on a combination of hydrodynamic focusing and sequential segmentation to reduce the mixing path and shorten mixing times. The sequential segmentation step divided the solvent and the solute into segments that usually occupied the whole channel width. Because of the additional segmentation step, the dispersion occurred even along the flow direction, leading to increased mixing efficiency.

Typically, focusing-enhanced micromixers focus the sample flow only in the horizontal dimension. Different authors have proposed microfabricated devices capable of focusing the sample horizontally and vertically [104–107]. Such devices add an additional dimension of focusing and are often referred to as 3D hydrodynamic focusing devices to distinguish them from traditional 2D focusing devices. Building these devices requires complex methods such as multistep photolithography, leading to an increase in fabrication cost. Recently, a novel fluid manipulation technique called "microfluidic drifting" was applied to obtain 3D focusing with a single-layer planar chip that is relatively easy to make [38, 108, 109] (Fig. 8).

The process of 3D focusing in this device can be divided into two steps. First, the sample stream is focused in the vertical direction using microfluidic drifting. The lateral drift of the sample flow is caused by the effect of the Dean vortices induced by the centrifugal effect of the curve, which transports the fluid in the opposite side of the channel. Second, classic horizontal focusing is obtained using two horizontal sheath streams. The result of these two steps is a stream focused in both the vertical and horizontal directions.

3.1.5 Chaotic Advection Micromixers

Advection is the transport of a substance within a moving fluid. In the micromixers discussed above, advection generally occurs in the direction of the flow, hence it has no effect on the transversal transport of the substance. However, advection in other directions, so-called chaotic advection [110], can generate a transverse

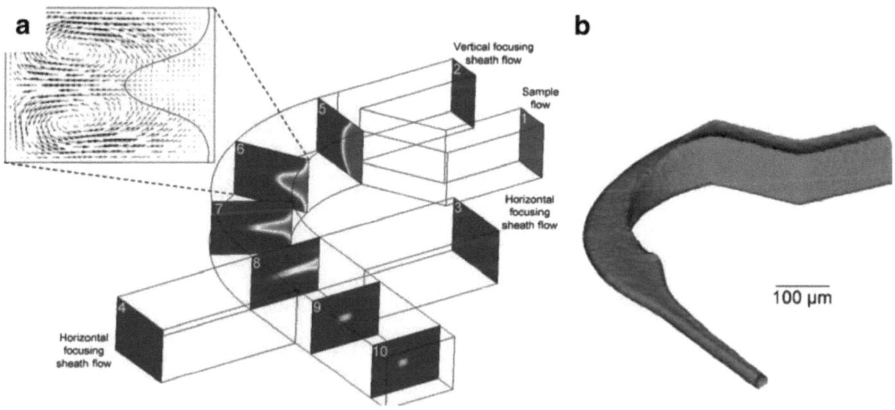

Fig. 8 The "microfluidic drifting" process. (**a**) Slices 1–10 are the cross-sectional profiles of the sample stream in the device; *inset* shows formation of Dean vortices in the 90° curve. (**b**) 3D microfluidic drifting focusing characterized by confocal microscopy (Reproduced from [109] by permission of The Royal Society of Chemistry)

component of flow [55]. These generated transverse flow components cause an exponential growth of the interfacial area and a corresponding decrease in the striation thickness, which can significantly improve mixing.

These "stirring" transverse flows can be generated by channel shapes that stretch, fold, break, and split the laminar flow over the cross-section of the channel. This effect can be achieved using 2D curved [111–113], or 3D convoluted channels [114–116] and by inserting obstacles [117] and bas-reliefs on the channel walls [54, 118, 119]. It must be noted that such type of chaotic flow could also be achieved by an active mixing strategy such as one using electrokinetic instability (EKI), as described in Sect. 3.2.2.

The simplest way to induce transverse flow is to insert an obstacle into the mixing channel. Obstacles can be inserted into the walls of the microchannel [62] or into the channel itself [117, 120–122]. The presence of obstacles alters the flow direction, and the streamlines induce whirl flow and recirculation that create the transversal mass transport. Generally obstacles in microchannel are not very efficient in creating transverse flow unless they are used at moderately high *Re* (typically more than 100) [120]. However, Bhagat et al. [121, 122] recently reported a micromixer with optimized cubic and rectangular structure capable of mixing with $Re < 1$ (Fig. 9).

Another efficient solution to induce transverse flows and chaotic advection at small *Re* (typically $Re \approx 1$) is to use a channel wall with a grooved pattern. SGM [55, 118] and staggered-herringbone micromixers (SHM) [55] subject the fluid to a repeated sequence of rotational and extensional local flow that, as result, produces a chaotic flow. The internal structures endow SHM with high mixing efficiency compared to the classic T-type mixer without them. In particular, a classic T-shaped mixer requires a mixing length (1–10 m) that is two order of magnitude larger than SHM mixers (1–1.5 cm) at *Pe* within 10^{-4} to 10^{-5}.

A series of improved grooved pattern micromixers has been proposed. A modified SHM micromixer that utilizes sequences of asymmetrical herringbone grooves

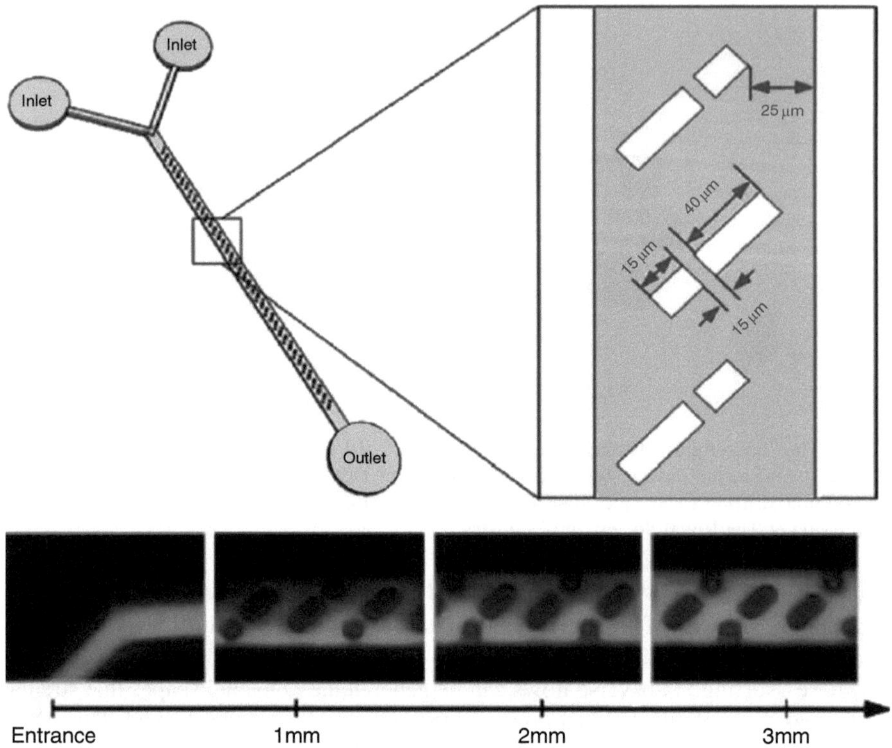

Fig. 9 Mixer with rectangular structure and *inset* reporting key features and dimension (*above*). Mixing extent at various portions downstream of the entrance at $Re = 0.05$ (*below*) (Adapted from [121] with permission. Copyright IOP Publishing)

was introduced and computationally studied by Hassel et al. [119]. Different authors have proposed micromixers in which the grooved pattern and zigzag barriers are not only applied on the bottom wall of the channel but also on the side and top walls to promote mixing, named respectively connected-groove micromixer (CGM) [123] (Fig. 10a) and circulation–disturbance micromixer (CDM) [124, 125] (Fig. 10b). Adding additional mixing elements to the side and top walls promotes lateral mass transport and assists the formation of advection patterns increasing mixing efficiency. In particular, CGM showed a mixing performance over 50% better than the classic SGM for *Re* ranging from 1 to 100 as a result of the intense transverse transport induced in the fluids [123].

Chaotic advection can be induced with a 2D alternatively curved microchannel (2D serpentine) [112, 113] or zigzag channel shape [111]. In the first case, the chaotic advection is induced in the curved microchannel by consecutive generation of Dean vortices (Fig. 11a). Typically such type of micromixer can provide an effective mixing only for high *Re* in the range of few hundreds. These micromixers are generally described using another dimensionless number, the Dean number (*De*):

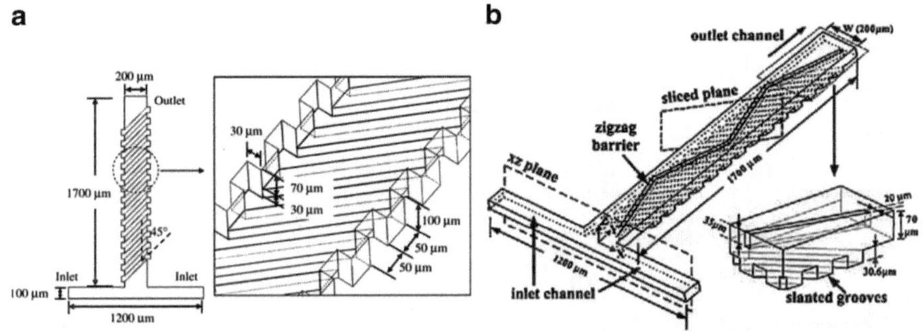

Fig. 10 Micromixers with grooved pattern: (a) Connected-groove micromixer (CGM) and inset reporting key features and dimension (Reprinted from [123] with permission from Elsevier. Copyright 2008). (b) Schematic representation of circulation–disturbance micromixer (CDM) and inset reporting key features and dimension (Adapted from [124] with permission. Copyright IOP Publishing)

$$De = Re\sqrt{\frac{D_h}{R}} \tag{14}$$

where R represent the channel curvature. Jiang et al. [112] demonstrated that in order to provide an efficient mixing De must be greater than 140.

Sundarsan et al. [95, 127] reported two improved 2D serpentine micromixers, namely the planar spiral micromixer [127] and the asymmetric serpentine micromixer (ASM) [95] (Fig. 11a). Both the micromixers were able to produce effective mixing at low Re number. The mixing enhancement was due to the synergistic effect of Dean and expansion vortices, where the latter were introduced by abrupt expansions of the microchannels.

In a zig-zag micromixer [111], mixing is provided by laminar recirculation that induces trasverse velocity components localized at each channel angle. The micromixer studied had a critical Re number ($Re = 80$), below which the mixing was solely due to molecular diffusion (see Fig. 11c).

Another interesting planar structure able to induce chaotic advection has been reported by Hong et al. [126] (Fig. 11d). This micromixer comprised a modified tesla structure that redirected the streams, by exploiting the Coanda effect. The authors demonstrated an efficient mixing at relative low Re number ($Re < 10$).

Based on the 2D twisted micromixers, the so-called 3D serpentine micromixers (or 3D twisted micromixers) have been developed. These micromixers have a complex 3D structure with a repetition mixing unit that induces the formation of secondary flows that stretch and fold the fluids. Different channel arrangements have been presented. Liu et al. [115] fabricated a 3D structure created by a series of C-shaped segments aligned in a perpendicular plane. The authors showed that the microreactor performed well at relatively high Re number, ($Re > 25$) and that the mixing efficiency increased with an increase of the Re number. A 3D structure comprising an L-shaped segment aligned in the perpendicular plane has also been presented (Fig. 12) [116].

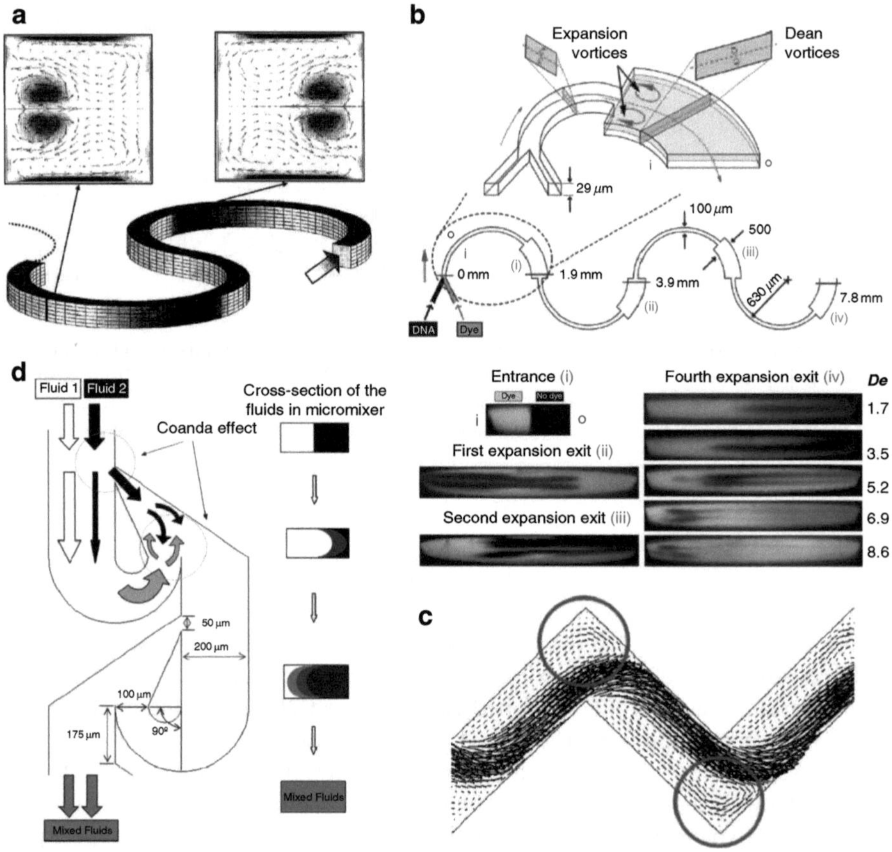

Fig. 11 Planar micromixer for chaotic advection: (**a**) 2D serpentine micromixer and *insets* showing the cross-section view of the channel and corresponding secondary Dean flows vortices (Adapted from [113] with permission. Copyright 2004 American Institute of Chemical Engineers). (**b**) Asymmetric serpentine micromixer (ASM) (*top*) and confocal cross section view showing synergistic effect of Dean vortices and expansion vortices at different section along the mixer and at different *De* number (Adapted from [95] with permission. Copyright 2006 National Academy of Sciences, USA). (**c**) Zigzag micromixer and recirculating pattern created at each zigzag (Adapted from [111] with permission). (d) Mixer with modified tesla structure and their effect on the liquid interface (Adapted from [126] with permission. Copyright 2002 American Chemical Society)

Chen et al. [114] reported a more complex structure derived from the connection of two helical flow channels with opposite chirality, and called it a topological mixer. By splitting, rotating and recombining the flow streams, the micromixer provided an effective and fast mixing at low *Re* between 0.1 and 2.

Park et al. [128] reported a structure that added the break-up effect in order to increase mixing efficiency. The break-up process enhances the mixing process by increasing the interfacial area as a result of the production of smaller fragments of blobs. Another interesting approach using a 3D structure was recently developed by

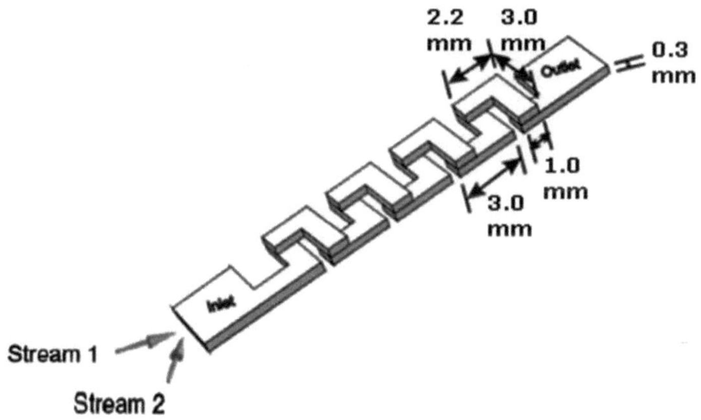

Fig. 12 3D serpentine micromixers with L-shaped segment. (Adapted from [116] with permission. Copyright 2003 American Chemical Society)

Long et al. [129]. They used an easily fabricated 3D structure comprising a circular chamber in which inlet and outlet channel are connect tangentially. When the two liquids to be mixed flow in the circular chamber, a vortical motion is generated. The vertical motion provides distortions and elongations of material interfaces and high mixing efficiency at relatively low *Re* ($Re > 21$).

Also, many micromixers combining chaotic advection and an SAR approach have been presented [85, 130–133]. Park et al. [130] and Kim et al. [131] presented a serpentine-laminating micromixer (Fig. 13a, b). These micromixers have a series of F-shaped mixing units that combine the effect of a 3D serpentine structure with a splitting/recombination mechanism. Wang et al. [132, 133] reported a micromixer that combines the effects of a grooved surface and a splitting/recombination process (Fig. 13c, d). This micromixer, called staggered overlapping crisscross micromixer (SOC μ-mixer), consists of an overlapping crisscross entrance and asymmetrical herringbone grooved surface channels. The author demonstrate 46% better mixing indices at the same longitudinal distance when compared with the SHM mixer. The authors claimed that the superior mixing characteristic was due to the induction of vertical tumbling near the intersections of the two crossing channels.

3.1.6 Multiphase Microfluidics and Microdroplet-Based Mixers

Microdroplets can be generated within microfluidic devices using different methods such as electric fields [134], micro-injectors [135] and needles [136]. However, the most widely used methods for droplet generation rely on flow instabilities between immiscible fluids that lead to the so-called multiphase flow. Any fluid flow consisting of more than one phase or component (e.g., emulsions and foams) are examples of multiphase fluids. Traditional emulsification methods are based on the agitation of immiscible fluids and result in the formation of a polydisperse collection of droplets. By

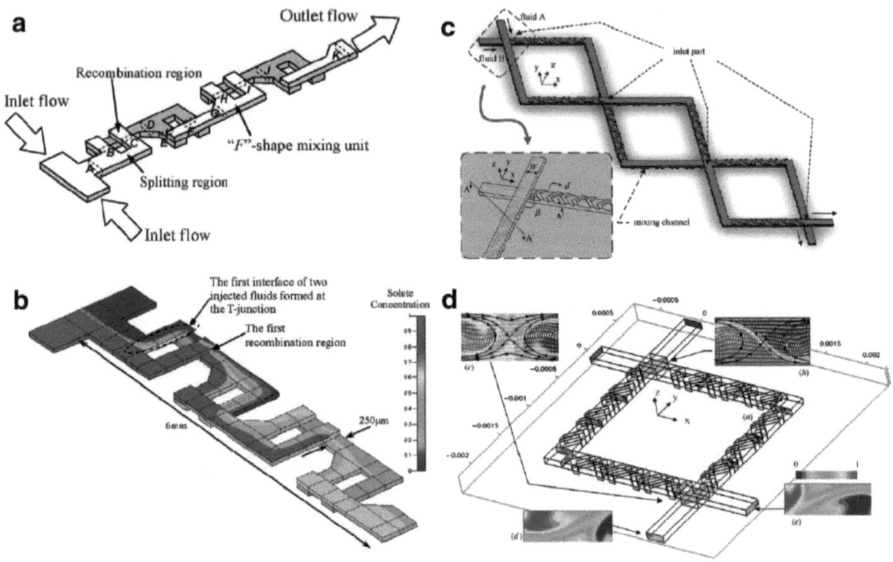

Fig. 13 Micromixer combining SAR and chaotic advection approaches: (a) Serpentine laminating micromixer (SLM) and (b) concentration contours along the mixers channels, (Reproduced from [131] by permission of The Royal Society of Chemistry). (c) Staggered overlapping crisscross micromixer (SOC μ-mixer) and (d) corresponding cross-section view showing concentration profiles after flowing through two junctions (Adapted from [132] with permission. Copyright IOP Publishing)

contrast, methods based on the use of microfluidic devices have been shown to produce highly monodisperse emulsions with a small size variation (e.g., $<1\%$) [53, 137–140].

In microfluidic devices, multiphase flows are created when two (or more) immiscible fluids come into contact. Depending on the interaction between the interfacial and viscous forces, the resulting multiphase flow can take different forms, such as suspended droplets, slugs (droplets occupying the whole channel) or stratified flow (parallel) [141, 142]. In addition to the forces exerted between the two liquids, the channel geometry and physical characteristics also play an important role in the process [143]. In this respect, the use of hydrophobic channels is suitable for the formation of water-in-oil emulsions, whereas hydrophilic channels favor the creation of oil-in-water emulsions [144].

Capillary number (*Ca*) is a parameter that expresses the competition between viscous and interfacial forces, and is generally used to describe multiphase flow behavior in micro- and nanochannels. *Ca* is defined as the ratio between the viscous and interfacial force:

$$Ca = \frac{\mu u}{\gamma}, \tag{15}$$

where μ is the viscosity of the continuous phase, u is the average flow velocity, and γ is the interfacial tension between the two fluid phases. The viscous forces mainly act tangentially to the fluid interface causing the elongation of it, whereas the interfacial force acts preferentially normal to the interface and leads to droplet or slug formation [145]. Generally, when viscous forces dominate interfacial forces, a stratified flow occurs, whereas capillary instability leads to the formation of segmented flow when the interfacial forces dominate.

The basic channel configurations used to generate multiphase flows include the T-junction and flow focusing configurations (Fig. 14a, b). In the T-junction configuration, the channel transporting the dispersed phase intersects perpendicularly with the continuous phase channel, and an emerging droplet is formed at the intersection of the two channels. The effect of the viscous force generated by the continuous phase flow, and the pressure gradient generated upstream of the junction, causes the narrowing of the neck and merging of the droplet, which eventually breaks, leaving the liquid plug (or droplet) flowing downstream. By varying the viscosity of the two phases, the relative flow rates, or the channel dimension it is possible to tune the dimensions of the produced microdroplets [146, 148, 149].

In flow focusing configuration (Fig. 14c, d), the dispersed phase flows in the middle channel, whereas the continuous phase flows in two lateral channels [147,

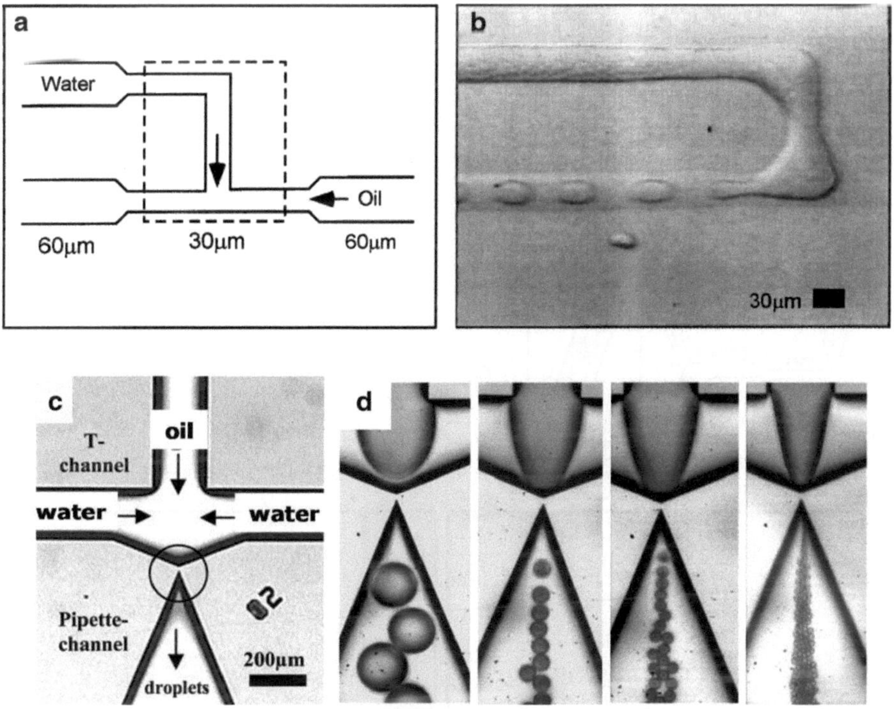

Fig. 14 Mechanism of droplet formation by flow instabilities between immiscible fluids: (**a**) T-junction droplet generator and (**b**) photomicrograph of water-in-oil emulsion formation (Reprinted from [146] with permission. Copyright 2001 by the American Physical Society). (**c**) Flow focusing configuration, and (**d**) formation of the water-in-oil droplets (Reproduced from [147] by permission of The Royal Society of Chemistry)

150]. The two phases are forced through a narrow region (orifice) located downstream of the three channels. The effects of pressure and shear stress exerted on the inner fluid cause the formation of a thin neck that eventually collapses, leading to the formation of a droplet. In this design, the flow rates of the two phases and their viscosity play crucial roles in controlling droplet generation [150].

Generally speaking, mixing inside microdroplets is enhanced by a reduction in diffusion length and by the intimate contact between the fluids to be mixed due to the geometrical confinement of the droplet itself. Furthermore, the contact between the droplet surface and the channel walls causes the generation of recirculating flow within the droplet fluid [137] (Fig. 15). When the droplet is transported through a straight channel, these flows are generated in the two halves of the channel. Each flow pattern consists of two counter-circulating flows. This flow pattern provides a mixing of the two halves; however, mass transport is not

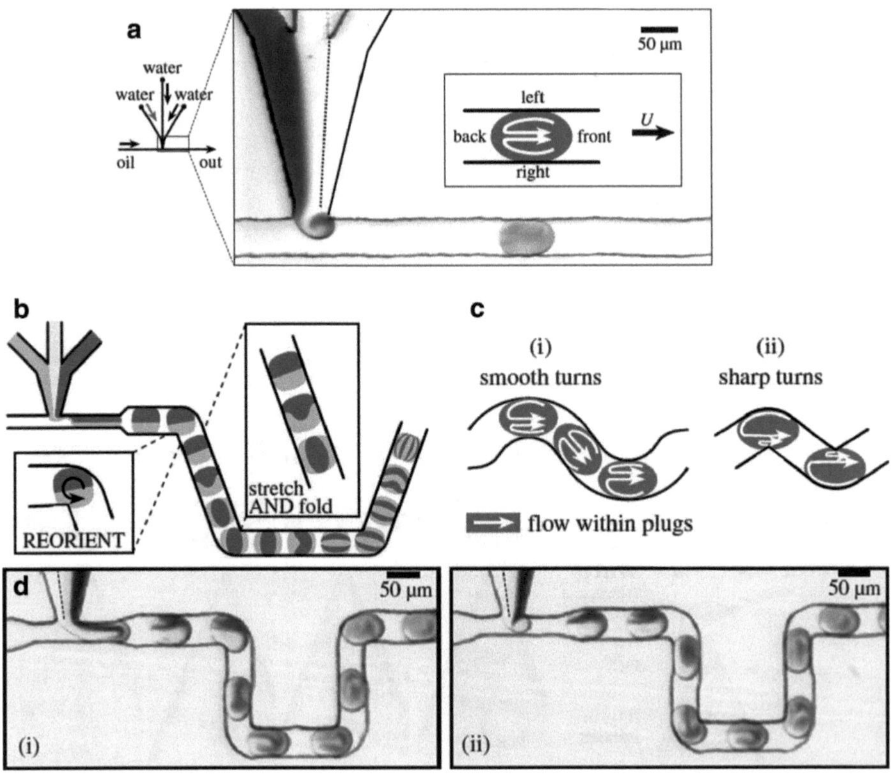

Fig. 15 Mixing in microdroplets flowing in a microchannel: (**a**) Recirculation flow generated in a straight channel. (**b**, **c**) Mixing patterns generated in winding channels that causes (**b**) stretch, fold, and reorientation of the fluids interface and (**c**) asymmetrical recirculation patterns in the droplet halves. (**d**) Experimental results showing the chaotic advection thus generated in microdroplets (Adapted from [151] with permission)

activated between the two halves that thus remain separated and unmixed [137] (Fig. 15a). In order to create chaotic advection within the whole volume of the microdroplets, the channel geometry must be varied in order to stretch and fold the liquid in the droplets [53].

A classic passive way to introduce chaotic advection relies on the introduction of turns and bends in the channel in order to introduce unequal recirculating flow in each half of the droplet. This type of micromixer is known as a planar serpentine micromixer (PSM). When a droplet is driven through a winding channel, each half is in contact with a different section of the turn. The half that is exposed to the inner arc is in contact with a shorter section, while the other is in contact with the larger section of the outer arc. Within the half exposed to the inner part, a smaller recirculating flow is generated compared with the other half. This asymmetrical distribution of the recirculating flows in combination with the alternate switching of them therefore causes chaos and crossing of fluid streams (Fig. 15c) [53, 139, 151, 152]. Furthermore, the turn in the winding channel causes the interface between the two halves of the plug to be reoriented from the direction of plug movement, leading to an exponentially thinner striation between the two fluids to be mixed (Fig. 15b). It must be noted that the extent of mixing can be controlled by controlling the number of turn in the microchannel.

Interestingly, has been reported that a combination of very high relative viscosity of disperse and continuous phases, together with the use of surface active molecules, caused a combination of slug flow and fine droplet dispersion, providing efficient mixing and increasing the interfacial area between the two phases [153]. This mechanism was particularly useful in enhancing the reaction rate of interfacial reactions, such as lipase-catalyzed acetyl isoamyl acetate synthesis [153].

Other channel geometries to induce chaotic advection in microdroplets have also been developed. Liau et al. [154] proposed the introduction of bumps on one side of the channel wall to promote droplet deformation. The authors proposed that the enhancement of mixing could be addressed to the thinning of the lubricant layer of dispersant fluid and by the interfacial stress induced by the bumps. A similar approach was presented by Liau et al. [154]. However, in this case the bumps were introduced in both the lateral channels walls. Similarly, Tung et al. [155] proposed the introduction of a nonuniform cross-section of the wall to deform the microdroplets and enhance the mixing.

3.2 *Active Micromixers*

As described previously, active micromixers rely on an external energy input to introduce perturbation within the fluid streamlines to achieve mixing. Therefore, they are categorized with respect to the type of external perturbation energy:

1. Pressure field
2. Elecrokinetic

3. Dielectrophoretic
4. Electrowetting
5. Magneto-hydrodynamic
6. Ultrasound

3.2.1 Pressure Field Disturbance

One of the simple ways to achieve active mixing is to induce a pressure field disturbance. Active micromixers relying on this strategy have been reported from different authors [43, 156–159]. Deshmuck et al. [156, 157] proposed a T-junction microfluidic chip with an integrated micropump that alternatively drives and stops the flow within the microdevice to create a segmented flow.

A similar approach was presented by Glasgow et al. [43] (Fig. 16) that proposed the use of a pulsing velocity fluid altering periodically the flow rate in the inlet channel from high to low. The author used a simple T-shaped mixer to demonstrate the effectiveness of this method at very low *Re* (from 0.30 to 2.55). They also demonstrated that when both inlets were pulsed simultaneously the interface between the two liquid was stretched through the confluence zone, leading to an enhanced mixing. The authors also showed the influence of the amount and periodicity of the pulsing on mixing efficiency, reporting that the best results were obtained when the pulsing had a phase difference of 180°. Lei et al. [160] reported a microfluidic mixer based on the same concept of fluid discretization and operated by two vortex micropumps. The discretized fluid, constituted of discrete volumes of liquids to be mixed, is then pumped into an expansion chamber to increase the interfacial surface area between the volumes. The flow in the micromixer had an *Re* of 30.

3.2.2 Electrokinetic Disturbance

Elecrokinetic instability (EKI) (or disturbance) micromixers take advantage of fluctuating electric field to induce mixing in microfluidic channels or chambers [161–163]. The fluctuating electric fields cause rapid stretching and folding of the fluids interfaces that are able to stir the fluid stream in highly laminar flow ($Re < 1$) [161]. Different mixing strategies that implement EKI have been presented. Oddy et al. [161] reported a pressure-driven micromixer (i.e., connected with syringe pump) in which oscillating electroosmotic flows were induced by an alternating current voltage (Fig. 17). A periodically alternated flow approach was also presented for an electroosmotic-driven flow device by using either a nonuniform zeta potential along the walls [164] or by varying the voltage with time [165, 166].

Recently, it has also been reported that EKI mixing effectiveness can be enhanced by combining its action with a passive micromixing strategy using channel geometries

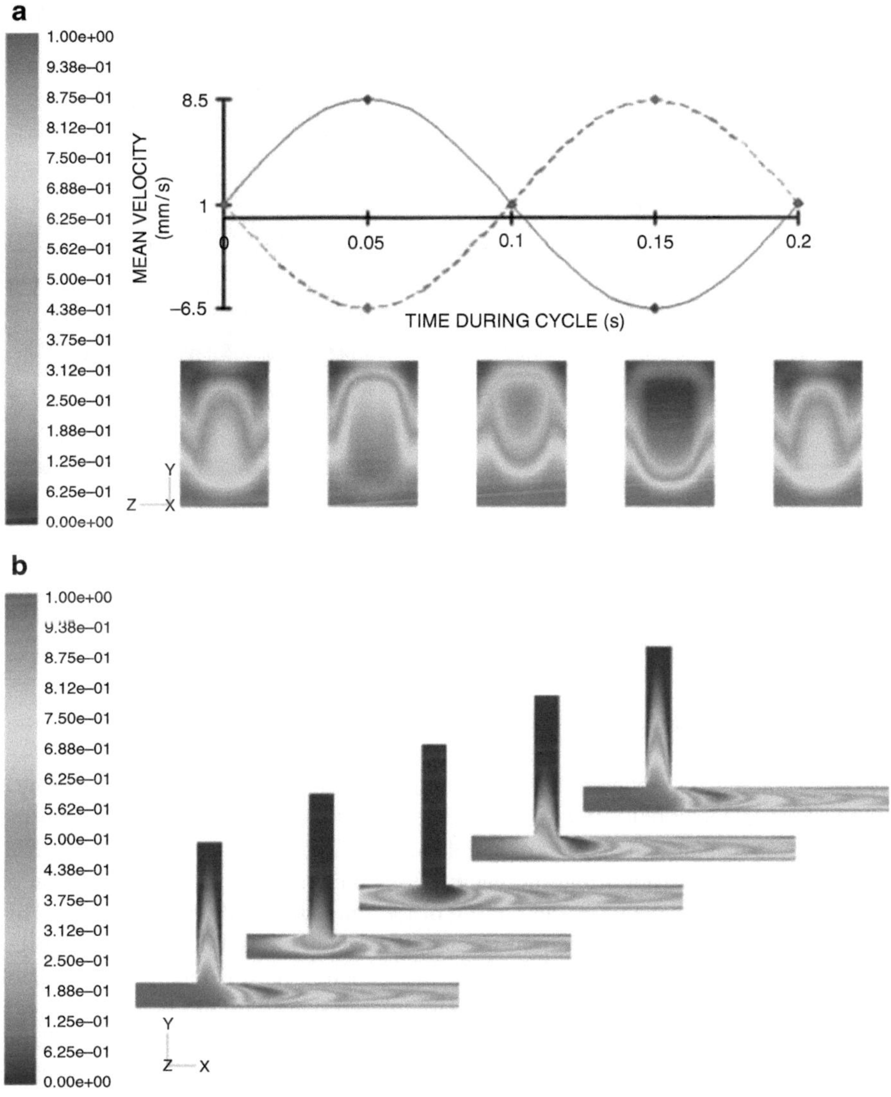

Fig. 16 Mixing by pressure field disturbance: (**a**) Mean fluid velocity along the channel as function of time for in-line inlet and perpendicular inlet. The fluids are pulsed with a simulated 180° phase difference. Contour levels of mass fraction of the fluids in the *Y–Z* plane are shown. (**b**) Contour levels in the *Y–X* plane as a function of time as expressed in the graph in (**a**). Alternate puffs of fluids are created as result of the pulsation introduced within the fluid stream (Reproduced from [43] by permission of The Royal Society of Chemistry)

that induce secondary flow [167]. Results show that for a 10-mm long T-type mixer combining active and passive mixing strategies, the mixing efficiency can be enhanced from 50% to 90% with respect to the solely active mixing strategy.

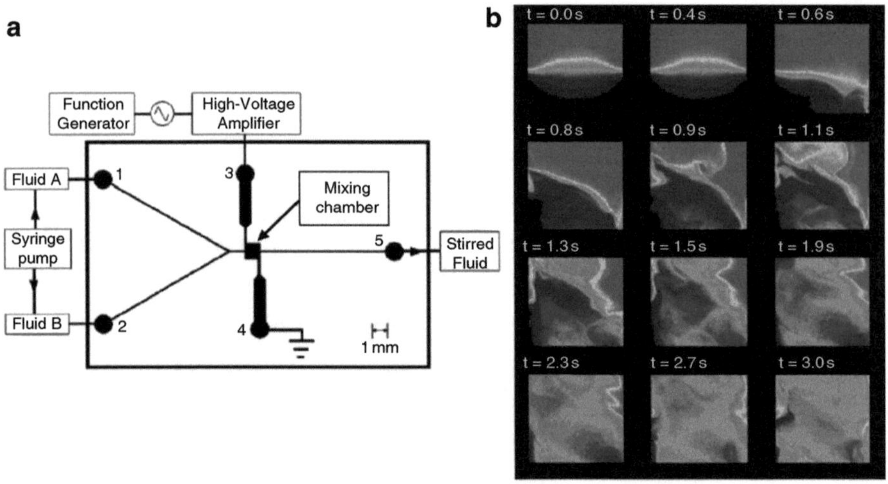

Fig. 17 Electrokinetic instability micromixer. (**a**) EKI micromixer. Fluids are pumped from inlets *1* and *2*, and flow toward outlet *5*, passing through the square mixing chamber. Side ports *3* and *4* allow for AC excitation. The mixing effect is confined within the mixing chamber. (**b**) Complex fluid motion generated within the mixing chamber after the application of the AC field causes rapid stretching and folding of the fluid interface, thus enhancing the mixing (Adapted from [161] with permission. Copyright 2001 American Chemical Society)

3.2.3 Dielectrophoretic Disturbance

Dielectrophoresis is a phenomenon in which polarization of particle is induced by a nonuniform electric field. Polarized particles can move towards or away from the electrodes in response to the electrical field applied. A synergistic effect between the movement of the particles and the geometry of the channel causes the creation of chaotic advection that causes the mixing of the fluid surrounding the particle. This approach was explored by different research groups in the last decade [168–170].

Recently, a similar approach based on isotachophoresis was reported in a microfluidic device, demonstrating its usefulness for micromixing purposes [170]. The author demonstrated that a small sample volume could be brought in contact in a controllable manner to trigger a fairly fast mixing. This type of mixer does not require complicated geometry and could be particular useful in the field of digital microfluidics.

3.2.4 Electrowetting Shaking

As described in the passive mixer section, movement of liquid droplets can generate flow patterns within the fluid and enhance the mixing of species inside the droplets. An active way to induce mixing in droplets is represented by electrowetting on dielectrics (EWOD), or simply electrowetting. EWOD relies on the control of the interfacial tension of a droplet by means of an electric field. Droplets containing different species can be electrically actuated to coalesce using electrowetting effect. After the coalescence, diffusion begins in the droplet and mixing of the two fluid species occurs. However, this passive-like mixing is rather slow [171]. To speed up the mixing process, different authors [171–173] have proposed the use of electrowetting to shake, split, and merge the droplets in order to create recirculating patterns that increase the interfacial area between the two liquids to be mixed. The droplets act as virtual mixing chambers, and mixing occurs by oscillating the droplet across a number of electrodes at various frequencies. The authors demonstrated an increase in mixing as the number of electrodes and transport velocity of the droplet increase [171]. Furthermore, it must be noted that EWOD can achieve mixing in a much more confined space than channel-based mixing.

3.2.5 Magneto-Hydrodynamic Disturbance

Magneto-hydrodynamic (MHD) disturbance relies on the induction of Lorentz body forces in an electrolyte solution [46, 174, 175]. MHD devices utilize an array of electrodes deposited in the channel walls to create current flows within the fluid to be mixed, in the presence of an alternate potential difference on the electrode pair. By coupling the generated electric field with a magnetic field, Lorentz body force could be generated. The complex flow field generated deformations and stretched the material interface, enhancing the mixing (Fig. 18).

3.2.6 Ultrasound Disturbance

Mixing can be achieved by means of acoustic stirring created by ultrasonic waves [42, 44, 176–181]. Ultrasounds are introduced into the channel by integrated piezoelectric ceramic transducers [42, 44, 176, 177]. The ultrasonic action causes an acoustic stirring of the fluid perpendicular to the flow direction and leads to an enhancement of the mixing inside the microfluidic channel [42] or chamber [44, 180]. A turbulent-like mixing was achieved at $Re < 1$.

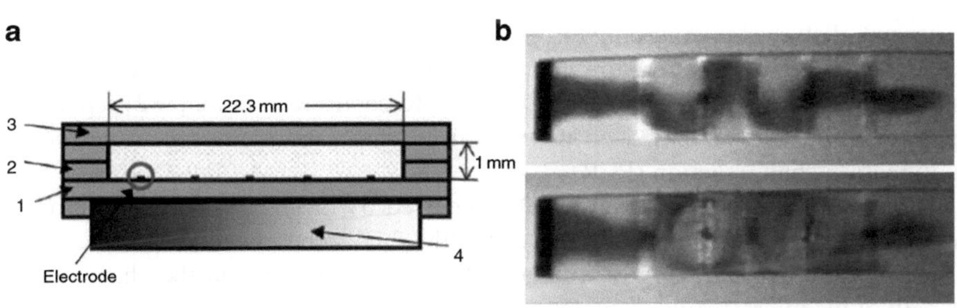

Fig. 18 Magneto-hydrodynamic disturbance (MHD) micromixer. (**a**) Cross-section view of an MHD mixer. MHD mixer comprises the following layers: (*1*) channel bottom wall containing the electrodes; (*2*) spacer layers that constitute the mixing chamber; (*3*) cover plates; (*4*) permanent magnet. (**b**) Deformation of fluid stream resulting from the application of a Lorentz body force (*upper panel*) and corresponding creation of eddies (*lower panel*) (Reprinted from [46] with permission. Copyright 2001 Elsevier)

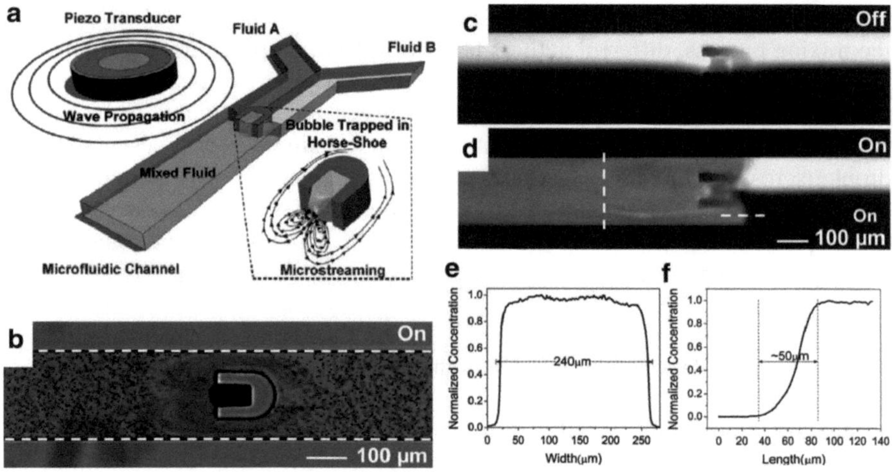

Fig. 19 Mixing by acoustic microstreaming. (**a**) Micromixer. An air bubble is trapped within the horseshoe structure and when activated by the piezo transducer generates a microstreaming around it, as shown in the *inset*. (**b**) Photomicrograph showing microstreaming. (**c**) No mixing effect was observed in the absence of acoustic waves. (**d**) Fast mixing was achieved in the presence of acoustic waves. Concentration plots (**e**) across the channel width and (**f**) near the bubble (Reproduced from [181] by permission of The Royal Society of Chemistry)

Air bubbles can be introduced in a microfluidic mixer in order to enhance the mixing process. The surface of an air bubble in a liquid medium exposed to a sound field can act as a vibrating membrane. The membrane vibration causes a bulk fluid movement at the air–liquid interface. This effect, known as cavitation microstreaming, has been applied in microfluidic micromixers using a single bubble [181] (Fig. 19) or an array of bubbles [178, 179].

3.2.7 Other Types of Active Micromixers

Thermal disturbance micromixers rely either on the increase in the diffusion coefficient due to a temperature increase, or on natural convection to enhance the mixing in the microfluidic channel [45, 182]. Recently Kim et al. [182] reported a pumpless micromixer based on thermal disturbance and applied it to perform a polymerase chain reaction (PCR).

Stirring in microdevices can also be achieved by means of microsized moving elements that create a turbulent flow within the fluid to be mixed [183–185]. These devices allow rapid mixing of a large volume of sample within the characteristic length of the microstirring elements and can be used with a wide range of fluids. Lu et al. [183] reported the use of a rotating bar made of ferromagnetic material driven by means of an external rotating magnetic field. A similar approach was developed by Huh et al. [185] and applied for sensing infectious viral disease. The system proposed used a rotating micromagnetic disk driven by a commercial rotating magnet.

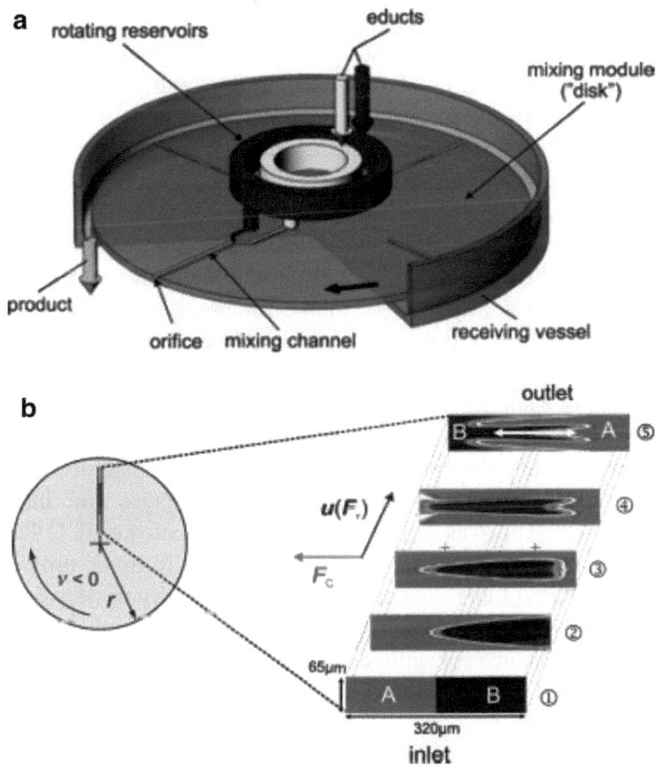

Fig. 20 Centrifugal micromixer. (**a**) Micromixer device and its components. (**b**) Simulation of the mixing process shows the effect of the Coriolis pseudoforce (F_c) in folding the interface of the two liquids. Simultaneously, the centrifugal force (F_v) drives the liquid toward the outlet (Reproduced from [186] with permission. Copyright Wiley-VCH)

Another novel micromixer was presented by Haeberle et al. [186]. It consisted of a modular centrifugal micromixer (Fig. 20) relying on the force generated by the rotating drive to provide both pumping and mixing energies. The mixing is due to the Coriolis force, which causes the interface between the two liquids to be mixed to folds and therefore increases the contact surface between them. The micromixer consists of a planar rotating disk hosting the mixing channel, a rotating unit (standard laboratory centrifuge), and a contact-free reservoir. This mixing system has a high volume throughput on the order of milliliters per minute.

4 Why Microfluidic Mixers?

The intense research over the last two decades on developing microfluidic-based mixers was driven by the possibility to leverage a number of advantages that stem from the unique fluid behavior in a microfluidic environment. At microscale level, fluid properties become increasingly controlled and provide the possibility of exerting control over the different processes carried out in a microfluidic environment.

First, the degree of mixing of reactants could greatly influence the product composition for very fast reactions, as demonstrated for both macro- [187, 188] and microreactors [74, 77]. In this respect, micromixers are appealing tools since they can provide a fast and controllable mixing process as a result of their small dimension and the omnipresence of a highly predictable laminar flow. These properties make micromixers particularly applicable for reactions with very fast kinetics or for dealing with unstable intermediate substances [56, 189]. The fast mixing process is also a valuable advantage for precipitation/crystallization processes such as production of colloidal systems or nanoparticles [190]. Fast mixing provided by micromixers can also be exploited to obtain a freeze-quenching process that allows trapping of metastable intermediates during fast chemical or biochemical reactions [117].

Microfluidic systems also provide the possibility to spatially and temporally monitor and control reactions by adding reagents at precise time intervals during the reaction progress. This was demonstrated by Shestopalov et al. [191] by carrying out a multistep synthesis of quantum dots using the microfluidic droplet reactor.

The small internal volume also provides an opportunity to decrease the amount of sample required for the analysis or reaction. This is particularly useful when rare and valuable substances or samples are used, as well as when a large number of samples in a limited small volume are available for biological and chemical analysis and screening [56].

Typically, microfluidic devices have channels with dimensions between 10 and 400 μm. This small dimension compared to conventional mixing systems results in an increase in the surface-to-volume ratio, to 10,000–50,000 $m^2\ m^{-3}$ compared with 100–2,000 $m^2\ m^{-3}$ of their macroscale counterparts. This characteristic endows micromixers with a superior heat transfer and control [192], allowing reactions to be performed in an isothermal manner [83, 193, 194]. Due to their heat transfer efficiency, micromixers can be conveniently applied for handling reactions in which fast heating and cooling of the reaction mixture are required, such as highly exothermal reactions [9, 195–199].

Recently, the development of microfluidic systems that are able to controllably generate liquid droplets has defined the possibility to perform a diverse range of chemical and biological processes, including the synthesis of biomolecules, drug delivery systems, and diagnostic testing [149, 200]. Microfluidic devices are able to generate highly monodisperse droplets [201–203] in a parallelized fashion, in line with industrial applications, and avoid cross contamination between the droplets by separating the droplets with an immiscible fluid or gas [201, 203]. Digital or droplet microfluidics is particular interesting because it allows the handling of each droplet as a singular microreactor that can be individually controlled and analyzed in an high-throughput fashion [204, 205]. The confinement of the substances in discrete droplets also allows the elimination of residence time variation due to the parabolic flow profile that frustrates single-phase microfluidic mixers [206].

In general, the small internal volume of the micromixers is beneficial in a safety-related point of view when handling hazardous substances and chemical reactions

[198]. The possibility of generation on-demand and in-situ of hazardous and toxic substances (e.g., H_2O_2) represents an important advantage [207].

Lastly, it must be noted that another important driving force in developing microfluidic based mixers and reactors is represented by their possible application as elements of more complicated and multifunctional μTAS.

5 Conclusions

Research on microfluidics and μTAS has been progressing rapidly in the last two decades. Micromixers represent one of the essential components in integrated microfluidic systems for chemical, biological, and medical applications. This chapter describes the characteristics of the microfluidic environment and the peculiar fluid behavior at the microscale. Among the different features of fluid at the microscale, one of the most relevant to mixing applications is the omnipresence of laminar flow where mixing can be dominantly accomplished by molecular diffusion. Nevertheless, this apparent disadvantage coupled with the reduced dimension of microfluidic devices has been leveraged to provide faster and controllable mixing. Design and characterization of various microfluidic mixers have been reported, and their operation conditions and implications for mixing at microscale have been discussed.

The mixing principles applied can be categorized into two groups, active and passive. Passive micromixers rely solely on pumping energy to manipulate the fluid interface and enhance mixing. Passive micromixers are then divided into different categories that reflect the way in which fluid streams are manipulated in order to increase mixing. Specifically, T- or Y-shaped micromixers, parallel lamination micromixers, focusing enhanced mixers, sequential lamination micromixers, chaotic advection micromixers, and droplet micromixers have been discussed.

Active micromixers use an external energy source to introduce perturbation in the fluid stream and are then categorized with respect to the type of energy source used. Pressure field, elecrokinetic, dielectrophoretic, electrowetting shaking, magneto-hydrodynamic, ultrasound, and thermal-assisted micromixers have been presented and discussed.

Finally, the practical advantages of mixing on the microscale have been discussed, showing that microfluidic technology provides a useful approach for solving a number of practical issues for chemical, biological and medical applications. Nevertheless, micromixing technology is still mostly focused on laboratory applications because the possible industrial applications are still frustrated by a lack of understanding of scalability, profitability, and operational flexibility.

Owing to the excellent capability of microfluidic devices and the presence of micromixer technology that has been studied in depth, it can be anticipated that in the near future more practical laboratory and industrial applications for micromixers will be developed.

References

1. Ståhl M, Åslund B, Rasmuson Å (2001) Reaction crystallization kinetics of benzoic acid. AIChE J 47(7):1544–1560
2. Schwarzer H, Schwertfirm F, Manhart M, Schmid H, Peukert W (2006) Predictive simulation of nanoparticle precipitation based on the population balance equation. Chem Eng Sci 61(1): 167–181
3. Mae K, Maki T, Hasegawa I, Eto U, Mizutani Y, Honda N (2004) Development of a new micromixer based on split/recombination for mass production and its application to soap free emulsifier. Chem Eng J 101(1–3):31–38
4. Okubo Y, Toma M, Ueda H, Maki T, Mae K (2004) Microchannel devices for the coalescence of dispersed droplets produced for use in rapid extraction processes. Chem Eng J 101 (1–3):39–48
5. Sprogies T, Köhler J, Gross G (2008) Evaluation of static micromixers for flow-through extraction by emulsification. Chem Eng J 135:S199–S202
6. Iwasaki T, Yoshida J (2005) Free radical polymerization in microreactors. Significant improvement in molecular weight distribution control. Macromolecules 38(4):1159–1163
7. Nagaki A, Kawamura K, Suga S, Ando T, Sawamoto M, Yoshida J (2004) Cation pool-initiated controlled/living polymerization using microsystems. J Am Chem Soc 126(45): 14702–14703
8. Nagaki A, Tomida Y, Yoshida J (2008) Microflow-system-controlled anionic polymerization of styrenes. Macromolecules 41(17):6322–6330
9. Wilms D, Klos J, Frey H (2008) Microstructured reactors for polymer synthesis: a renaissance of continuous flow processes for tailor-made macromolecules? Macromol Chem Phys 209(4):343–356
10. Haswell S, Middleton R, O'Sullivan B, Skelton V, Watts P, Styring P (2001) The application of micro reactors to synthetic chemistry. Chem Commun (5):391–398
11. Hessel V, Hofmann C, Löb P, Löhndorf J, Löwe H, Ziogas A (2005) Aqueous Kolbe- Schmitt synthesis using resorcinol in a microreactor laboratory rig under high-p, T conditions. Org Process Res Dev 9(4):479–489
12. Kee S, Gavriiliidis A (2007) Batch versus continuous mg-scale synthesis of chalcone epoxide with soluble polyethylene glycol poly-L-leucine catalyst. J Mol Catal A Chem 263(1–2):156–162
13. Miller E, Wheeler A (2008) A digital microfluidic approach to homogeneous enzyme assays. Anal Chem 80(5):1614–1619
14. Ukita Y, Asano T, Fujiwara K, Matsui K, Takeo M, Negoro S, Kanie T, Katayama M, Utsumi Y (2008) Application of vertical microreactor stack with polystylene microbeads to immunoassay. Sens Actuators A Phys 145:449–455
15. Bilsel O, Kayatekin C, Wallace L, Matthews C (2005) A microchannel solution mixer for studying microsecond protein folding reactions. Rev Sci Instrum 76:014302
16. Park T, Lee S, Seong G, Choo J, Lee E, Kim Y, Ji W, Hwang S, Gweon D (2005) Highly sensitive signal detection of duplex dye-labelled DNA oligonucleotides in a PDMS microfluidic chip: confocal surface-enhanced Raman spectroscopic study. Lab Chip 5(4):437–442
17. Maerkl S (2009) Integration column: microfluidic high-throughput screening. Integr Biol 1(1):19–29
18. Chen X, Cui D, Liu C, Li H, Chen J (2007) Continuous flow microfluidic device for cell separation, cell lysis and DNA purification. Anal Chim Acta 584(2):237–243
19. Zhang C, Xing D, Li Y (2007) Micropumps, microvalves, and micromixers within PCR microfluidic chips: advances and trends. Biotechnol Adv 25(5):483–514
20. Sato K, Hibara A, Tokeshi M, Hisamoto H, Kitamori T (2003) Microchip-based chemical and biochemical analysis systems. Adv Drug Deliv Rev 55(3):379–391
21. Malic L, Herrmann M, Hoa X, Tabrizian M (2007) Current state of intellectual property in microfluidic nucleic acid analysis. Recent Pat Eng 1(1):71–88

22. Ingham C, Vlieg J (2008) MEMS and the microbe. Lab Chip 8(10):1604–1616
23. Micheletti M, Lye G (2006) Microscale bioprocess optimisation. Curr Opin Biotechnol 17(6):611–618
24. Schäpper D, Alam M, Szita N, Eliasson Lantz A, Gernaey K (2009) Application of microbioreactors in fermentation process development: a review. Anal Bioanal Chem 395(3):679–695
25. Schulte T, Bardell R, Weigl B (2002) Microfluidic technologies in clinical diagnostics. Clin Chim Acta 321(1–2):1–10
26. Rapp B, Gruhl F, Länge K. (2010) Biosensors with label-free detection designed for diagnostic applications. Anal Bioanal Chem 398:2403–2412
27. Zafar Razzacki S, Thwar P, Yang M, Ugaz V, Burns M (2004) Integrated microsystems for controlled drug delivery. Adv Drug Deliv Rev 56(2):185–198
28. Reyes D, Iossifidis D, Auroux P, Manz A (2002) Micro total analysis systems. 1. Introduction, theory, and technology. Anal Chem 74(12):2623–2636
29. Lee S, Lee S (2004) Micro total analysis system (μ-TAS) in biotechnology. Appl Microbiol Biotechnol 64(3):289–299
30. Dittrich P, Tachikawa K, Manz A (2006) Micro total analysis systems. Latest advancements and trends. Anal Chem 78(12):3887–3908
31. Zhang Y, Ozdemir P (2009) Microfluidic DNA amplification – a review. Anal Chim Acta 638(2):115–125
32. Doku G, Verboom W, Reinhoudt D, van den Berg A (2005) On-microchip multiphase chemistry – a review of microreactor design principles and reagent contacting modes. Tetrahedron 61(11):2733–2742
33. Naher S, Orpen D, Brabazon D, Morshed M (2010) An overview of microfluidic mixing application. Adv Mat Res 83:931–939
34. Dittrich P, Manz A (2006) Lab-on-a-chip: microfluidics in drug discovery. Nat Rev Drug Discov 5(3):210–218
35. Yoshida J, Nagaki A, Iwasaki T, Suga S (2005) Enhancement of chemical selectivity by microreactors. Chem Eng Technol 28(3):259–266
36. Baldyga J, Pohorecki R (1995) Turbulent micromixing in chemical reactors – a review. Chem Eng J Biochem Eng J 58(2):183–195
37. Beebe DJ, Mensing GA, Walker GM (2002) Physics and application of microfluidic in biology. Annu Rev Biomed Eng 4:261–286
38. Weigl B, Bardell R, Cabrera C (2003) Lab-on-a-chip for drug development. Adv Drug Deliv Rev 55(3):349–377
39. Nguyen N, Wereley S (2002) Fundamentals and applications of microfluidics. Artech House, Norwood
40. Einstein A, Fürth R (1956) Investigations on the theory of the Brownian movement. Dover Publications, New York
41. Zhang Z, Zhao P, Xiao G, Lin M, Cao X (2008) Focusing-enhanced mixing in microfluidic channels. Biomicrofluidics 2:014101
42. Yaralioglu G, Wygant I, Marentis T, Khuri-Yakub B (2004) Ultrasonic mixing in microfluidic channels using integrated transducers. Anal Chem 76(13):3694–3698
43. Glasgow I, Aubry N (2003) Enhancement of microfluidic mixing using time pulsing. Lab Chip 3(2):114–120. http://dx.doi.org/10.1039/B302569A
44. Yang Z, Matsumoto S, Goto H, Matsumoto M, Maeda R (2001) Ultrasonic micromixer for microfluidic systems. Sens Actuators A Phys 93(3):266–272
45. Tsai H Jr, Lin L (2002) Active microfluidic mixer and gas bubble filter driven by thermal bubble micropump* 1. Sens Actuators A Phys 97:665–671
46. Bau H, Zhong J, Yi M (2001) A minute magneto hydro dynamic (MHD) mixer. Sens Actuators B Chem 79(2–3):207–215
47. Wu Z, Nguyen N (2005) Convective–diffusive transport in parallel lamination micromixers. Microfluid Nanofluid 1(3):208–217
48. Nguyen N, Wu Z (2005) Micromixers – a review. J Micromech Microeng 15:R1

49. Kamholz A, Yager P (2002) Molecular diffusive scaling laws in pressure-driven microfluidic channels: deviation from one-dimensional Einstein approximations. Sens Actuators B Chem 82(1):117–121
50. Schwesinger N, Frank T, Wurmus H (1996) A modular microfluid system with an integrated micromixer. J Micromech Microeng 6:99
51. Knight J, Vishwanath A, Brody J, Austin R (1998) Hydrodynamic focusing on a silicon chip: mixing nanoliters in microseconds. Phys Rev Lett 80(17):3863–3866
52. Günther A, Jhunjhunwala M, Thalmann M, Schmidt M, Jensen K (2005) Micromixing of miscible liquids in segmented gas- liquid flow. Langmuir 21(4):1547–1555
53. Song H, Tice J, Ismagilov R (2003) A microfluidic system for controlling reaction networks in time. Angew Chem 115(7):792–796
54. Johnson T, Ross D, Locascio L (2002) Rapid microfluidic mixing. Anal Chem 74(1):45–51
55. Stroock A, Dertinger S, Ajdari A, Mezic I, Stone H, Whitesides G (2002) Chaotic mixer for microchannels. Science 295(5555):647
56. Hessel V, Löwe H, Schönfeld F (2005) Micromixers – a review on passive and active mixing principles. Chem Eng Sci 60(8–9):2479–2501
57. Jeong G, Chung S, Kim C, Lee S (2010) Applications of micromixing technology. Analyst 135(3):460–473
58. Kamholz A, Weigl B, Finlayson B, Yager P (1999) Quantitative analysis of molecular interaction in a microfluidic channel: the T-sensor. Anal Chem 71(23):5340–5347
59. Gobby D, Angeli P, Gavriilidis A (2001) Mixing characteristics of T-type microfluidic mixers. J Micromech Microeng 11:126
60. Ismagilov R, Stroock A, Kenis P, Whitesides G, Stone H (2000) Experimental and theoretical scaling laws for transverse diffusive broadening in two-phase laminar flows in microchannels. Appl Phys Lett 76:2376
61. Soleymani A, Kolehmainen E, Turunen I (2008) Numerical and experimental investigations of liquid mixing in T-type micromixers. Chem Eng J 135:S219–S228
62. Wong S, Bryant P, Ward M, Wharton C (2003) Investigation of mixing in a cross-shaped micromixer with static mixing elements for reaction kinetics studies. Sens Actuators B Chem 95(1–3):414–424
63. Wong S, Ward M, Wharton C (2004) Micro T-mixer as a rapid mixing micromixer. Sens Actuators B Chem 100(3):359–379
64. Hoffmann M, Raebiger N, Schlueter M, Blazy S, Bothe D, Stemich C, Warnecke A (2003) Experimental and numerical investigations of T-shaped micromixers. In: Proceedings of the 11th European conference on mixing, Bamberg, Germany, 14–17 October 2003, pp 269–276
65. Veenstra T, Lammerink T, Elwenspoek M, Berg A (1999) Characterization method for a new diffusion mixer applicable in micro flow injection analysis systems. J Micromech Microeng 9:199
66. Erbacher C, Bessoth F, Busch M, Verpoorte E, Manz A (1999) Towards integrated continuous-flow chemical reactors. Microchim Acta 131(1):19–24
67. Lob P, Drese K, Hessel V, Hardt S, Hofmann C, Lowe H, Schenk R, Schonfeld F, Werner B (2004) Steering of liquid mixing speed in interdigital micro mixers-from very fast to deliberately slow mixing. Chem Eng Technol 27(3):340–345
68. Cha J, Kim J, Ryu S, Park J, Jeong Y, Park S, Kim H, Chun K (2006) A highly efficient 3D micromixer using soft PDMS bonding. J Micromech Microeng 16:1778. http://dx.doi.org/10.1088/0960-1317/16/9/004
69. Che-Hsin L, Chien-Hsiung T, Lung-Ming F (2005) A rapid three-dimensional vortex micromixer utilizing self-rotation effect under low Reynolds number conditions. J Micromech Microeng 15:935. http://dx.doi.org/10.1088/0960-1317/15/5/006
70. Bessoth F, deMello A, Manz A (1999) Microstructure for efficient continuous flow mixing. Anal Commun 36(6):213–215
71. Floyd T, Schmidt M, Jensen K (2005) Silicon micromixers with infrared detection for studies of liquid-phase reactions. Ind Eng Chem Res 44(8):2351–2358

72. Jackman R, Floyd T, Ghodssi R, Schmidt M, Jensen K (2001) Microfluidic systems with on-line UV detection fabricated in photodefinable epoxy. J Micromech Microeng 11:263
73. Drese K (2004) Optimization of interdigital micromixers via analytical modeling – exemplified with the SuperFocus mixer. Chem Eng J 101(1–3):403–407
74. Ehlers S, Elgeti K, Menzel T, Wießmeier G (2000) Mixing in the offstream of a microchannel system* 1. Chem Eng Process 39(4):291–298
75. Hardt S, Schönfeld F (2003) Laminar mixing in different interdigital micromixers: II. Numerical simulations. AIChE J 49(3):578–584
76. Hessel V, Hardt S, Löwe H, Schönfeld F (2003) Laminar mixing in different interdigital micromixers: I. Experimental characterization. AIChE J 49(3):566–577
77. Ehrfeld W, Golbig K, Hessel V, Lowe H, Richter T (1999) Characterization of mixing in micromixers by a test reaction: single mixing units and mixer arrays. Ind Eng Chem Res 38(3):1075–1082
78. Rosenfeld C, Serra C, Brochon C, Hadziioannou G (2007) High-temperature nitroxide-mediated radical polymerization in a continuous microtube reactor: towards a better control of the polymerization reaction. Chem Eng Sci 62(18–20):5245–5250
79. Rosenfeld C, Serra C, Brochon C, Hessel V, Hadziioannou G (2008) Use of micromixers to control the molecular weight distribution in continuous two-stage nitroxide-mediated copolymerizations. Chem Eng J 135:S242–S246
80. Chung Y, Hsu Y, Jen C, Lu M, Lin Y (2004) Design of passive mixers utilizing microfluidic self-circulation in the mixing chamber. Lab Chip 4(1):70–77
81. Lin C, Tsai C, Fu L (2005) A rapid three-dimensional vortex micromixer utilizing self-rotation effects under low Reynolds number conditions. J Micromech Microeng 15:935
82. Ehrfeld W, Hessel V, Haverkamp V (2000) Microreactors. In: Ullmann's encyclopedia of industrial chemistry. Wiley-VCH, Weinheim. doi: 10.1002/14356007.b16_b37
83. Jähnisch K, Hessel V, Löwe H, Baerns M (2004) Chemistry in microstructured reactors. Angew Chem Int Ed 43(4):406–446
84. Lee S, Kim D, Lee S, Kwon T (2006) A split and recombination micromixer fabricated in a PDMS three-dimensional structure. J Micromech Microeng 16:1067. http://dx.doi.org/10.1088/0960-1317/16/5/027
85. Fang W, Yang J (2009) A novel microreactor with 3D rotating flow to boost fluid reaction and mixing of viscous fluids. Sens Actuators B Chem 140(2):629–642
86. Branebjerg J, Gravesen P, Krog J, Nielsen C (2002) Fast mixing by lamination. In: Proceedings IEEE 9th annual international workshop on micro electro mechanical systems (MEMS'96), San Diego, CA, 11–15 Febuary 2002, pp 441–446
87. Hardt S, Pennemann H, Schönfeld F (2006) Theoretical and experimental characterization of a low-Reynolds number split-and-recombine mixer. Microfluid Nanofluid 2(3):237–248
88. Munson M, Yager P (2004) Simple quantitative optical method for monitoring the extent of mixing applied to a novel microfluidic mixer. Anal Chim Acta 507(1):63–71
89. Radadia A, Cao L, Jeong H, Shannon M, Masel R (2008) A 3D micromixer fabricated with dry film resist. In: Proceedings IEEE 20th international conference on micro mechanical systems (MEMS'07), Hyogo, Japan, 21–25 January 2007, pp 361–364
90. Schönfeld F, Hessel V, Hofmann C (2004) An optimised split-and-recombine micro-mixer with uniform 'chaotic'mixing. Lab Chip 4(1):65–69
91. Bertsch A, Heimgartner S, Cousseau P, Renaud P (2001) Static micromixers based on large-scale industrial mixer geometry. Lab Chip 1(1):56–60. http://dx.doi.org/10.1039/DOI/B103848F.
92. Lim T, Son Y, Jeong Y, Yang D, Kong H, Lee K, Kim D (2010) Three-dimensionally crossing manifold micro-mixer for fast mixing in a short channel length. Lab Chip 11 (1):100–103. http://dx.doi.org/10.1039/DOI/C005325M
93. Melin J, Giménez G, Roxhed N, Wijngaart W, Stemme G (2004) A fast passive and planar liquid sample micromixer. Lab Chip 4(3):214–219
94. He B, Burke B, Zhang X, Zhang R, Regnier F (2001) A picoliter-volume mixer for microfluidic analytical systems. Anal Chem 73(9):1942–1947

95. Sudarsan A, Ugaz V (2006) Multivortex micromixing. Proc Natl Acad Sci USA 103: 7228–7233
96. Vestad T, Marr D, Oakey J (2004) Flow control for capillary-pumped microfluidic systems. J Micromech Microeng 14:1503
97. Stiles T, Fallon R, Vestad T, Oakey J, Marr D, Squier J, Jimenez R (2005) Hydrodynamic focusing for vacuum-pumped microfluidics. Microfluid Nanofluid 1(3):280–283
98. Lee G, Chang C, Huang S, Yang R (2006) The hydrodynamic focusing effect inside rectangular microchannels. J Micromech Microeng 16:1024
99. Lee G, Hung C, Ke B, Huang G, Hwei B, Lai H (2001) Hydrodynamic focusing for a micromachined flow cytometer. Trans ASME J Fluid Eng 123(3):672–679
100. Wu Z, Nguyen N (2005) Hydrodynamic focusing in microchannels under consideration of diffusive dispersion: theories and experiments. Sens Actuators B Chem 107(2):965–974
101. Wu Z, Nguyen N (2005) Rapid mixing using two-phase hydraulic focusing in microchannels. Biomed Microdevices 7(1):13–20
102. Park H, Qiu X, Rhoades E, Korlach J, Kwok L, Zipfel W, Webb W, Pollack L (2006) Achieving uniform mixing in a microfluidic device: hydrodynamic focusing prior to mixing. Anal Chem 78(13):4465–4473
103. Nguyen N, Huang X (2005) Mixing in microchannels based on hydrodynamic focusing and time-interleaved segmentation: modelling and experiment. Lab Chip 5(11):1320–1326
104. Chang C, Huang Z, Yang R (2007) Three-dimensional hydrodynamic focusing in two-layer polydimethylsiloxane (PDMS) microchannels. J Micromech Microeng 17:1479
105. Simonnet C, Groisman A (2005) Two-dimensional hydrodynamic focusing in a simple microfluidic device. Appl Phys Lett 87:114104
106. Sundararajan N, Pio M, Lee L, Berlin A (2004) Three-dimensional hydrodynamic focusing in polydimethylsiloxane (PDMS) microchannels. J Microelectromech Syst 13(4):559–567
107. Yang R, Feeback D, Wang W (2005) Microfabrication and test of a three-dimensional polymer hydro-focusing unit for flow cytometry applications. Sens Actuators A Phys 118(2):259–267
108. Mao X, Lin S, Dong C, Huang T (2009) Single-layer planar on-chip flow cytometer using microfluidic drifting based three-dimensional (3D) hydrodynamic focusing. Lab Chip 9(11): 1583–1589
109. Mao X, Waldeisen J, Huang T (2007) "Microfluidic drifting" – implementing three-dimensional hydrodynamic focusing with a single-layer planar microfluidic device. Lab Chip 7(10): 1260–1262. http://dx.doi.org/10.1039/B711155J
110. Ottino J (1989) The kinematics of mixing: stretching, chaos, and transport. Cambridge University Press, Cambridge
111. Mengeaud V, Josserand J, Girault H (2002) Mixing processes in a zigzag microchannel: finite element simulations and optical study. Anal Chem 74(16):4279–4286
112. Jiang F, Drese K, Hardt S, Küpper M, Schönfeld F (2004) Helical flows and chaotic mixing in curved micro channels. AIChE J 50(9):2297–2305
113. Schönfeld F, Hardt S (2004) Simulation of helical flows in microchannels. AIChE J 50(4): 771–778
114. Chen H, Meiners J (2004) Topologic mixing on a microfluidic chip. Appl Phys Lett 84:2193
115. Liu R, Stremler M, Sharp K, Olsen M, Santiago J, Adrian R, Aref H, Beebe D (2000) Passive mixing in a three-dimensional serpentine microchannel. J Microelectromech Syst 9(2): 190–197
116. Vijayendran R, Motsegood K, Beebe D, Leckband D (2003) Evaluation of a three-dimensional micromixer in a surface-based biosensor†. Langmuir 19(5):1824–1828
117. Lin Y, Gerfen G, Rousseau D, Yeh S (2003) Ultrafast microfluidic mixer and freeze-quenching device. Anal Chem 75(20):5381–5386
118. Stroock A, Dertinger S, Whitesides G, Ajdari A (2002) Patterning flows using grooved surfaces. Anal Chem 74(20):5306–5312

119. Hassell D, Zimmerman W (2006) Investigation of the convective motion through a staggered herringbone micromixer at low Reynolds number flow. Chem Eng Sci 61(9):2977–2985
120. Wang H, Iovenitti P, Harvey E, Masood S (2002) Optimizing layout of obstacles for enhanced mixing in microchannels. Smart Mater Struct 11:662
121. Bhagat A, Papautsky I (2008) Enhancing particle dispersion in a passive planar micromixer using rectangular obstacles. J Micromech Microeng 18:085005. http://dx.doi.org/10.1088/0960-1317/18/8/085005
122. Bhagat A, Peterson E, Papautsky I (2007) A passive planar micromixer with obstructions for mixing at low Reynolds numbers. J Micromech Microeng 17:1017
123. Yang J, Fang W, Tung K (2008) Fluids mixing in devices with connected-groove channels. Chem Eng Sci 63(7):1871–1881
124. Yang J, Huang K, Tung K, Hu I (2007) A chaotic micromixer modulated by constructive vortex agitation. J Micromech Microeng 17:2084. http://dx.doi.org/10.1088/0960-1317/17/10/021
125. Howell P, Mott D, Fertig S, Kaplan C, Golden J, Oran E, Ligler F (2005) A microfluidic mixer with grooves placed on the top and bottom of the channel. Lab Chip 5(5):524–530
126. Hong C, Choi J, Ahn C (2004) A novel in-plane passive microfluidic mixer with modified Tesla structures. Lab Chip 4(2):109–113
127. Sudarsan A, Ugaz V (2006) Fluid mixing in planar spiral microchannels. Lab Chip 6(1):74–82
128. Park S, Kim J, Park J, Chung S, Chung C, Chang J (2004) Rapid three-dimensional passive rotation micromixer using the breakup process. J Micromech Microeng 14:6
129. Long M, Sprague M, Grimes A, Rich B, Khine M (2009) A simple three-dimensional vortex micromixer. Appl Phys Lett 94:133501
130. Park J, Kim D, Kang T, Kwon T (2008) Improved serpentine laminating micromixer with enhanced local advection. Microfluid Nanofluid 4(6):513–523
131. Kim D, Lee S, Kwon T, Ahn C (2005) A serpentine laminating micromixer combining splitting/recombination and advection. Lab Chip 5(7):739–747. http://dx.doi.org/10.1039/B418314B
132. Wang L, Yang J (2006) An overlapping crisscross micromixer using chaotic mixing principles. J Micromech Microeng 16:2684. http://dx.doi.org/10.1088/0960-1317/16/12/022
133. Wang L, Yang J, Lyu P (2007) An overlapping crisscross micromixer. Chem Eng Sci 62(3):711–720
134. Kim S, Song Y, Skipper P, Han J (2006) Electrohydrodynamic generation and delivery of monodisperse picoliter droplets using a poly (dimethylsiloxane) microchip. Anal Chem 78(23):8011–8019
135. Lorenz R, Edgar J, Jeffries G, Chiu D (2006) Microfluidic and optical systems for the on-demand generation and manipulation of single femtoliter-volume aqueous droplets. Anal Chem 78(18):6433–6439
136. Quevedo E, Steinbacher J, McQuade D (2005) Interfacial polymerization within a simplified microfluidic device: capturing capsules. J Am Chem Soc 127(30):10498–10499
137. Handique K, Burns M (2001) Mathematical modeling of drop mixing in a slit-type microchannel. J Micromech Microeng 11:548
138. Nisisako T, Torii T, Takahashi T, Takizawa Y (2006) Synthesis of monodisperse bicolored janus particles with electrical anisotropy using a microfluidic Co flow system. Adv Mater 18(9):1152–1156
139. Song H, Bringer M, Tice J, Gerdts C, Ismagilov R (2009) Experimental test of scaling of mixing by chaotic advection in droplets moving through microfluidic channels. Appl Phys Lett 83(22):4664–4666
140. Song H, Ismagilov R (2003) Millisecond kinetics on a microfluidic chip using nanoliters of reagents. J Am Chem Soc 125(47):14613–14619
141. Günther A, Jensen K (2006) Multiphase microfluidics: from flow characteristics to chemical and materials synthesis. Lab Chip 6(12):1487–1503

142. Shui L, Eijkel J, van den Berg A (2007) Multiphase flow in micro-and nanochannels. Sens Actuators B Chem 121(1):263–276
143. Sugiura S, Nakajima M, Seki M (2002) Effect of channel structure on microchannel emulsification. Langmuir 18(15):5708–5712
144. Okushima S, Nisisako T, Torii T, Higuchi T (2004) Controlled production of monodisperse double emulsions by two-step droplet breakup in microfluidic devices. Langmuir 20(23): 9905–9908
145. Shui L, Eijkel J, van den Berg A (2007) Multiphase flow in microfluidic systems-control and applications of droplets and interfaces. Adv Colloid Interface Sci 133(1):35–49
146. Thorsen T, Roberts R, Arnold F, Quake S (2001) Dynamic pattern formation in a vesicle-generating microfluidic device. Phys Rev Lett 86(18):4163–4166
147. Yobas L, Martens S, Ong W, Ranganathan N (2006) High-performance flow-focusing geometry for spontaneous generation of monodispersed droplets. Lab Chip 6(8): 1073–1079. http://dx.doi.org/10.1039/B602240E
148. Garstecki P, Fuerstman M, Stone H, Whitesides G (2006) Formation of droplets and bubbles in a microfluidic T-junction – scaling and mechanism of break-up. Lab Chip 6(3):437–446
149. Teh S, Lin R, Hung L, Lee A (2008) Droplet microfluidics. Lab Chip 8(2):198–220
150. Garstecki P, Fuerstman M, Whitesides G (2005) Nonlinear dynamics of a flow-focusing bubble generator: an inverted dripping faucet. Phys Rev Lett 94(23):234502
151. Bringer M, Gerdts C, Song H, Tice J, Ismagilov R (1818) Microfluidic systems for chemical kinetics that rely on chaotic mixing in droplets. Philos Trans R Soc Lond A Math Phys Eng Sci 362:1087
152. Tice J, Lyon A, Ismagilov R (2004) Effects of viscosity on droplet formation and mixing in microfluidic channels. Anal Chim Acta 507(1):73–77
153. Pohar A, Plazl I, Žnidarši-Plazl P (2009) Lipase-catalyzed synthesis of isoamyl acetate in an ionic liquid/n-heptane two-phase system at the microreactor scale. Lab Chip 9(23): 3385–3390
154. Liau A, Karnik R, Majumdar A, Cate J (2005) Mixing crowded biological solutions in milliseconds. Anal Chem 77(23):7618–7625
155. Tung K, Li C, Yang J (2009) Mixing and hydrodynamic analysis of a droplet in a planar serpentine micromixer. Microfluid Nanofluid 7(4):545–557
156. Deshmukh A, Liepmann D, Pisano A (2000) Characterization of a micro-mixing, pumping, and valving system. In: Proceedings 11th international conference on solid-state sensor and actuators (Transducers'01), Munich, Germany, 10–14 June 2001, pp 950–953
157. Deshmukh A, Liepmann D, Pisano A (2001) Continuous micromixer with pulsatile micro-pumps. In: Proceedings IEEE solid-state sensor and actuator workshop, Hilton Head Island, SC, 4–8 June 2000, pp 73–76
158. Fujii T, Sando Y, Higashino K, Fujii Y (2003) A plug and play microfluidic device. Lab Chip 3(3):193–197
159. Lim C, Lam Y, Yang C (2010) Mixing enhancement in microfluidic channel with a constriction under periodic electro-osmotic flow. Biomicrofluidics 4:014101
160. Lei K, Li W (2008) A novel in-plane microfluidic mixer using vortex pumps for fluidic discretization. J Assoc Lab Automation 13(4):227–236
161. Oddy M, Santiago J, Mikkelsen J (2001) Electrokinetic instability micromixing. Anal Chem 73(24):5822–5832
162. Yang R, Wu C, Tseng T, Huang S, Lee G (2005) Enhancement of electrokinetically-driven flow mixing in microchannel with added side channels. Jpn J Appl Phys 1 44(10):7634
163. Posner J, Santiago J (2006) Convective instability of electrokinetic flows in a cross-shaped microchannel. J Fluid Mech 555:1–42
164. Qian S, Bau H (2002) A chaotic electroosmotic stirrer. Anal Chem 74(15):3616–3625
165. Glasgow I, Batton J, Aubry N (2004) Electroosmotic mixing in microchannels. Lab Chip 4(6):558–562

166. Tang Z, Hong S, Djukic D, Modi V, West A, Yardley J, Osgood R (2002) Electrokinetic flow control for composition modulation in a microchannel. J Micromech Microeng 12: 870
167. Yan DG, Yang C, Miao JM, Lam YC, Huang XY (2009) Enhancement of electrokinetically driven microfluidic T-mixer using frequency modulated electric field and channel geometry effects. Electrophoresis 30(18):3144–3152
168. Lee H, Voldman J (2007) Optimizing micromixer design for enhancing dielectrophoretic microconcentrator performance. Anal Chem 79(5):1833–1839
169. Deval J, Tabeling P, Ho C (2002) A dielectrophoretic chaotic mixer. In: Proceedings 15th international conference on micro electro mechanical systems, Las Vegas, NV, 20–24 January 2002, pp 36–39
170. Goet G, Baier T, Hardt S (2009) Micro contactor based on isotachophoretic sample transport. Lab Chip 9(24):3586–3593. doi:10.1039/b914466h
171. Paik P, Pamula V, Pollack M, Fair R (2003) Electrowetting-based droplet mixers for microfluidic systems. Lab Chip 3(1):28–33
172. Paik P, Pamula V, Fair R (2003) Rapid droplet mixers for digital microfluidic systems. Lab Chip 3(4):253–259
173. Fowler J, Moon H, Kim C (2002) Enhancement of mixing by droplet-based microfluidics. In: Proceedings IEEE 15th international conference on micro electro mechanical systems, Las Vegas, NV, 20–24 January 2002, pp 97–100
174. West J, Karamata B, Lillis B, Gleeson J, Alderman J, Collins J, Lane W, Mathewson A, Berney H (2002) Application of magnetohydrodynamic actuation to continuous flow chemistry. Lab Chip 2(4):224–230
175. Oh D, Jin J, Choi J, Kim H, Lee J (2007) A microfluidic chaotic mixer using ferrofluid. J Micromech Microeng 17:2077
176. Yang Z, Goto H, Matsumoto M, Maeda R (2000) Active micromixer for microfluidic systems using lead-zirconate-titanate (PZT)-generated ultrasonic vibration. Electrophoresis 21(1):116–119
177. Woias P, Hauser K, Yacoub-George E (2000) An active silicon micromixer for mTAS applications. In: van den Berg A, Olthuis W, Bergveld P (eds) Proceedings micro total analysis systems symposium (μTAS2000), Enschede, The Netherlands, 14–18 May 2000, pp 277–282
178. Liu R, Yang J, Pindera M, Athavale M, Grodzinski P (2002) Bubble-induced acoustic micromixing. Lab Chip 2(3):151–157
179. Liu R, Lenigk R, Druyor-Sanchez R, Yang J, Grodzinski P (2003) Hybridization enhancement using cavitation microstreaming. Anal Chem 75(8):1911–1917
180. Jang L, Chao S, Holl M, Meldrum D (2007) Resonant mode-hopping micromixing. Sens Actuators A Phys 138(1):179–186
181. Ahmed D, Mao X, Shi J, Juluri B, Huang T (2009) A millisecond micromixer via single-bubble-based acoustic streaming. Lab Chip 9(18):2738–2741. http://dx.doi.org/10.1039/B903687C
182. Kim S, Wang F, Burns M, Kurabayashi K (2009) Temperature-programmed natural convection for micromixing and biochemical reaction in a single microfluidic chamber. Anal Chem 81(11):4510–4516
183. Lu L, Ryu K, Liu C (2002) A magnetic microstirrer and array for microfluidic mixing. J Microelectromech Syst 11(5):462–469
184. Mensing G, Pearce T, Graham M, Beebe D (1818) An externally driven magnetic microstirrer. Philos Trans R Soc Lond A Math Phys Eng Sci 362:1059
185. Huh Y, Park T, Lee E, Hong W, Lee S (2008) Development of a fully integrated microfluidic system for sensing infectious viral disease. Electrophoresis 29(14):2960–2969
186. Haeberle S, Brenner T, Schlosser H, Zengerle R, Ducrée J (2005) Centrifugal micromixery. Chem Eng Technol 28(5):613–616
187. Chakraborty S, Balakotaiah V (2003) A novel approach for describing mixing effects in homogeneous reactors. Chem Eng Sci 58(3–6):1053–1061

188. Baldyga J, Bourne J, Hearn S (1997) Interaction between chemical reactions and mixing on various scales. Chem Eng Sci 52(4):457–466
189. Hessel V, Hardt S, Löwe H (2004) Chemical micro process engineering: fundamentals, modelling and reactions. Wiley-Vch, Weinheim
190. Karnik R, Gu F, Basto P, Cannizzaro C, Dean L, Kyei-Manu W, Langer R, Farokhzad O (2008) Microfluidic platform for controlled synthesis of polymeric nanoparticles. Nano Lett 8(9):2906–2912
191. Shestopalov I, Tice J, Ismagilov R (2004) Multi-step synthesis of nanoparticles performed on millisecond time scale in a microfluidic droplet-based system. Lab Chip 4(4):316–321
192. Chow A (2002) Lab on a chip: opportunities for chemical engineering. AIChE J 48(8): 1590–1595
193. Jensen K (2001) Microreaction engineering – is small better? Chem Eng Sci 56(2):293–303
194. Mason B, Price K, Steinbacher J, Bogdan A, McQuade D (2007) Greener approaches to organic synthesis using microreactor technology. Chem Rev 107(6):2300–2318
195. Aoki N, Hasebe S, Mae K (2004) Mixing in microreactors: effectiveness of lamination segments as a form of feed on product distribution for multiple reactions. Chem Eng J 101 (1–3):323–331
196. Chambers R, Fox M, Sandford G (2005) Elemental fluorinePart 18. Selective direct fluorination of 1, 3-ketoesters and 1, 3-diketones using gas/liquid microreactor technology. Lab Chip 5(10):1132–1139
197. Demello A (2006) Control and detection of chemical reactions in microfluidic systems. Nature 442(7101):394–402
198. Kestenbaum H, Lange de Oliveira A, Schmidt W, Schüth F, Ehrfeld W, Gebauer K, Löwe H, Richter T (2000) Synthesis of ethylene oxide in a catalytic microreactor system. Stud Surf Sci Catal 130:2741–2746
199. Surangalikar H, Ouyang X, Besser R (2003) Experimental study of hydrocarbon hydrogenation and dehydrogenation reactions in silicon microfabricated reactors of two different geometries. Chem Eng J 93(3):217–224
200. Huebner A, Sharma S, Srisa-Art M, Hollfelder F, Edel J, demello A (2008) Microdroplets: a sea of applications? Lab Chip 8(8):1244–1254
201. Li W, Young E, Seo M, Nie Z, Garstecki P, Simmons C, Kumacheva E (2008) Simultaneous generation of droplets with different dimensions in parallel integrated microfluidic droplet generators. Soft matter 4(2):258–262
202. Capretto L, Mazzitelli S, Balestra C, Tosi A, Nastruzzi C (2008) Effect of the gelation process on the production of alginate microbeads by microfluidic chip technology. Lab Chip 8(4):617–621
203. Nisisako T, Torii T (2008) Microfluidic large-scale integration on a chip for mass production of monodisperse droplets and particles. Lab Chip 8(2):287–293
204. Fair R (2007) Digital microfluidics: is a true lab-on-a-chip possible? Microfluid Nanofluid 3(3):245–281
205. Link D, Grasland Mongrain E, Duri A, Sarrazin F, Cheng Z, Cristobal G, Marquez M, Weitz D (2006) Electric control of droplets in microfluidic devices. Angew Chem 118(16): 2618–2622
206. Krishnadasan S, Brown R, Demello A, Demello J (2007) Intelligent routes to the controlled synthesis of nanoparticles. Lab Chip 7(11):1434–1441
207. Voloshin Y, Halder R, Lawal A (2007) Kinetics of hydrogen peroxide synthesis by direct combination of H2 and O2 in a microreactor. Catal Today 125(1–2):40–47

Top Curr Chem (2011) 304: 69–90
DOI: 10.1007/128_2011_149

Published online: 20 May 2011

Basic Technologies for Droplet Microfluidics

Shaojiang Zeng, Xin Liu, Hua Xie, and Bingcheng Lin

Abstract In recent years droplet microfluidics has become a quickly evolving research field. The availability of a wide range of technologies for droplet generation and manipulation has enabled the applications of droplet microfluidics in a wide variety of fields, from single cell analysis to material synthesis. In this review we summarize the main technologies for droplet microfluidics and discuss the recent advances in technologies that enable droplets as microreactors with complete functions. Applications of microdroplets in chemical reactions, particle synthesis, and single cell analysis are also briefly reviewed.

Keywords Microdroplet · Microfluidics

Contents

1 Introduction 70
2 Technologies for Droplet Microfluidics 71
  2.1 Droplet Generation 71
  2.2 Droplet Fusion 73
  2.3 Droplet Splitting 74
  2.4 Mixing Inside Droplets 75
  2.5 Droplet Sorting 76
  2.6 Trapping and Incubation of Droplets 78
  2.7 Droplet Content Characterization 79
  2.8 Hyphenation of Droplets to CE and HPLC 80

S. Zeng, H. Xie, and B. Lin (✉)
Dalian Institute of Chemical Physics, Chinese Academy of Sciences, 457 Zhongshan Road, Dalian 116023, China
e-mail: bclin@dicp.ac.cn

X. Liu
Department of Chemistry, University of Cambridge, Lensfield Road, Cambridge, Cambridgeshire CB2 1EW, UK

3 Applications 81
3.1 Chemical Reactions 82
3.2 Microparticle Synthesis 83
3.3 Single Cell Analysis 85
4 Conclusions and Outlook 86
References 87

Abbreviations

PDMS	Poly(dimethyl) siloxane
PCR	Polymerase chain reaction
AC	Alternating current
CE	Capillary electrophoresis
HPLC	High performance liquid chromatography

1 Introduction

Droplet microfluidics deals with generation, manipulation, and applications of droplets with dimensions, typically, in the range of several micrometers to hundreds of micrometers in diameter (femtoliter to nanoliter in volume) in microfluidic devices. Droplet microfluidics has defined a new experimental platform for a diverse range of biological and chemical processes. Essentially, droplets partition experiments into individual tiny compartments, each serving as a discrete microreactor and separated from each other by an immiscible carrier fluid. This concept of compartmentalization is particularly appealing to perform chemistry and biology since reactants are isolated within individual droplets. Since its advent in the early 2000s [1, 2], droplet microfluidics has become an increasingly popular and quickly evolving research field. The rapid proliferation of droplet microfluidics has been driven by (1) the need to understand new fluid dynamics phenomena involved in multiphase microfluidic systems and (2) the growing interest in using droplets as microreactors for applications in chemistry and biology.

Early works demonstrated the use of microchannels with prescribed dimensions and geometries to generate streams of microdroplets (disperse phase) dispersed in an immiscible continuous phase [1, 2]. The sizes and production rates of the droplets were shown to be able to be tuned by simply varying the flow rates of the disperse phase and continuous phase. Later on theoretical interests arose on the characterization and exploration of complex dynamic phenomena involved in droplet generation and trafficking within microfluidic channel networks [3, 4]. Meanwhile, a wealth of technologies for improved droplet generation, intra-droplet content manipulation, and methods for controlled trafficking, all enabled a wide range of appealing applications [5–8] from chemical kinetics [9] and protein crystallization [10, 11] to material synthesis [12, 13], single cell analysis [14, 15], polymerase chain reaction (PCR) [16], protein engineering, and high throughput screening technologies [17].

This review will discuss the recent advances in technologies that enable droplets as microreactors with complete functions and the applications of droplets in chemistry and biology. Discussions will be focused on various technologies for generation, fusion, splitting, and detection of droplets in microfluidic devices. Applications of droplets in chemical reactions, particle synthesis, and single cell analysis will also be briefly reviewed. The readers are also addressed to some recent excellent reviews on various aspects of droplet microfluidics [5–9, 14–17]. Studies from fluid mechanics perspectives, which focus on the understanding of the basic fluid dynamic phenomena involved in droplet or multiphase microfluidics in general, have been summarized in recent reviews [18–21] and are not detailed here. Digital microfluidics [22] that generate and manipulate relatively bigger droplets based on electro-wetting will not be covered in this review.

2 Technologies for Droplet Microfluidics

2.1 Droplet Generation

Conventional methods for making droplets involve manual or mechanical agitation of multiphase fluids. Since there is no way to control the uniformity of the shear or impact stresses involved in droplet breakup, droplets formed in these ways are highly polydisperse in size. In contrast, using microfluidic devices with prescribed channel dimensions and geometries, and by controlling the flow rates of two immiscible liquids, the flow conditions in droplet breakup can be highly repeatable and thus it is feasible to generate microdroplets of uniform size distributions (1–3% dispersity).

In droplet microfluidic devices, T-shaped junction [1, 23] and "flow-focusing" [2, 24] are the most commonly used channel geometries for generating droplets. Droplet formation in a T-junction was first proposed by Thorsen et al. [1] to form water droplets in a set of oil phases. In the T-junction device, the disperse phase and continuous phase flowed out from two perpendicular channels and formed an interface at the junction (Fig. 1a, b). Due to the shear force exerted by the continuous phase, the disperse phase thinned gradually and eventually broke into droplets. Later on, Anna et al. developed a "flow-focusing" device [2], where the disperse and continuous phases flowed in a central channel and two outside channels respectively. The two phases were then impelled to flow through a small orifice located downstream the convergent region and the droplets formed inside or downstream the orifice (Fig. 1c, d). Both types of devices were shown to be able to change the sizes of droplets by varying the relative flow rates of the two immiscible phases. "Flow-focusing" configuration allows generating droplets smaller than the orifice size [25]. This feature will be attractive when small sized droplets are required. A T-junction structured device has a simpler channel

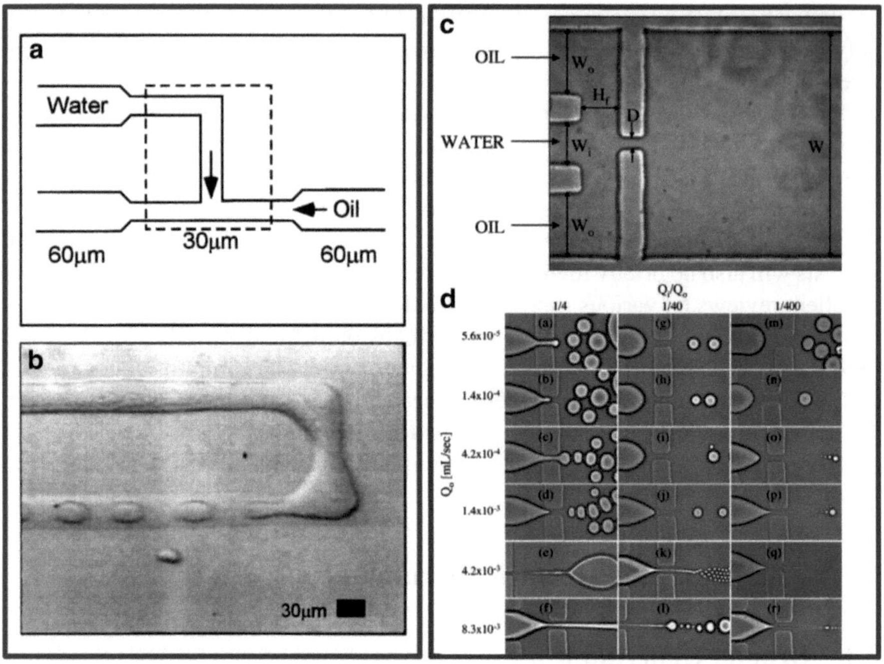

Fig. 1 Droplets generation in microfluidic devices. (**a**) Schematic channel layout of T-junction device. (**b**) Micrograph showing the formation of droplets in a T-junction device. Reproduced with permission from [1]. (**c**) Micrograph of the "flow-focusing" device. (**d**) Droplets generation at different flow conditions in "flow-focusing" device. Reproduced with permission from [2]

structure and is relatively easy to operate. Detailed discussions on the hydrodynamic features of the two devices have been summarized in recent reviews [18–21].

In both types of devices, the surface properties of the microchannels were found to be critically important for droplets generation [26]. The channel surface should preferentially be wetted by the continuous phase. Poly(dimethyl) siloxane (PDMS) is the most widely used material to fabricate microfluidic devices for use in droplet research. PDMS is inherently hydrophobic and can be directly used to generate water-in-oil droplets. Oil-in-water droplets can also be generated using PDMS microfluidic device after hydrophilic patterning of the channel wall. The particularly appealing advantages of PDMS include the low cost of fabrication and rapid and easy prototyping process. However, PDMS will undergo swelling and deformation in the presence of organic solvents, and other materials with higher solvent resistance, such as glass, will be chosen. Glass is hydrophilic and can easily be modified to be hydrophobic, and both water-in-oil and oil-in-water droplets can be produced using glass microfluidic devices. The time-consuming process of device fabrication and relatively high cost, however, limit its wide use in droplet research.

Other methods for generating droplets are also available, for instance, using capillary systems [27], axisymmetric "flow-focusing" structured devices [28, 29]. In these devices the disperse phase is shielded by the continuous phase from touching the channel wall, and the surface problems could be alleviated.

One limitation of the aforementioned technologies is that parameters, such as droplet size or generation rate, cannot be independently controlled. One demands methods [30–35] that can actively control the generation of droplets and various methods have been reported, including electro-wetting [30], centrifugal [31], electric control [32], thermal capillary valving [33], and pneumatic microvalving-based methods [34, 35].

2.2 Droplet Fusion

Controlled fusion of droplets is a fundamental operation unit for the droplets to be used as microreactors. To initiate a reaction in a droplet, two sets of droplets are separately generated and then brought into contact and merged through either active or passive processes. Under the occasions requiring multistep reactions or controlling the reaction kinetics, fusion steps are demanded to concentrate or dilute the reactants in droplets or to add new reagents into droplets [36–39].

Ahn et al. proposed an electrocoalescence device (Fig. 2a) [40] that relied on pair-wise synchronizing droplets by their size-dependent flow velocities in the

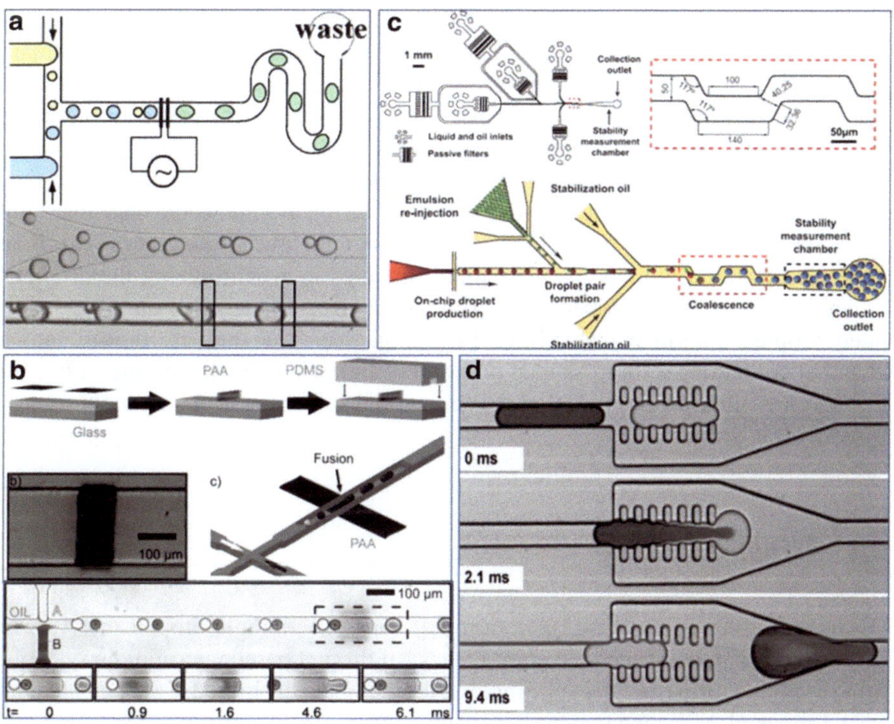

Fig. 2 Droplet fusion devices. (**a**) Fusion based on electrocoalescence. Reproduced with permission from [40]. (**b**) Droplet fusion based on surface energy patterning. Reproduced with permission from [41]. (**c**) Fusion based on changing the concentration of surfactant in continuous phase. Reproduced with permission from [42]. (**d**) Pillars assisted droplet fusion. Reproduced with permission from [43]

microchannels, then using an electric field to fuse the paired droplets. Smaller droplets moved faster due to the Poiseuille flow, allowing pairs of surfactant-stabilized droplets to be brought into contact and then fused under an electric field. Takeuchi and co-workers reported a similar approach for droplet fusion [44]. In their approach, they incorporated a widening chamber in the droplet flow pathway, where the precedent droplet would slow down, providing a chance for a second droplet to catch up. An alternating current (AC) electric field was applied across the chamber to facilitate the droplet fusion. The mechanism of electrocoalescence is still a subject of study. Yet it was observed that the phenomenon is related to electric field induced interface instability [37]. Link et al. proposed another way of using electric force to fuse droplets [32]. In their method, opposite charges were applied to the interfaces of aqueous and oil phases, and two streams of oppositely charged droplets were produced simultaneously. These droplets would immediately coalesce once they came into contact with each other.

Aside from methods that rely on using active (electric) force to trigger droplet coalescence, passive droplet fusion methods based on tuning channel surface energy, geometry, or liquid interfacial tension have also been proposed [41–43, 45, 46]. For instance, Fidalgo et al. proposed a fusion strategy based on patterning the wettability of the microchannel [41]. As shown in Fig. 2b, they fabricated a hydrophilic patch within a hydrophobic PDMS channel. Water-in-oil droplets containing different chemicals were generated with double T-junction upstream patches, flowed downstream, and were trapped and fused at the hydrophilic patch. Mazutix et al. proposed another strategy to fuse pair-wise synchronized droplets [42] (Fig. 2c). They diluted the concentration of surfactant in the continuous phase, hence de-stabilizing the droplets, allowing them to coalesce upon touching each other. One interesting observation in their system was that one big droplet would only fuse with one small droplet, even if multiple small droplets might touch the big ones. This is an appealing feature that may alleviate the difficulties in synchronizing the droplets. Recently, Niu et al. developed a novel fusion device, which relied on pillar facilitated droplets interface instabilities [43] (Fig. 2d). The droplets flowed into a chamber containing a comb-shaped pillar array and, while the first droplet squeezed through the pillar array, it slowed down, and the following droplet caught up and fused with it.

2.3 Droplet Splitting

Droplet splitting is a critical operation that can split a single droplet into two or more droplets, and enhance the effectiveness of droplet based microfluidic systems. Song et al. first proposed the splitting of a droplet in a constricted T-shaped channel [47]. Link et al. [48] then comprehensively investigated the process of splitting of a droplet in a T-shaped channel (Fig. 3a). They found that when a droplet flowed toward the bifurcation section, it was stretched by the predominant elongation flow in that region. If the droplet was large enough initially, it would thin at its center and

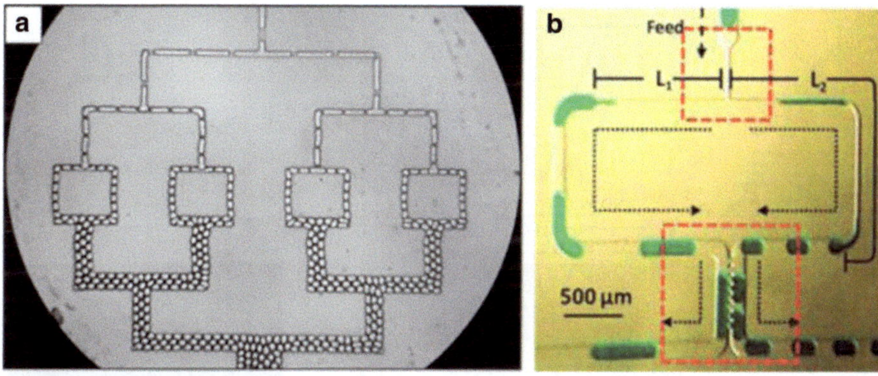

Fig. 3 Droplet splitting devices. (**a**) Sequential droplets splitting. Reproduced with permission from [48]. (**b**) Asymmetrical droplets splitting. Reproduced with permission from [49]

split into two daughter droplets as it passed through the junction. They demonstrated that asymmetrical splitting was also attainable by altering the lengths of the two branch channels.

More active control of the splitting, where the size ratio of the two daughter droplets could be actively controlled, was also demonstrated [49–51]. Nie et al. recently reported using asymmetrical droplet splitting as a method for sampling small amounts of the contents from larger nanoliter plugs [49]. In their device, plugs were pumped into a loop where they split according to the flow resistance of each arm of the loop (Fig. 3b). The splitting ratio depended on both the dimensions of the loop channels and the frequency of plugs entering the loop. Splitting ratios from 1:1 to 34:1 were achieved for samples from 1.7 to 3.3 nL. The system showed promise in performing multiple assays on a single sample, preserving the sample while some was removed for analysis, and splitting reagents or test compounds for use in multiple assays [49].

2.4 *Mixing Inside Droplets*

Rapid mixing of reagents is necessary to determine accurately the starting time of a reaction. The control of mixing is also important for carrying out and studying the kinetics of chemical and biological reactions. In continuous flow microfluidic systems, fluids commonly flow laminarly and mixing is achieved by diffusion, which is commonly a very slow process. Droplet-based microfluidic systems allow rapid mixing simply by implementing special channel configuration [47, 52–55] (Fig. 4a–c). When a droplet moves along a microchannel and touches the channel wall, two recirculation flows are generated inside the droplet due to the shear interaction of the channel wall. In a straight channel, two equal recirculation flows are generated in each half of the droplet that touches the channel wall. Fluids

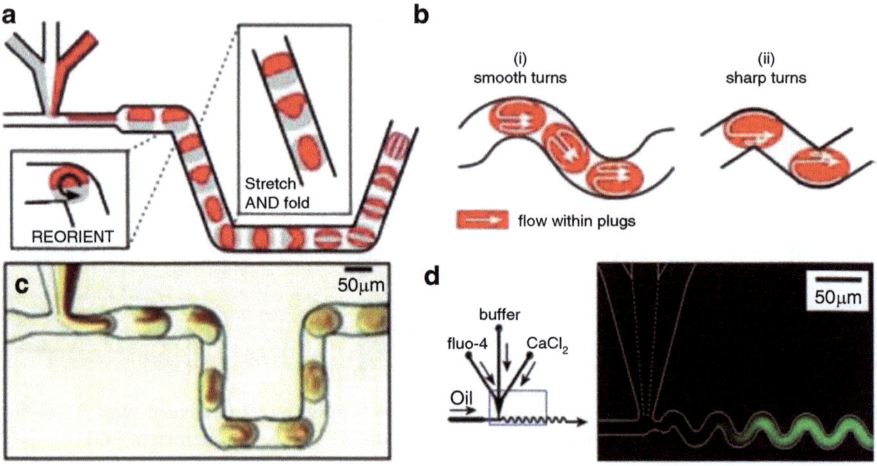

Fig. 4 The process of mixing within a droplet moving through a winding microfluidic channel. (**a**) Schematic of stretching and folding of the recirculation flows inside droplets. (**b**) Schematic of recirculating flows in droplets moving through smooth and sharp turns. (**c**) Pictures showing the flowing patterns inside droplets at different positions in a winding microchannel. Reproduced with permission from [55]. (**d**) Fluorescence images of droplets moving through winding channel. Reproduced with permission from [53]

within each half of the droplet are mixed by advection; however, mixing between the two halves is still dominated by diffusion and is slow. To enhance the internal mixing within droplets, a winding channel is introduced after the droplet generation section. As a droplet traverses through a winding channel, the recirculation flows in the two halves of the droplet are unequal. In the half exposed to the inner arc of the winding channel, a smaller recirculation flow is generated and, in the other half of the droplet, a larger recirculation flow is generated. This irregular motion of droplet contents along the channel repeatedly folds and stretches the two fluids to achieve fluids striations that become exponentially thinner, the process of which is called chaotic advection, and decreases the distance of diffusive mixing. As the diffusive mixing time t_{diff} (s) can be expressed as $t_{\mathrm{diff}} = d^2/2D$, where d (m) is the striation thickness and $D(\mathrm{m}^2\mathrm{s}^{-1})$ is the diffusion coefficient, mixing by diffusion would occur rapidly when the striation thickness is on the sub-micrometer scale. The striation thickness and the extent of mixing are dependent on the number of winding turns that the droplet has experienced. By increasing the flow rates of oil and aqueous phases, complete mixing within droplets on a sub-millisecond time scale has been achieved. This rapid mixing has been used to investigate the fast reaction kinetics of enzyme [53] (Fig. 4d).

2.5 *Droplet Sorting*

Droplet sorting enables separation of specific droplets of interest from a large population of droplets, which is an essential module for high-throughput screening applications. In the process of droplet sorting, a bias is applied to a specific droplet when some criteria are satisfied such that it can be picked out. Various approaches have been proposed to generate the bias for droplet sorting, including electric field, surface acoustic wave, and so on.

Electric field has been the most developed method for use in droplet sorting. Link et al. developed a set of tools for droplet manipulation by electric field [32] and demonstrated the ability to steer actively the direction of droplets flow into branching channels. Ahn et al. [56] developed a dielectrophoresis-based technology for high-speed sorting of water droplets in microfluidic devices (Fig. 5). They fabricated microfluidic devices containing microelectrodes underneath PDMS channels that produced a force of more than 10 nN on a water droplet in oil phase, resulting in a sorting rate of higher than 1.6 kHz. Baret et al. [57] further developed a fluorescence activated droplet sorting device, which depended on laser induced fluorescence to detect the chemical contents within a train of droplets continuously passing the excitation light spot. The fluorescence signal was coupled to a downstream electronic sorting unit, which would selectively steer the droplets into either of the two branching channels based on their fluorescence signal intensity. They demonstrated sorting single cells compartmentalized in droplets at rates up to 2,000 droplets per second.

Surface acoustic wave is another useful method that can direct the motion of droplets in microfluidic channels and be used for droplet sorting [58]. The surface acoustic wave device is assembled on top of the surface of a piezoelectric substrate, on which an interdigital transducer of gold electrodes is fabricated. A surface acoustic wave, known as Rayleigh wave, is excited and propagates along the piezoelectric substrate when an alternating RF voltage signal is applied to the interdigital transducer. This surface acoustic wave will propagate into the continuous phase and push the droplet along the direction of wave propagation. Since acoustic streaming only depends on the compressibility of the continuous phase, two especially attractive features of this method are that it is label-free and material independent, and is thus easy to integrate into complex systems.

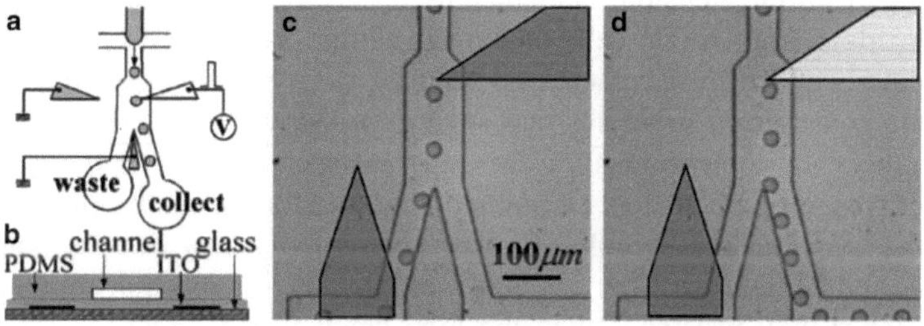

Fig. 5 Dielectrophoretic droplets sorting. (**a**) Schematic view of the device. (**b**) Schematic cross section of the device. (**c**) In the absence of an electric field, water droplets flow into the waste channel. (**d**) Applying an electric field, the droplets are attracted toward the energized electrode and flow into the collection channel. Reproduced with permission from [56]

Other methods, such as using the magnetic field, have also been employed to manipulate droplets loaded with magnetic particles [59]. Laser-induced localized heating, which alters the interfacial tension, offers another way to sort droplets [60]. However, these methods either need complex operation or lack controllability, which may limit their applications.

2.6 Trapping and Incubation of Droplets

Typically, due to the limited length of a microchannel, the residence time of droplets in microfluidic devices is in the range of seconds to a few minutes. However, in many circumstances, it will be necessary to store droplets for up to several hours or even days, after or between different stages of droplets processing to allow a particular reaction to take place. The simplest and most straightforward method to increase the residence time of droplets in microfluidic devices is to increase the length and width of the channel. Frenz fabricated a microfluidic device with a delay line that is wider and deeper than the microchannel at the droplet generation section [61]. In order to make sure that all droplets would have a similar residence time in the delay line, periodic constriction structures were added to it. They then used this device to measure the reaction kinetics of enzyme β-lactamase over several minutes. However, this device was not suitable for applications where longer storage times of droplets were required. Courtois et al. [62] developed a microfluidic device containing a reservoir that can hold up to 10^6 droplets. Droplets could reside in the reservoir for at least 6 h and in vitro expression of green fluorescent proteins was conducted in this device. The disadvantage of this method was that the original sequence of droplets was lost in large reservoirs. One way to solve this problem is to trap the droplets in the reservoir. Schmitz et al. [63] proposed the "dropspot" device, a microfluidic device in which arrays of round chambers were connected by narrow constrictions. When droplets flowed into the chamber arrays, they had to squeeze through the constrictions. With appropriate droplet size, and in the absence of flow, a chamber would contain only a single droplet due to surface tension. They then used the device to monitor the activity of β-galactosidase from single cells contained within droplets. Takeuchi et al. [64] designed a novel flow resistance-based trapping device in which a straight channel containing constriction structures was superimposed with loop channels (Fig. 6a). The length of the loop channels were designed such that, when the constriction structures were empty, the flow resistance along the straight channel was lower than that of the loop channel. Droplets would preferentially flow into the constriction structures and be trapped. After trapping a droplet, the constriction structure would have a higher flow resistance than the loop channel and direct the droplet flow through the loop channel. Subsequent droplets would bypass the constriction structure and be trapped by the next constriction structure. Shi et al. [65] adopted this design to trap droplets containing *Caenorhabditis elegans* and characterized the behavior of *C. elegans* in response to neurotoxin (Fig. 6b). Later, Huebner et al. [66]

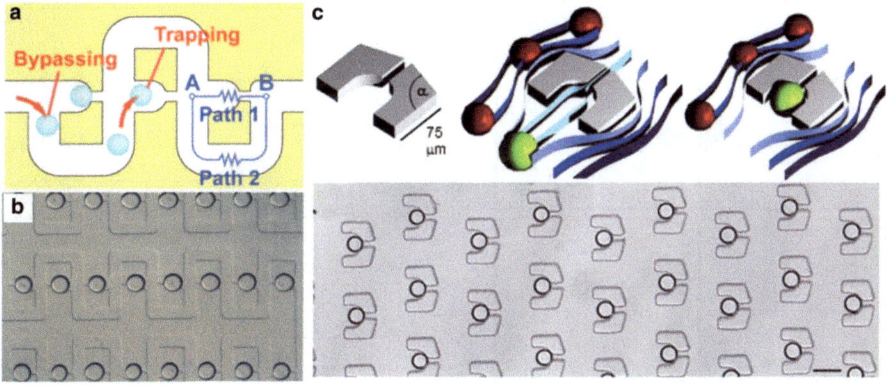

Fig. 6 Trapping of droplets inside microfluidic devices. (**a**) The design and working mechanism of the flow resistance-based droplet-trapping microfluidic device. Reproduced with permission from [64]. (**b**) Image of arrays of trapped droplets encapsulating *C. elegans*. Reproduced with permission from [65]. (**c**) Geometry-based droplet trapping device (*top*) and image of trapped droplet array (*bottom*). Reproduced with permission from [66]

developed a similar design and investigated the enzymatic reactions and cell behaviors in the trapped droplets (Fig. 6c). Although droplets can be trapped by devices with trapping structures, droplets can still not be incubated for long term, days to months, due to the gas permeability of PDMS. One solution to this problem is to collect the droplets in a glass capillary and this method has been applied to investigate protein crystallization [67].

2.7 *Droplet Content Characterization*

Once the reactions have completed inside droplets, the reaction products have to be detected and characterized. Various methods have been employed for the detection and characterization of droplet contents. Since most microfluidic devices are fabricated using transparent materials such as PDMS and glass, optical systems, including fluorescence microscopy [53, 68, 69] and Raman spectroscopy [70], can be directly used to detect the reaction products inside droplets. Up until now, fluorescence measurement has been the most successful and prevalent method to detect the droplet contents. The advantage of fluorescence detection is that it can simultaneously monitor the fluorescence signal from multiple droplets and has been applied to characterize enzyme kinetics [53] and protein–protein interactions [68] in continuously moving droplets. However, it requires the reaction products to be fluorescent, which limits its applications. Besides optical methods, electrochemical detection can also be used to gain information on droplet contents and has been used to measure the K_m and V_{max} in conversion of H_2O_2 by catalase [71].

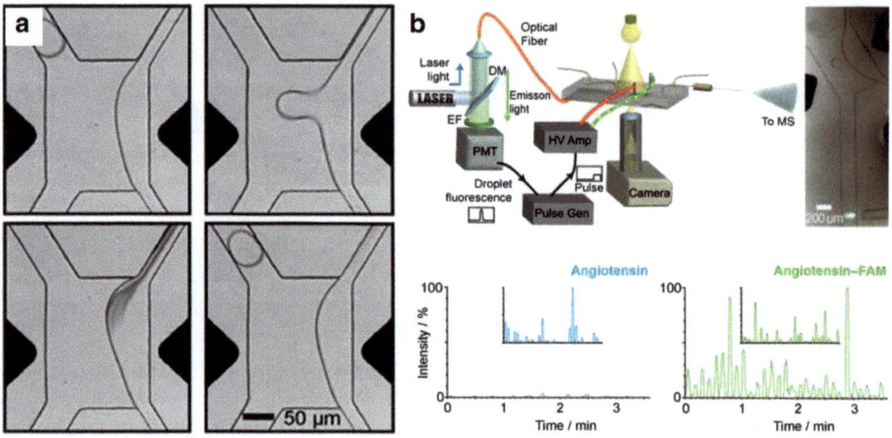

Fig. 7 Characterization of droplet contents by mass spectroscopy. (**a**) Pictures showing the process of extracting droplets from oil into aqueous stream. (**b**) Schematic of the experimental setup for combining fluorescence detection and mass analysis of droplets. Reproduced with permission from [74, 75]

In the above methods, the droplet contents to be detected and characterized are relatively simple. However, in many cases, the composition of droplet contents is complex, and then separation will be necessary. Electrophoresis has been the most developed separation method in microfluidic devices. Recently, Roman successfully extracted the contents of individual droplets from oil and interfaced it to electrophoresis separation [72]. Later, Wang et al. [73] improved the method and applied it to in vivo chemical monitoring of the amino acids from brain after microdialysis. As a label free and universal method, mass spectroscopy has also been successfully applied to study the reaction products in droplets. Since droplets are surrounded by carrier oil, the droplet contents cannot be directly characterized by mass spectroscopy. Fidalgo et al. [74, 75] employed a high voltage to extract a droplet from carrier oil into an aqueous buffer stream and interfaced the buffer into ESI-MS (Fig. 7). When combined with fluorescence signal actuated droplet extraction, this method can be used to carry out advanced experiments in which the chemical identity of droplet contents is determined depending on their biological or chemical activity.

2.8 Hyphenation of Droplets to CE and HPLC

One unique characteristic of droplets is their isolated nature by immiscible carrier oil. This feature, in combination with the tiny size of droplets, is particularly useful for preserving the separation bands of high resolution separation methods, such as capillary electrophoresis (CE) and microscale high performance liquid chromatography (HPLC), in which the separated components are usually difficult to collect

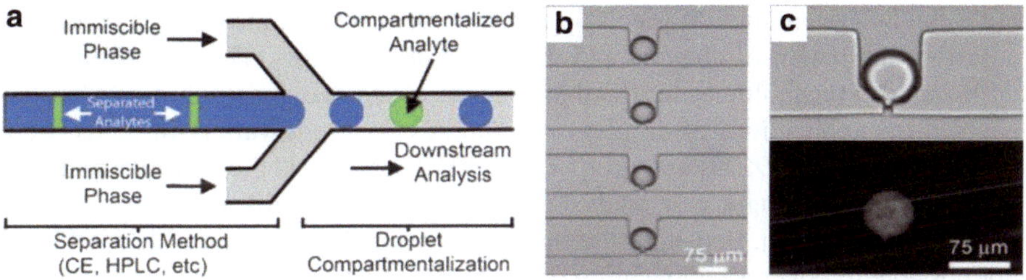

Fig. 8 Compartmentalization of the separation components of capillary electrophoresis. (**a**) Schematic of the device and process of compartmentalizing the separation bands of capillary electrophoresis into individual droplets. (**b**) Trapped droplets generated by electroosmotic flow. (**c**) Bright-field (*top*) and fluorescence (*bottom*) images of trapped droplets containing the band of fluorescein after capillary electrophoresis. Reproduced with permission from [76]

owing to the extremely small volume involved and molecular diffusion. Edgar et al. [76] first proposed the use of droplets for the compartmentalization of separated components of microchip capillary electrophoresis (Fig. 8). In their device, electroosmotic flow was used to drive the aqueous flow and generate droplets. They demonstrated that the droplet generation would not interfere with the electrophoresis separation process. Further processing of the droplets can be performed after encapsulation of the separation components into droplets. In another example, Theberge et al. [77] described the generation of droplets with defined contents and concentration gradients from the separation components of high performance liquid chromatography. Niu et al. [78] then utilized droplets to couple two-dimensional separations, with capillary HPLC as the first dimension and electrophoresis as the second dimension.

3 Applications

In general, several features of droplets make droplet-based microfluidics an appealing platform for chemistry and biology. (1) The compartmentalization of reactions into individual droplets is particularly effective for eliminating the dispersion of reactants along the microchannel encountered in continuous flow microfluidic systems. (2) Evaporation of solvent is also prevented and reaction condition inside a droplet is very stable. (3) As a result of the tiny size of a droplet, the consumption of reagents and sample will be extremely low and the enclosed space of a droplet can be used to compartmentalize single cells or even single molecules. (4) Furthermore, droplets generated by microfluidic devices are highly monodisperse, which allows for quantitative measurements and analysis. (5) The high surface area-to-volume ratio of droplets and the differential flow speeds between carrier fluid and droplet offers extremely efficient mass and heat transfer between the carrier fluid and the droplet. (6) The chaotic advection inside droplets enables complete mixing on a millisecond scale. Other advantages of droplet-based microfluidic system include precisely controllable reaction time inside a droplet by adjusting the length of channel and the flow rates of fluids. The availability of a wide range of technologies for flexible generation and manipulation of droplets has enabled the applications of droplet microfluidics in a wide variety of fields, from chemical reactions [9, 79] and protein crystallization [10, 11] to material synthesis [12, 13, 80–83], single cell analysis [84–91], DNA amplification [16], protein engineering [62, 92], and high-throughput screening technologies [17, 93]. It is not possible to cover all application areas in this review. For the demonstration of the use of droplets as microreactors, applications of droplet microfluidics will be briefly reviewed in chemical reactions, material synthesis and single cell analysis.

3.1 Chemical Reactions

In conventional single phase microfluidic systems, flow in the microchannel is laminar; a parabolic velocity profile is established with fluid velocity zero at the channel walls and maximum at the channel center [18, 21] (Fig. 9a). There are two implications to this behavior: (1) a reagent sample plug will constantly dissipate along the microchannel, and (2) the mixing of samples could be very slow in co-flow streams.

In contrast, these problems could be easily solved in droplet based microfluidic systems. Song et al. reported a droplet-based reaction system that could be used to control networks of many chemical reactions on the millisecond scale [47]. They reported that the droplet reactor, which was defined by the enclosed water–oil interface, would confine the reactants within the droplet and henceforth eliminate sample dissipation problems; furthermore, they observed that, by allowing micro-droplets to flow through winding channels, the flowfield inside the droplet could be transformed from recirculating to chaotic, which greatly enhanced the mixing inside the droplets (Fig. 9a). These two features of droplet microreactor allowed one to control when each reaction began, for how long each reaction evolved before it was separated or combined with other reactions, and when each reaction was analyzed or quenched [47]. Later on, the same group demonstrated using a droplet microfluidic system to probe millisecond reaction kinetics using nanoliters of reagents (Fig. 9b) [53]. A similar concept has recently been adopted by Churski et al. who developed a droplet on demand technique and an integrated system for scanning arbitrary combinations of three miscible solutions in droplets at 3 Hz and used them to scan up to 10,000 conditions of chemical and biochemical reactions per hour using 10 mL of solutions in total [95].

Through years of development, the portfolio of applications of droplet-based microreactors has expanded from chemical kinetics to a wide spectrum of applications including protein crystallization [10, 67, 96–99] and modeling complex reaction networks [100–104]. Interfacial reaction at the oil/water interface has also been explored for chemical synthesis [105]. Another interesting area is using droplets as highly effective reaction system to prepare nanoparticles [94, 106–108].

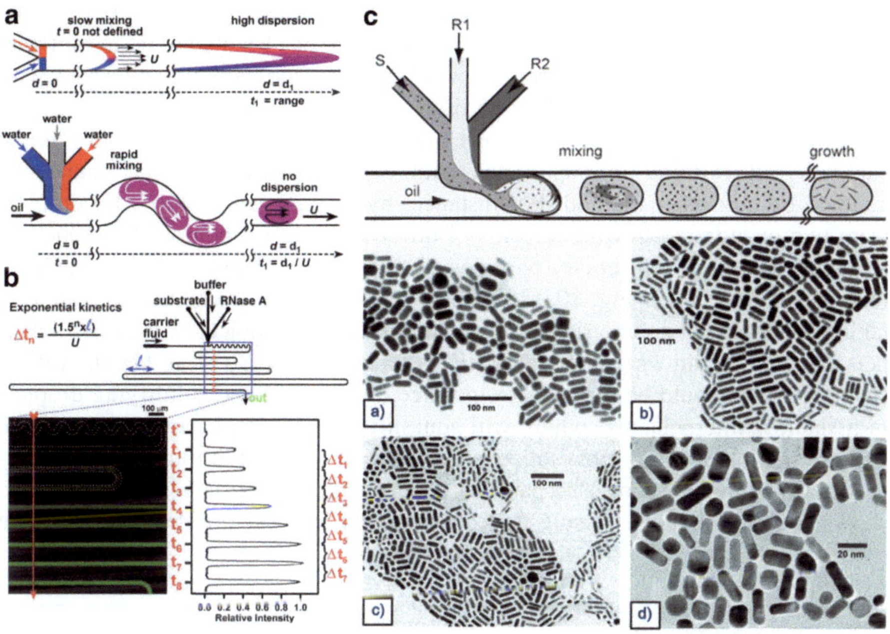

Fig. 9 Reactions in microdroplets. (**a**) Schematic showing enhanced mixing and suppressed dispersion in droplets. Reproduced with permission from [47]. (**b**) Millisecond kinetics on droplet microfluidics. Reproduced with permission from [53]. (**c**) Droplet-based microfluidic synthesis of anisotropic metal nanocrystals. Reproduced with permission from [94]

For instance, Duraiswamy et al. [94] recently reported a droplet-based microfluidic method for the preparation of anisotropic gold nanocrystal dispersions. Gold nanoparticle seeds and growth reagents were dispensed into monodisperse picoliter droplets within a microchannel. Confinement of reaction solutions within small droplets prevents contact between the growing nanocrystals and the microchannel walls. The critical factors in translating macroscale flask-based methods to a droplet-based microfluidic method were highlighted and approaches were demonstrated to fine-tune nanoparticle shapes into three broad classes: spheres/spheroids, rods, and extended sharp-edged structures, thus varying the optical resonances in the visible–near-infrared spectral range (Fig. 9c) [94].

3.2 *Microparticle Synthesis*

One apparent feature of droplets generated in microfluidic devices is their monodispersity in size. Formation of droplets containing precursor solutions dispersed in a continuous phase and then initiation of crosslinking, polymerization or phase separation produces monodisperse microparticles of defined compositions.

The size of the particles could be tuned by simply controlling the flow rates of the disperse phase and continuous phase. This approach has been adopted to fabricate microparticles of gels [80, 81], polymers [13, 82, 83], and inorganic materials with designed physical and chemical properties.

A more appealing feature of droplet microfluidics for particle synthesis is the capability of forming microparticles with complex structures that cannot be attained by any other method. For instance, by forming and solidifying droplets from two or more co-flowing streams of different chemical composition, one can fabricate Janus or even ternary particles of asymmetric chemical or geographical feature [82, 109, 110] (Fig. 10a). Aspherical particles could also be fabricated by taking advantage of geometric confinement of the microdroplets by the microchannels. Figure 10b shows aspherical particles fabricated by Xu et al. [111]. A disk-shaped particle could be formed, for instance, by forming and solidifying droplets in a shallow microchannel which will squash the microdroplet into a disk shape upon formation. Several groups also exploited the formation of multiple emulsions [113, 114] to produce core-shell particles or capsules [27, 112, 115–117]. Figure 10c shows a composite gel capsule that contains multiple oil droplets [112]. These

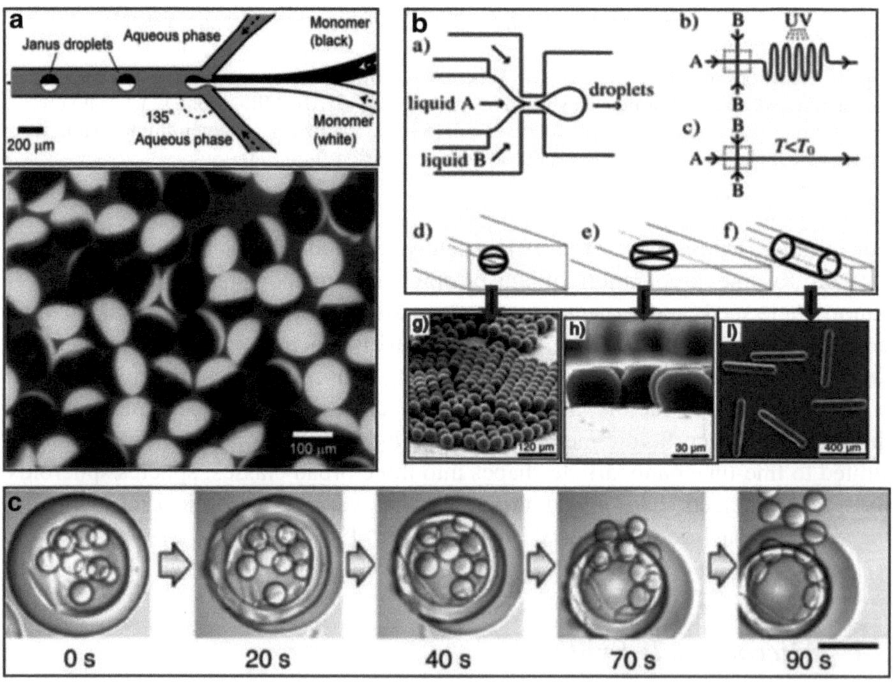

Fig. 10 Droplet microfluidics for microparticle synthesis. (**a**) Janus particle synthesis by co-flowing two monomer streams. Reproduced with permission from [82]. (**b**) Channel geometry facilitated aspherical particles synthesis. Reproduced with permission from [111]. (**c**) Multiple emulsion templated composite particles synthesis. Reproduced with permission from [112]

particles were fabricated by forming triple emulsions from gel precursor solutions and then allowing polymerization of the hydrogel.

3.3 Single Cell Analysis

Technologies for high-throughput, quantifiable single cell analysis is highly demanding in modern biological research [15, 118]. The capability of single cell analysis would allow for (1) spotting stochastic events among a cell population, (2) tracing spatiotemporal changes on cells to elucidate mechanistically cellular functions, and (3) developing high throughput technology for identifying and purifying rare cells from a cell population. In general, technologies with low sample consumption, capable of quantitative analysis, ease of operation, and scalable to different throughput levels, will be highly appealing. In this scene, droplet microfluidics is potentially a very competitive technical platform for high-throughput single cell analysis. Technologies for encapsulating individual cells into separated water in oil droplets [119–121], which allow for confining [122, 123], content analysis [87, 88, 90, 119, 124], screening [36, 57], and temporal changes tracing [63] of the cells, have become available recently, indicating promising applications for droplet microfluidics in this field.

The first technical step for droplet-based single cell analysis is to encapsulate single cells into droplets. Typical practice is to use a cell suspension as the disperse phase to form droplets. The distribution of the number of cells in droplets obeys Poisson statistics. To ensure that most droplets contain only one cell, one has to compromise on forming a large amount of empty droplets [87]. To overcome this limitation, several technologies have been developed to enhance encapsulation efficiency [119–121]. For instance, Edd et al. [119] reported a method that induced cells to self-organize into two evenly-spaced streams whose longitudinal order is shifted by half the particle–particle spacing. This occurred when a high density suspension of cells or particles is forced to travel rapidly through a high aspect-ratio microchannel, where particle diameter is a large fraction of the channel's narrow dimension (Fig. 11a). This phenomenon provides a purely passive method to load single cells controllably into droplets, overcoming the intrinsic limitation set by Poisson statistics and ensuring that virtually every drop contains exactly one cell.

One straightforward consequence of compartmentalizing cells into the water in oil droplets is that the cells will be confined in a small space. Thus molecules secreted by the cells will not dissipate into the bulk but keep accumulating around the cell. This feature provides opportunities to increase the sensitivity in detecting cellular events [87, 125]. For instance, Joensson et al. [125] developed an assay in droplets based on enzymatic amplification of low-abundance cell-surface biomarkers (Fig. 11b). This method is an effective miniaturized ELISA (enzyme-linked immunosorbent assay). Cells bearing specific biomarkers were labeled with an antibody linked to β-galactosidase and were co-encapsulated with a fluorogenic substrate. The increase in fluorescence was measured by laser detection after incubation

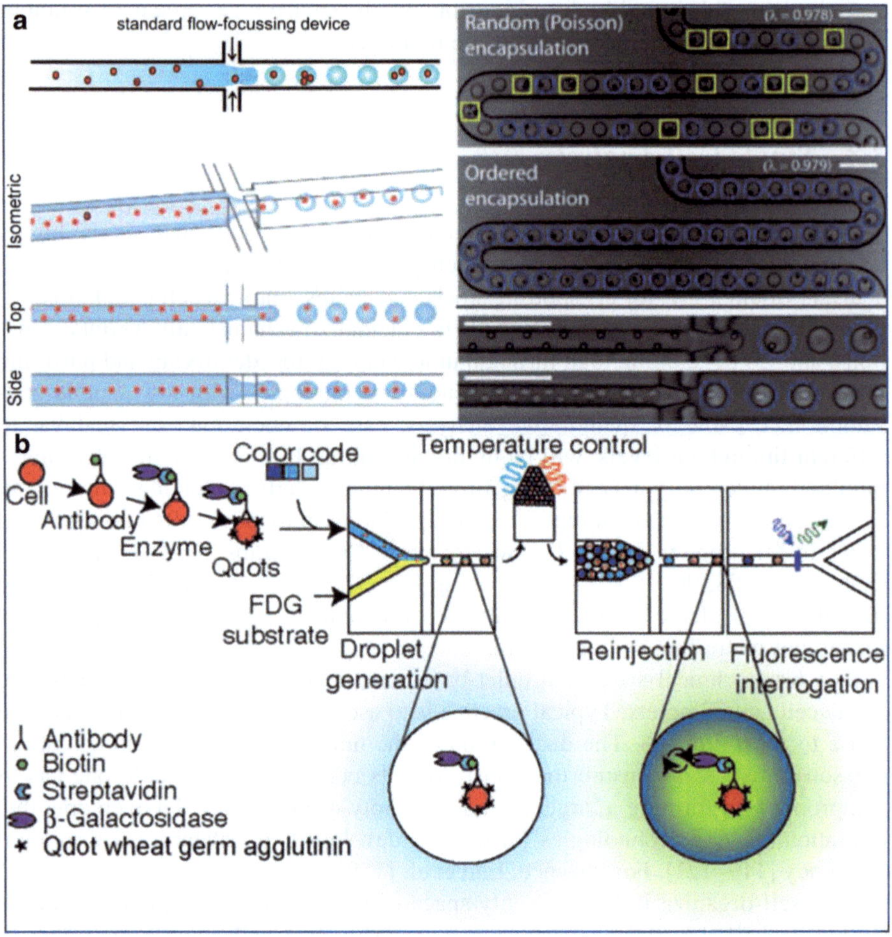

Fig. 11 Droplet-based single cell analysis. (**a**) Highly efficient encapsulation of single cells into microdroplets. Reproduced with permission from [119]. (**b**) Detection and analysis of low-abundance cell-surface biomarkers using enzymatic amplification in microfluidic droplets. Reproduced with permission from [125]

offline and reinjection of the droplets. Increased signal discrimination compared to standard fluorescence activated cell sorting (FACS) methods will enable sorting of cells on the basis of these low abundance markers.

4 Conclusions and Outlook

Droplet microfluidics has been a quickly evolving research field in recent years. In this review we summarize the main technologies for droplet microfluidics. Several applications are also discussed as a showcase of the wide range applications of droplet microfluidics. Since an increasing community of scientists and engineers is working on both theoretical and application aspects of droplet microfluidics, we would expect even further proliferation of this research field, marked by the emergence of both new technologies and new applications.

Acknowledgements This work was supported by 973 program (No. 2007CB714505 and 2007CB714507), Knowledge Innovation Program of the Chinese Academy of Sciences (KJCX2-YW-H18), and Instrument Research and Development Program of the Chinese Academy of Sciences (YZ200908).

References

1. Thorsen T, Roberts RW, Arnold FH, Quake SR (2001) Phys Rev Lett 86:4163
2. Anna SL, Bontoux N, Stone HA (2003) Appl Phys Lett 82:364
3. Garstecki P, Fuerstman MJ, Whitesides GM (2005) Nat Phys 1:168
4. Fuerstman MJ, Garstecki P, Whitesides GM (2007) Science 315:828
5. Chiu DT, Lorenz RM (2009) Acc Chem Res 42:649
6. Schaerli Y, Hollfelder F (2009) Mol Biosyst 5:1392
7. Huebner A, Sharma S, Srisa-Art M, Hollfelder F, Edel JB, Demello AJ (2008) Lab Chip 8:1244
8. Kelly BT, Baret JC, Taly V, Griffiths AD (2007) Chem Commun 1773
9. Song H, Chen DL, Ismagilov RF (2006) Angew Chem Int Ed 45:7336
10. Li L, Ismagilov RF (2010) Annu Rev Biophys 39:139
11. Leng J, Salmon JB (2009) Lab Chip 9:24
12. Hung LH, Lee AP (2007) J Med Biol Eng 27:1
13. Park JI, Saffari A, Kumar S, Günther A, Kumacheva E (2010) Annu Rev Mater Res 40:415
14. Zare RN, Kim S (2010) Annu Rev Biomed Eng 12:187
15. Borland LM, Kottegoda S, Phillips KS, Allbritton NL (2008) Annu Rev Anal Chem 1:191
16. Zhang Y, Ozdemir P (2009) Anal Chim Acta 638:115
17. Hong J, Edel JB, deMello AJ (2009) Drug Discov Today 14:134
18. Squires TM, Quake SR (2005) Rev Mod Phys 77:977
19. Christopher GF, Anna SL (2007) J Phys D Appl Phys 40:R319
20. Baroud CN, Gallaire F, Dangla R (2010) Lab Chip 10:2032
21. Stone HA, Stroock AD, Ajdari A (2004) Annu Rev Fluid Mech 36:381
22. Abdelgawad M, Wheeler AR (2009) Adv Mater 21:920
23. Garstecki P, Fuerstman MJ, Stone HA, Whitesides GM (2006) Lab Chip 6:437
24. Garstecki P, Stone HA, Whitesides GM (2005) Phys Rev Lett 94:164501
25. Anna SL, Mayer HC (2006) Phys Fluids 18:121512
26. Dreyfus R, Tabeling P, Willaime H (2003) Phys Rev Lett 90:144505
27. Shah RK, Shum HC, Rowat AC, Lee D, Agresti JJ, Utada AS, Chu LY, Kim JW, Fernandez-Nieves A, Martinez CJ, Weitz DA (2008) Mater Today 11:18
28. Takeuchi S, Garstecki P, Weibel DB, Whitesides GM (2005) Adv Mater 17:1067
29. Morimoto Y, Tan WH, Takeuchi S (2009) Biomed Microdevices 11:369
30. Malloggi F, Vanapalli SA, Gu H, Van Den Ende D, Mugele F (2007) J Phys Condens Matter 19:462101
31. Haeberle S, Zengerle R, Ducree J (2007) Microfluid Nanofluid 3:65
32. Link DR, Grasland-Mongrain E, Duri A, Sarrazin F, Cheng ZD, Cristobal G, Marquez M, Weitz DA (2006) Angew Chem Int Ed 45:2556
33. Baroud CN, Delville JP, Gallaire F, Wunenburger R (2007) Phys Rev E 75:046302
34. Abate AR, Romanowsky MB, Agresti JJ, Weitz DA (2009) Appl Phys Lett 94:023503

35. Zeng SJ, Li BW, Su XO, Qin JH, Lin BC (2009) Lab Chip 9:1340
36. Brouzes E, Medkova M, Savenelli N, Marran D, Twardowski M, Hutchison JB, Rothberg JM, Link DR, Perrimon N, Samuels ML (2009) Proc Natl Acad Sci USA 106:14195
37. Priest C, Herminghaus S, Seemann R (2006) Appl Phys Lett 89:3
38. Mazutis L, Baret JC, Treacy P, Skhiri Y, Araghi AF, Ryckelynck M, Taly V, Griffiths AD (2009) Lab Chip 9:2902
39. Mazutis L, Araghi AF, Miller OJ, Baret JC, Frenz L, Janoshazi A, Taly V, Miller BJ, Hutchison JB, Link D, Griffiths AD, Ryckelynck M (2009) Anal Chem 81:4813
40. Ahn K, Agresti J, Chong H, Marquez M, Weitz DA (2006) Appl Phys Lett 88:3
41. Fidalgo LM, Abell C, Huck WTS (2007) Lab Chip 7:984
42. Mazutis L, Baret JC, Griffiths AD (2009) Lab Chip 9:2665
43. Niu X, Gulati S, Edel JB, deMello AJ (2008) Lab Chip 8:1837
44. Tan WH, Takeuchi S (2006) Lab Chip 6:757
45. Hung LH, Choi KM, Tseng WY, Tan YC, Shea KJ, Lee AP (2006) Lab Chip 6:174
46. Tan YC, Fisher JS, Lee AI, Cristini V, Lee AP (2004) Lab Chip 4:292
47. Song H, Tice JD, Ismagilov RF (2003) Angew Chem Int Ed 42:768
48. Link DR, Anna SL, Weitz DA, Stone HA (2004) Phys Rev Lett 92:054503
49. Nie J, Kennedy RT (2010) Anal Chem 82:7852
50. Yamada M, Doi S, Maenaka H, Yasuda M, Seki M (2008) J Colloid Interface Sci 321:401
51. Christopher GF, Anna SL (2009) J Rheol 53:663
52. Song H, Bringer MR, Tice JD, Gerdts CJ, Ismagilov RF (2003) Appl Phys Lett 83:4664
53. Song H, Ismagilov RF (2003) J Am Chem Soc 125:14613
54. Tice JD, Song H, Lyon AD, Ismagilov RF (2003) Langmuir 19:9127
55. Bringer MR, Gerdts CJ, Song H, Tice JD, Ismagilov RF (2004) Philos Trans R Soc Lond A Math Phys Eng Sci 362:1087
56. Ahn K, Kerbage C, Hunt TP, Westervelt RM, Link DR, Weitz DA (2006) Appl Phys Lett 88:024104
57. Baret JC, Miller OJ, Taly V, Ryckelynck M, El-Harrak A, Frenz L, Rick C, Samuels ML, Hutchison JB, Agresti JJ, Link DR, Weitz DA, Griffiths AD (2009) Lab Chip 9:1850
58. Franke T, Abate AR, Weitz DA, Wixforth A (2009) Lab Chip 9:2625
59. Zhang K, Liang QL, Ma S, Mu XA, Hu P, Wang YM, Luo GA (2009) Lab Chip 9:2992
60. Baroud CN, de Saint Vincent MR, Delville JP (2007) Lab Chip 7:1029
61. Frenz L, Blank K, Brouzes E, Griffiths AD (2009) Lab Chip 9:1344
62. Courtois F, Olguin LF, Whyte G, Bratton D, Huck WTS, Abell C, Hollfelder F (2008) Chembiochem 9:439
63. Schmitz CHJ, Rowat AC, Koster S, Weitz DA (2009) Lab Chip 9:44
64. Tan WH, Takeuchi S (2007) Proc Natl Acad Sci USA 104:1146
65. Shi WW, Qin JH, Ye NN, Lin BC (2008) Lab Chip 8:1432
66. Huebner A, Bratton D, Whyte G, Yang M, deMello AJ, Abell C, Hollfelder F (2009) Lab Chip 9:692
67. Zheng B, Tice JD, Roach LS, Ismagilov RF (2004) Angew Chem Int Ed 43:2508
68. Srisa-Art M, Kang DK, Hong J, Park H, Leatherbarrow RJ, Edel JB, Chang SI, de Mello AJ (2009) Chembiochem 10:1605
69. Srisa-Art M, deMello AJ, Edel JB (2008) Phys Rev Lett 101:4
70. Barnes SE, Cygan ZT, Yates JK, Beers KL, Amis EJ (2006) Analyst 131:1027
71. Liu SJ, Gu YF, Le Roux RB, Matthews SM, Bratton D, Yunus K, Fisher AC, Huck WTS (2008) Lab Chip 8:1937
72. Roman GT, Wang M, Shultz KN, Jennings C, Kennedy RT (2008) Anal Chem 80:8231
73. Wang M, Roman GT, Schultz K, Jennings C, Kennedy RT (2008) Anal Chem 80:5607
74. Fidalgo LM, Whyte G, Bratton D, Kaminski CF, Abell C, Huck WTS (2008) Angew Chem Int Ed 47:2042
75. Fidalgo L, Whyte G, Ruotolo B, Benesch J, Stengel F, Abell C, Robinson C, Huck W (2009) Angew Chem Int Ed 48:3665

76. Edgar JS, Milne G, Zhao YQ, Pabbati CP, Lim DSW, Chiu DT (2009) Angew Chem Int Ed 48:2719
77. Theberge AB, Whyte G, Huck WTS (2010) Anal Chem 82:3449
78. Niu XZ, Zhang B, Marszalek RT, Ces O, Edel JB, Klug DR, Demello AJ (2009) Chem Commun 6159
79. McMullen JP, Jensen KF (2009) Annu Rev Anal Chem 3:19–42
80. Tumarkin E, Kumacheva E (2009) Chem Soc Rev 38:2161
81. Um E, Lee DS, Pyo HB, Park JK (2008) Microfluid Nanofluid 5:541
82. Nisisako T, Torii T, Takahashi T, Takizawa Y (2006) Adv Mater 18:1152
83. Li W, Greener J, Voicu D, Kumacheva E (2009) Lab Chip 9:2715
84. Chiu DT (2010) Anal Bioanal Chem 397:3179–3183
85. Di Carlo D, Aghdam N, Lee LP (2006) Anal Chem 78:4925
86. He MY, Edgar JS, Jeffries GDM, Lorenz RM, Shelby JP, Chiu DT (2005) Anal Chem 77:1539
87. Huebner A, Olguin LF, Bratton D, Whyte G, Huck WTS, de Mello AJ, Edel JB, Abell C, Hollfelder F (2008) Anal Chem 80:3890
88. Kumaresan P, Yang CJ, Cronier SA, Blazei RG, Mathies RA (2008) Anal Chem 80:3522
89. Sasuga Y, Iwasawa T, Terada K, Oe Y, Sorimachi H, Ohara O, Harada Y (2008) Anal Chem 80:9141
90. Yong Z, Novak R, Shuga J, Smith MT, Mathies RA (2010) Anal Chem 82:3183
91. Wu HK, Wheeler A, Zare RN (2004) Proc Natl Acad Sci USA 101:12809
92. Dittrich PS, Jahnz M, Schwille P (2005) Chembiochem 6:811
93. Agresti JJ, Antipov E, Abate AR, Ahn K, Rowat AC, Baret JC, Marquez M, Klibanov AM, Griffiths AD, Weitz DA (2010) Proc Natl Acad Sci USA 107:4004
94. Duraiswamy S, Khan SA (2009) Small 5:2828
95. Churski K, Korczyk P, Garstecki P (2010) Lab Chip Miniaturisation Chem Biol 10:816
96. Chen DL, Gerdts CJ, Ismagilov RF (2005) J Am Chem Soc 127:9672
97. Zheng B, Gerdts CJ, Ismagilov RF (2005) Curr Opin Struct Biol 15:548
98. Yadav MK, Gerdts CJ, Sanishvili R, Smith WW, Roach LS, Ismagilov RF, Kuhn P, Stevens RC (2005) J Appl Crystallogr 38:900
99. Li L, Fu Q, Kors CA, Stewart L, Nollert P, Laible PD, Ismagilov RF (2010) Microfluid Nanofluid 8:789
100. Kastrup CJ, Runyon MK, Lucchetta EM, Price JM, Ismagilov RF (2008) Acc Chem Res 41:549
101. Meier M, Kennedy-Darling J, Choi SH, Norstrom EM, Sisodia SS, Ismagilov RF (2009) Angew Chem Int Ed 48:1487
102. Shen F, Kastrup CJ, Liu Y, Ismagilov RF (2008) Arterioscler Thromb Vasc Biol 28:2035
103. Kastrup CJ, Ismagilov RF (2007) J Phys Org Chem 20:711
104. Gerdts CJ, Sharoyan DE, Ismagilov RF (2004) J Am Chem Soc 126:6327
105. Theberge AB, Whyte G, Frenzel M, Fidalgo LM, Wootton RCR, Huck WTS (2009) Chem Commun 6225
106. deMello AJ (2006) Nature 442:394
107. Marre S, Park J, Rempel J, Guan J, Bawendi MG, Jensen KF (2008) Adv Mater 20:4830
108. Marre S, Jensen KF (2010) Chem Soc Rev 39:1183
109. Seiffert S, Romanowsky MB, Weitz DA (2010) Langmuir 26:14842
110. Nie ZH, Li W, Seo M, Xu SQ, Kumacheva E (2006) J Am Chem Soc 128:9408
111. Xu SQ, Nie ZH, Seo M, Lewis P, Kumacheva E, Stone HA, Garstecki P, Weibel DB, Gitlin I, Whitesides GM (2005) Angew Chem Int Ed 44:724
112. Chu LY, Utada AS, Shah RK, Kim JW, Weitz DA (2007) Angew Chem Int Ed 46:8970
113. Utada AS, Lorenceau E, Link DR, Kaplan PD, Stone HA, Weitz DA (2005) Science 308:537
114. Seo M, Paquet C, Nie ZH, Xu SQ, Kumacheva E (2007) Soft Matter 3:986
115. Shum HC, Abate AR, Lee D, Studart AR, Wang BG, Chen CH, Thiele J, Shah RK, Krummel A, Weitz DA (2010) Macromol Rapid Commun 31:108
116. Thiele J, Abate AR, Shum HC, Bachtler S, Forster S, Weitz DA (2010) Small 6:1723

117. Seiffert S, Thiele J, Abate AR, Weitz DA (2010) J Am Chem Soc 132:6606
118. Schmid A, Kortmann H, Dittrich PS, Blank LM (2010) Curr Opin Biotechnol 21:12
119. Edd JF, Di Carlo D, Humphry KJ, Koster S, Irimia D, Weitz DA, Toner M (2008) Lab Chip 8:1262
120. Chabert M, Viovy JL (2008) Proc Natl Acad Sci USA 105:3191
121. Koster S, Angile FE, Duan H, Agresti JJ, Wintner A, Schmitz C, Rowat AC, Merten CA, Pisignano D, Griffiths AD, Weitz DA (2008) Lab Chip 8:1110
122. Vincent ME, Liu WS, Haney EB, Ismagilov RF (2010) Chem Soc Rev 39:974
123. Boedicker JQ, Vincent ME, Ismagilov RF (2009) Angew Chem Int Ed 48:5908
124. Baret JC, Beck Y, Billas-Massobrio I, Moras D, Griffiths AD (2010) Chem Biol 17:528
125. Joensson HN, Samuels ML, Brouzes ER, Medkova M, Uhlen M, Link DR, Andersson-Svahn H (2009) Angew Chem Int Ed 48:2518

Top Curr Chem (2011) 304: 91–115
DOI: 10.1007/128_2011_148

Published online: 29 April 2011

Electrorheological Fluid and Its Applications in Microfluidics

Limu Wang, Xiuqing Gong, and Weijia Wen

Abstract Microfluidics is a low-cost technique for fast-diagnosis and microsynthesis. Within a decade it might become the foundation of point-of-care and lab-on-a-chip applications. With microfluidic chips, high-throughput sample screening and information processing are made possible. The picoliter droplet runs in microfluidic chips are ideal miniaturized vessels for microdetection and microsynthesis. Meanwhile, individual manipulation of microdroplets remains a challenge: the shortcomings in automatic, reliable, and scalable methods for logic control prevent further integration of microfluidic applications. The giant electrorheological fluid (GERF), which is a kind of "smart" colloid, has tunable viscosity under the influence of external electric field. Therefore, GERF is introduced as the active controlling medium, with real-time response in on-chip fluid control. This review article introduces the working principles and fabrication methods of different types of electrorheological fluid, and extensively describes the strategies of GERF-assisted microfluidic controlling schemes.

Keywords Electrorheological fluid · Logic control · Microfluidics · Microdroplet

Contents

1 Introduction 92
2 Electrorheological Fluid 93
  2.1 Inorganic ER Particles 94
  2.2 Inorganic–Polymer Hybrid ER Particles 97
  2.3 Conducting-Polymer-Based ER Particles 98
  2.4 Dispersing Phase 98
  2.5 Future Research Directions 98
3 ERF-Based Precise Microfluidic Control System 99
  3.1 Soft Conducting Electrodes for Droplet Detection and ERF On-Chip Control 101

L. Wang, X. Gong, and W. Wen (✉)
Department of Physics and KAUST-HKUST Micro/Nano-Fluidics Joint Laboratory, The Hong Kong University of Science and Technology, Clear Water Bay, Kowloon, Hong Kong
e-mail: phwen@ust.hk

3.2 GERF Microvalves ... 102
3.3 Integratable Microfluidic Components Based on GERF Microvalves ... 103
3.4 "Smart" Droplet Control by GERF ... 105
3.5 Droplet Logic ... 108
4 Family Tree and Development of GERF-Based Microdevices ... 111
References ... 113

Abbreviation

AgPDMS	Silver-PDMS composite
CPDMS	Carbon-PDMS composite
CPU	Central processing unit
CTP	Calcium and titanium precipitate
DNA	Deoxyribonucleic acid
ERF	Electrorheological fluid
EWOD	Electrowetting on dielectric
GERF	Giant electrorheological fluid
LOC	Lab-on-a-chip
MCM-41	Mobil composition of matter no. 41
MWNT	Multiwall-nanotube
PANI	Polyaniline
PCR	Polymerase chain reaction
PDMS	Polydimethylsiloxane
PM	Polar-molecule
PMMA	Poly(methyl metharcylate)
POC	Point of care
PPY	Polypyrrole
PS	Polystyrene
SBA-15	Santa Barbara amorphous no.15

1 Introduction

During the last decade, microfluidics emerged as a successfully revitalized, "slimmer" version of fluidics.[1, 2] Through microfluidics, new ground in point-of-care (POC) [3] and lab-on-chip (LOC) [4–6] applications has been carved out. Picoliter droplet runs in microfluidic channels are perfectly suitable for micro-synthesis, drug screening, and chemical tracing; droplets in their discrete nature, moreover, lend themselves analogously to the binary 0/1 coding system, implying, thus, an arithmetic application for microfluidics. Here an interesting question arises: can droplets and bubbles think? [7] What besides the binary logic operations could it provide?

Droplets, in any case, extend the concept of "information" from the binary world characterized by 0/1 to "information as a real entity," such as a DNA sequence and its chemical composition. As Marshall McLuhan famously announced: "The medium is the message". With microfluidic chips, high-throughput sample screening and information processing are made possible, and thousands of individual droplet "message carriers" can be individually dispensed [8]. As a result, high-density microfluidic control unions [9] are required. Despite the off-chip macroscaled solenoid arrays controlled by peripheral equipments, on-chip control components have spurred interest and attracted enormous attention owing to their scalability and cascadability. Among the proposed on-chip control schemes, the electrowetting-on-dielectric (EWOD) system [10, 11] is famous for its fine "digital" control of droplets, yet its predefined round-trip control, reflecting its electronic rather than fluidic nature, diminishes its flexibility. Likewise, individual manipulation of microdroplets remains a challenge: the shortcomings in automatic, reliable, and scalable methods for logic control prevent further integration of microfluidic applications [12]. By employing giant electrorheological fluid (GERF) [13], researchers have realized a series of fully chip-embedded soft-valves and have made strides in fluidic-based automatic droplet control systems.

Electrorheological fluid (ERF), composed of dielectric particles suspended in insulating oil, is a type of smart material that can be utilized as a two-phase system: fluid phase or solid phase. The viscosity of ERF can vary by a few orders of magnitude under the application of an external electric field. If the field is sufficiently strong, ERF can "solidify" into an anisotropic solid boasting a yield stress befitting its strength. The change in rheological characteristics usually is accomplished within 10 ms and is reversible. Hence, ERF has utility as an electrical–mechanical interface [14–16] for potential active-control clutch, damper, and valve applications [17–19], and is denoted "smart".

In this review article, we look at current topics in electrorheological (ER) material design and effect enhancement, including dispersed particles and the dispersing phase, and explore their applications in microfluidic devices for precise manipulation of fluids and droplets.

2 Electrorheological Fluid

The ER phenomenon was first discovered by Winslow in the 1940s; appropriately enough, the "smart" property of this material was initially named "the Winslow effect" [20]. Key experiments since then have established that certain kinds of particles suspended in insulating oils tend to form a fibrous structure in the direction of an applied electric field. Those particles found to be responsive to the electric field usually contain semiconductive solids of high dielectric constants. Early ERF were mostly hydrous, with cellulose, corn starch, silica gel, or zeolite as the dispersed particles [21–23]. The problems with hydrous ERF include weak yield stress under shearing, high corrosion, narrow working temperature ranges, low suspensibility, and low stability. By the 1980s, renewed experimental and theoretical interests advanced understanding of both the ER mechanism and the fabrication of novel materials with enhanced ER effects [24, 25]. Anhydrous ERF, which is to say, ERF without water or polar solvent in the dispersing phase, has become mainstream and boasts higher yield stress, an increased working temperature, improved stability, and lower zero-field viscosity, and other advantages. Considerable efforts have gone into the preparation of new and various particle suspensions with an eye to improving the ER yield stress. These suspensions include semiconducting organic polymers, inorganic solids with high dielectric constants, and organic–inorganic hybrid material. Particles of differing shape, surface morphology, and dispersing phase also have been closely scrutinized, specifically to increase local surface polarizations.

2.1 Inorganic ER Particles

According to well-known theoretical models, inorganic solids with high dielectric constant have a strong electroresponsive fibrous structure. Typical examples include TiO_2, calcium, strontium or barium titanate precipitates [26], zeolite or clay-type minerals, and polar-molecule-dominated ERFs (PM-ERFs) [27].

TiO_2 is among the most intriguing and well-studied ER materials [24, 28]. Given its high dielectric constant, it has been presumed to be a potential ER-active substrate; however, due to a low active intrinsic structure, it shows only very weak ER activity. Yin et al. doped TiO_2 with Cr ions to improve its ER activity, thereby increasing the defect and charge states and improving the local interfacial polarization effect; they found that the effective yield stress of a typical Cr-doped TiO_2 suspension at 3 kV/mm is 18 times higher that of a pure TiO_2 suspension [29, 30]. Also, assuming that a larger surface area and active interface would further enhance ER activity [31], they employed a block-copolymer-templated sol–gel method to obtain mesoporous TiO_2 of a different porosity. The effective yield stress of their mesoporous, Cr-doped TiO_2 ER suspension was three times higher than that of nonporous Cr-doped TiO_2 ERF, seven times higher than that of mesoporous pure TiO_2 ER suspension, and 20 times higher than that of nonporous pure TiO_2 ER suspension. Moreover, the ER effect of the mesoporous Cr-doped TiO_2 was found to be dependent on the surface area or porosity, and indeed, a high surface area or porosity sample was subsequently demonstrated to have a higher ER activity. Recently, Yin et al. modified the surface morphology of mesoporous Cr-doped TiO_2 to form sea-urchin-like TiO_2, which further improved ER performance. The hierarchical Cr-doped titania (TiO_2) particles consisted of high-density rutile Cr-doped titania nanorods of 20–30 nm diameter assembled radially on the surfaces of the particles. The specific surface area of the particles was close to 65 m^2/g, which was 13 times higher than that of smooth Cr-doped TiO_2 particles, the crystal structure and composition of which, significantly, were similar to those of the hierarchical ones. Their yield stress was approximately twice as high as that

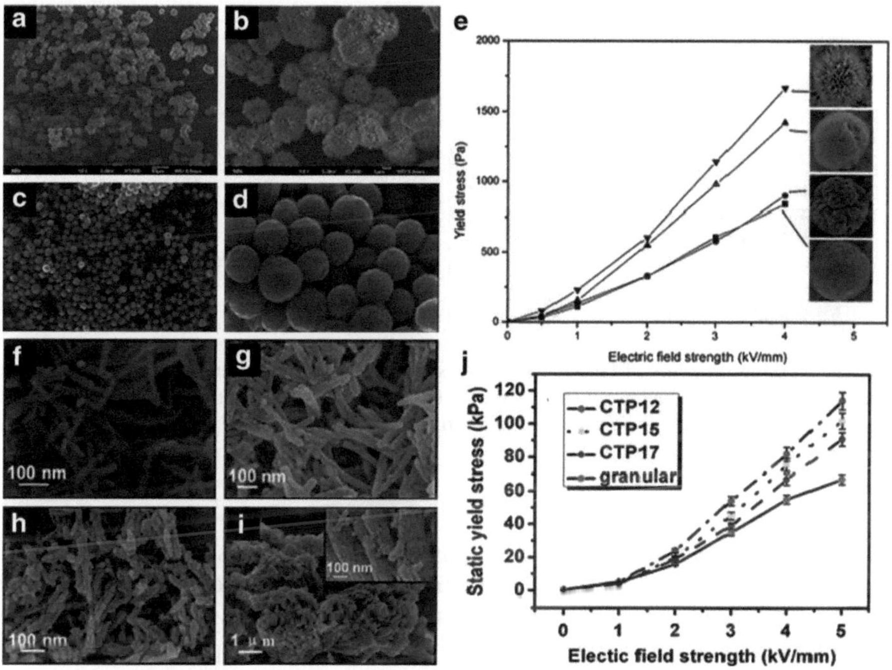

Fig. 1 (**a**, **b**) SEM images of sea-urchin-like hierarchical Cr-doped titania (TiO_2) particles, and (**c**, **d**) smooth Cr-doped TiO_2 particles. (**e**) Yield stress as function of electric field for ERFs of Cr-doped TiO_2 particles and differing surface morphology (*filled square* smooth Cr-doped TiO_2; *filled circle* surface-coarsened Cr-doped TiO_2 without sea-urchin-like nanostructure; *filled triangle* hierarchical Cr-doped TiO_2 with less well-developed sea-urchin-like nanostructure; *filled inverted triangle* hierarchical Cr-doped TiO_2) [32]. (**f–i**) 1D Ca–Ti–O-type nanorods of differing aspect ratio. (**j**) Yield stress as function of electric field for ERFs with CTP nanorods of differing aspect ratio (*dashed lines*) and with a granular CTP suspension (*solid line*) [26]

of the smooth Cr-doped TiO_2 suspension, 3.5 times as high as that of the hierarchical pure TiO_2 suspension, and 13 times as high as that of the smooth pure TiO_2 suspension (Fig. 1) [32].

Since the discovery of the giant ER (GER) effect of Ba–Ti–O-type nanoparticles (Fig. 2) [13, 33, 34], Ca–Ti–O- or Sr–Ti–O-type composites have been exploited for use as ER particles. These particles offer good wetting ability with silicon oil, and their ER effect surpasses the highest reported yield stress for conventional ERFs. To increase the suspension stability, Li et al. used carbon nanotubes to connect Ba–Ti–O particles together in a network structure. Carbon nanotubes prohibit the aggregation and sedimentation of particles, doubling the suspending time without deteriorating the GER effect [35]. Cheng et al. [26], using the co-precipitation method, synthesized one-dimensional (1D) nanorods of calcium and titanium precipitates (CTPs) for a high ER effect. The main components of CTP have been demonstrated to be polycrystalline $CaC_2O_4{\cdot}H_2O$, $TiOC_2O_4(H_2O)_2$, and

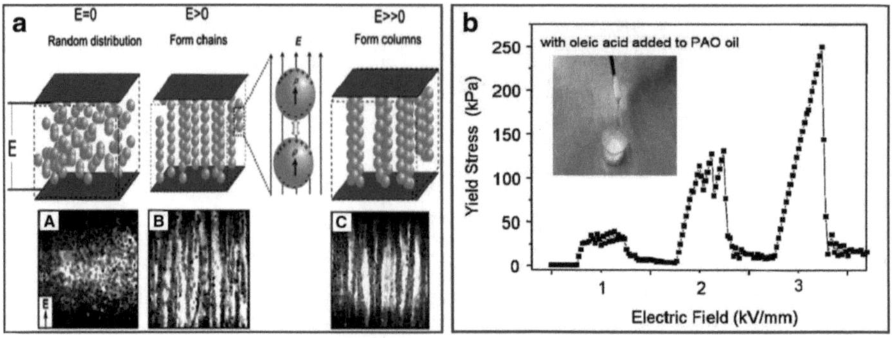

Fig. 2 Working principle of GERF. (**a**) GERF stays in liquid state when no electric field (*E*) is applied, begins to form chains under a moderate field of 500 V/mm, and then grows into columns as electric field increases. (**b**) Yield stress of hydrocarbon oil-based GERF with oleic acid additive, as function of applied electric field [33]

$TiO(OH)_2$. These rod-like particles have highly uniform widths and tunable lengths. Although a high aspect ratio of CTP particles was designed for a high static yield stress, experiment indicates that the effective static yield stress is a function of the electric field, and a moderate aspect ratio of CTP particles provides the highest effective yield stress. The effective static yield stress is defined as the static yield stress at electric field minuses the yield stress at zero electric field, and aspect ratio will also improve the yield stress at zero electric field. Through experiment, it was found that the yield stress of rod-like particles was 3.8 times that of a granular CTP suspension.

Besides TiO_2- and Ba–Ti–O-type materials, mesoporous molecular sieves and clay-type ionic crystalline materials also show admirable ER effects. Taking laponite-type ERFs as an example [36], a minimum electric field of 0.6 kV/mm can trigger laponite polarization and induce chain formation. Governed by the dielectric constant and the external electric field, laponite's rheology is similar to that of spherical-particle-based ER systems. Although the highest static yield stress obtained is lower than 1 kPa under an electric field, the rheology shows Newtonian-like behavior with relatively low shearing stress under a zero electric field.

Shen et al. [27] developed PM-ERFs as an effective means to enhance the ER effect of both TiO_2- and Ba–Ti–O-type materials. Their method adds urea (dipole moment = 4.56 D) or C=O and O–H polar molecules (2.3–2.7 D and 1.51 D) to TiO_2, and adds C=O and O–H polar molecules to Ca–Ti–O and Sr–Ti–O nano-particles, during their formation. The results showed that the yield stress of the PM-ERFs was greatly enhanced over that of the conventional ERFs. This effect of polar molecules can be verified by heating the particles at high temperature (500–800°C) and removing the absorbed polar groups (confirmed by IR and differential scanning calorimeter measurement), which leads to diminished ER yield stress comparable to that of the traditional dielectric ERFs. Shen et al. also derived a new ERF yield stress measurement method that roughens the surfaces of electrodes, preventing ERF from sliding under an applied electric field.

2.2 *Inorganic–Polymer Hybrid ER Particles*

The application of inorganic particles in ERF preparation accrues profitable advantages, including a high dielectric constant, size and shape variability, an accessible interfacial area, and versatility in surface modification. However, the relatively high particle density and low interfacial polarization (when suspending in dispersing fluid) bring about the instability of suspension and low yield stress, which inhibit the practical applications of inorganic ER particles. Hybrid ER particles, such as clay-type particles intercalated by semi-conductive polymers, carbon nanotube hybrid polymers, and metal-doped polymers, embody the organic–inorganic synergistic effect. A variety of polymers, including polystyrene (PS), poly(methyl metharcylate) (PMMA), polyaniline (PANI) and its derivatives, polypyrrole (PPY), and styrene-acrylonitrile, have been hybridized with inorganic materials such as carbon nanotubes, kaolinite, montmorillonite, laponite, mesoporous molecular sieves, and others [37].

Conducting polymers such as PANI- or PPY-intercalated clay-type minerals are another type of ER particle material. Kim et al. synthesized PPY-intercalated montmorillonite nanocomposites through inverted emulsion pathway polymerization, and characterized its ER effect under an electric field [38]. The nanocomposites showed not only a typical ER behavior under electric fields but also the existence of a critical electrical field when the yield stress was plotted as a function of the electric field.

Conducting polymers also can be utilized to form core–shell structures with high dielectric constant particles. Fang et al. used PANI to encapsulate barium titanate via in situ oxidative polymerization. They examined the influence of the fraction of $BaTiO_3$ particles on the ER behavior, and found that the PANI/ $BaTiO_3$ composites-based ERFs exhibit a better ER effect than does pure PANI, which result might be due to the unique ferroelectric properties as well as the high dielectric constant of $BaTiO_3$ nanoparticles.

Cho et al. reported ERFs with conducting PANI and a silica-based mesoporous molecular sieve (MCM-41). The PANI was confined in the channels of the mesoporous MCM-41, partially filling them. It induced enhanced dipole moments in the longitude direction, thereby increasing the ER effect relative to pure MCM-41 [39]. The same results were found for PANI-inserted particles of the mesoporous molecular sieve SBA-15 [40].

Recently, Jin et al. developed carbon-nanotube-absorbed polymeric microspheres and studied their potential application in ERF. Nanotubes were incorporated into the surface of PS and PMMA microspheres. Later, Park et al. investigated the enhanced ER effect of multiwall-nanotubes (MWNTs) on the insulating PMMA [41].

2.3 Conducting-Polymer-Based ER Particles

Semi-conducting polymers with the π-conjugated electron system have a fine electron-donating property but a low ionic potential, and thus can potentially be used as ER materials. These polymers include PANI, PPY, polythiophene, poly (phenylenediamine) and poly(*p*-phenylene), their derivatives, and copolymers. Among these polymers, PANI and its derivatives have been the most commonly reported, owing to their ease of preparation via chemical oxidation polymerization of aniline. PANI can be utilized also as an encapsulating or encapsulated material in the formation of core–shell structures. PANI has various derivatives and thus can be grafted onto inorganic or other organic materials to adjust particle conductivity or density. The core–shell PMMA/PANI microspheres employed as dispersed materials in ERFs represent an example [42].

2.4 Dispersing Phase

ERFs are a type of smart material composed of dielectric particles suspended in insulating oil, and can be utilized as a two-phase system. The flow characteristics of this system depend on both the properties of the dispersed particles and the oil (Fig. 3). The chain length of the oil and the functional groups in the chain can influence the interfacial adhesion strength between the two phases and, consequently, the viscoelastic properties. Such an effect can favor the agglomeration of particles into large clusters by interparticle electrostatic and van der Waals forces. To investigate the oil effect, oils with the same functional groups but different viscosities, and oils with same viscosity but different functional groups, have been studied and the results indicate effectively enhanced yield stresses. The choice of hydroxyl-terminated silicon oil of 25 cSt viscosity resulted in the highest yield stress, 300 kPa [43].

2.5 Future Research Directions

In order to synthesize effective nanosized ER particles, current experimental devices and processes should be modified. This involves controlling the reaction time and solvent temperature, as well as other conditions during the chemical synthesis. It is also crucial to prepare nanoparticles with various dehydration methods under suitable conditions. For example, for the synthesis of inorganic GER particles, researchers have found that pure $BaTiO_3$ nanoparticles do not form a promising ER suspension, even though such particles possess a very high dielectric constant; however, surface modification of such ceramic particles can dramatically improve the ER yield stress. Although water can also greatly enhance the ER yield stress, it suffers from an inherent drawback in that water can evaporate. Therefore, the ER effect is intensely associated with the water molecule concentration, leading

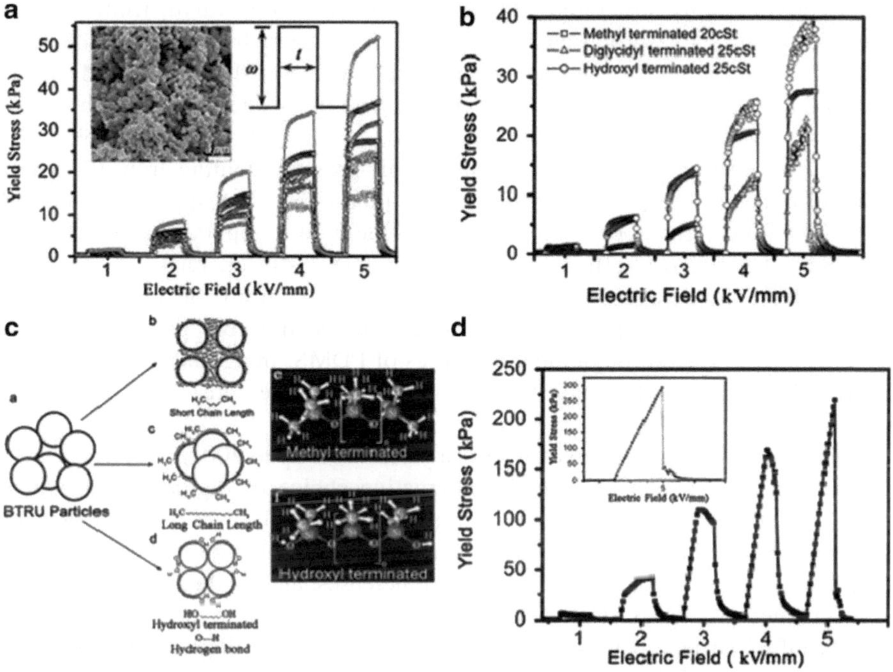

Fig. 3 (**a**) ER effects of seven GERFs constituted of different silicone oils. *Inset*: SEM photograph of the as-prepared BTRU (urea-coated barium titanyl oxalate) particles of around 100 nm diameter. (**b**) Yield stresses measured as function of applied electric field of GERFs constituted of hydroxyl-, methyl-, and diglycidyl-group-terminated silicone oils. (**c**) ERF structures with methyl-terminated silicone oil and hydroxyl-terminated silicone oil. (**d**) Yield stress variation of GERF constituted of hydroxyl-group-terminated silicone oil, measured as function of applied electric field. The concentration is 0.25. The *inset* shows the limiting yield stress value of 300 kPa obtained by direct application of a 5 kV/mm electric field [43]

to unstable and sometimes non-repeatable experimental results. Therefore, the role of water in the ER effect needs to be further explored, and optimization of coating processes to obtain a uniform surface coating layer is also necessary. In addition, the ER effect might also be increased by employing various surfactants with different molecular dipole moments.

It is also found that the liquid phase is crucial in obtaining a good ER suspension and intensive ER effect. The role of silicone oils and other nonconductive liquids should be examined from both experimental and theoretical aspects.

3 ERF-Based Precise Microfluidic Control System

Of all the microfluidic issues, microvalve and droplet logic are among the most basic and crucial areas of research for both one-step fluid actuation and multistep precise fluid control [44–47]. Researchers have never stopped pursuing an effective, simply structured and easy-to-fabricate microvalve; from the first thin membrane valve proposed by Unger et al. [9] to hydrogel valves [48, 49], heat-control valves [50], and screw-pneumatic valves [51–53], this effort has retained its significance through the decades. The first-generation polydimethylsiloxane (PDMS)-based microfluidic valves comprised two cross-channels separated by a thin PDMS membrane; whereby air pressure coming from the lower channel deflects the thin membrane, which then closes the upper channel and blocks the flow in the upper channel [9]. This membrane-separated cross-channel design has been proved effective, and its principle is still operative to date [54]. Nonetheless, the design is subject to flaws resulting from the laxly effective response of air (a compressible medium), and the unstable pressure brought about by air leakage through the micropores of PDMS. Integratable and digitally addressable microfluidic valves with fast response are rare in microanalysis systems.

GERF (discovered by Wen et.al.) has a yield stress up to 300 kPa in response to an electric field, which provides an alternative choice of digitally controllable microvalve that can respond within 10 μm [13, 43]. Its solid-like behavior sustains shear in the direction perpendicular to the applied electric field, the shear stress can be enhanced when the applied electric field increases, and its rheological variation is reversible upon removal of the electric field (Fig. 4). These marvelous features qualify GERF as an electric-fluid-mechanical interface for digital fluid control in microfluidics [55, 56].

As shown in Fig. 4, when there is no electric field applied to the GERF in the microfluidic channels, it keeps flowing; but when an electric field is applied, the GERF starts to form columns, effecting high yield stress to balance the pumping pressure, leading to flow blockage if the electric field is sufficiently strong. The channel size in microfluidic systems typically is around 10~100 μm. Rather than the high voltage (typically in kilovolts) required in macroscopic applications, the GERF control voltage in microfluidics is limited to ~200 V, a value both feasible and safe in daily life.

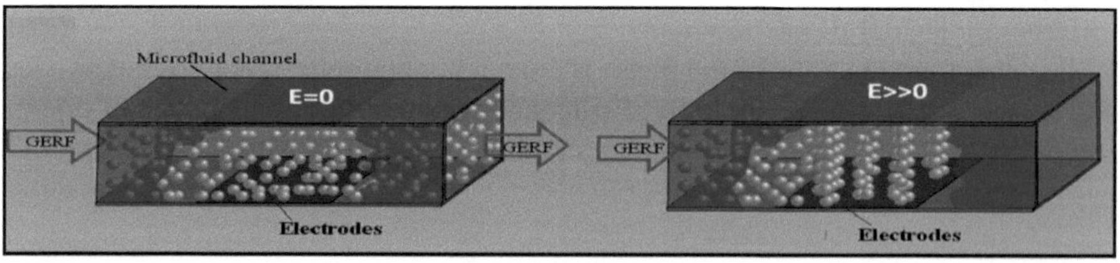

Fig. 4 When no electric field (E) is exerted, GERF keeps flowing because it is pumped, and when electric field is applied on GERF by embedded parallel plate electrodes, GERF solidifies. A sufficiently strong electric field will cause a very high GERF yield stress, which will result in the fluid suspending effect

3.1 *Soft Conducting Electrodes for Droplet Detection and ERF On-Chip Control*

PDMS is a silicon-based organic polymer that has been widely used in microfluidic chip fabrication owing to its good elastic property, nontoxicity, biocompatibility, optical transparence, non-inflammability, chemical inertness and conformability, among other attributes [57–61]. Regarding the demand for electrical signal detection and device control in microfluidic chips, integration of conducting structures into bulk PDMS has been a crucial issue. However, PDMS is an inert polymer lacking conductive and magnetic properties. In addition, due to the weak adhesion between PDMS and metal, it is problematic to pattern metallic structures onto its surface or into bulk PDMS for microdevice fabrication.

Recently, Niu et al. developed a method of patterning conductive structures using PDMS-based conducting composites, which are synthesized by uniformly mixing conductive micrometer silver or nanometer carbon particles with PDMS gel [62]; the resultant mixtures, respectively, were denoted AgPDMS and CPDMS. Silver and carbon-black particles, thanks to their desirable wetting characteristics, are easy to mix with PDMS gel. To fabricate the soft conducting composite into PDMS-based chips (Fig. 5a), Niu et al. embedded AgPDMS or CPDMS gel into a photoresist mold (Fig. 5b) on a glass substrate for patterning of conductive composites. After baking, the gel was cured into a solid, and the conducting composite pattern was retained on the substrate by removing the photoresist mold. (The complete fabrication process is described in [62].) The experimental results indicated that two-dimensional (2D) and three-dimensional (3D) conducting microstructures, the dimensions of which can range from tens to hundreds of micrometers, had been

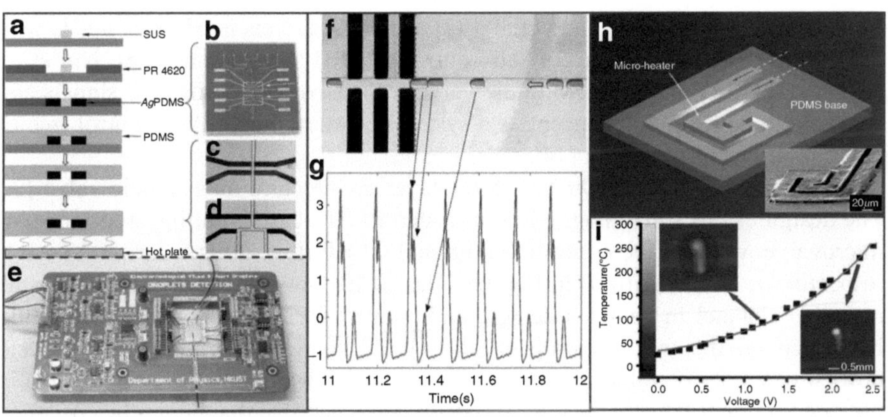

Fig. 5 (**a–f**) Method of fabrication of chip-embedded electrode (AgPDMS) [62]. (**e**) On-chip electric detection module. (**f**) Optical image of group of water droplets of different sizes. A small amount of dye was added for labeling [63]. (**g**) Detected signals. (**h**) Microheater fabricated from AgPDMS. (**i**) Response according to applied voltage [64]

successfully constructed and integrated into PDMS bulk material (Fig. 5c, d). The advantage of using PDMS-based conducting composites is the ease of microstructure bonding and embedding into PDMS-based microchips, along with the formation of electric signal connections in a 2D or 3D microstructure, thereby effecting great enhancements in the potential functionalities of microfluidic chips. With the help of AgPDMS or CPDMS, channel-wall-embedded electrodes, leading to highly integrated on-chip droplet detection (Fig. 5e–g) and GERF control, is achievable.

Electrical input, sensing, and detection of microfluidic signals is the alternative to pneumatic and optical methods. A PDMS/glass-based electroporation [65–68] or analytical microsystem [6, 59, 69] requires chip-patterned electrodes. For the purposes of sensing and detecting non-continuous droplet volumes [70, 71], Niu et al. considered that droplet size and distance can vary both spatially and temporally and introduced a capacity detection method for determining droplet size, shape, and composition [63]. By means of a pair of parallel electrodes installed across the microfluidic channel, very small capacitance variations can be detected when a droplet passes through. Thus, due to the electrode's design and feedback electronic circuit (Fig. 5e), real-time and accurate determination of the size, shape, and composition of droplets can be achieved (Fig. 5f, g). Soft conductive patterns can serve not only as electrodes but also as chip-embedded soft microresistors, microheaters (Fig. 5h) [64, 72], and micropressure sensors [72]. Figure 5i shows a plot of the voltage (or current)-dependent temperature control of such a microheater.

3.2 GERF Microvalves

Yoshida et al. [55] and Nakano et al. [73] tested ER effects in SU-8 channels with indium tin oxide electrodes, for self-control of ERF in hard substrates. Niu et al. designed a GERF-based microfluidic valve responsive to external DC signals and was able to develop this concept as a system for microfluidic flow control free of any limitations of flow type [74].

This approach follows Quake's three-layer-architecture microfluidic valve [9]. The design of this soft-lithography-fabricated PDMS-based multilayered four-port microvalve, wherein the controlled fluid and GERF flow along the x- and y-axes, respectively, is schematized in Fig. 6a. The GERF flows in layer 1, replacing the air-valve channel in Quake's design, and parallel CPDMS electrodes are tightly integrated on the two sidewalls of the GERF channel, forming upstream and downstream control valves. The controlled liquid flows in layer 2. As indicated in the intersectional scheme (Fig. 6a), a 35 μm-thick flexible diaphragm lies between the two layers, separating the cross-channels. Figure 6b shows an optical microscopy image of such a microvalve chip. GERF is continuously pushed at a constant pressure into the chip by a syringe pump (Fig. 6c). With an adequate DC electric field applied alternately to two electrode pairs, the pressure in the GER channel between the two valves can be modulated as the two valves are alternately opened

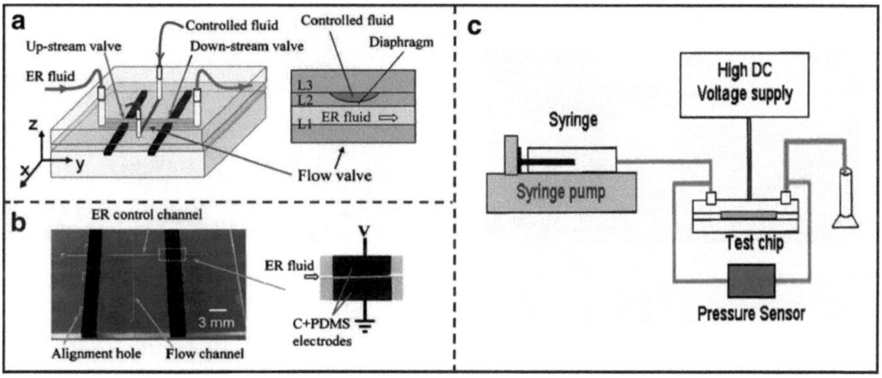

Fig. 6 Design and fabrication of four-port microGER valve. (**a**) Design of four-port valve chip. *Right*: cross-section of the different layers forming the flow valve; *L1* is the GER channel layer, *L2* the controlled flow channel layer, and *L3* the cover layer. (**b**) Optical photograph of fabricated microGER valve chip. *Right*: top-view image of the planner electrodes and the GER channel. (**c**) Experimental setup for microvalve testing [74]

and closed. Such a pressure change within the ER channel will eventually result in the deformation of the flexible diaphragm with a vertical pull-and-push movement. In this way, the liquid flow in flow channel layer 2 is controlled by the pressure changes in the GERF channel.

3.3 Integratable Microfluidic Components Based on GERF Microvalves

This microvalve approach, providing a variable pressure source by means of a chip-embedded microchannel, is a simplified on-chip pressure-control scheme that minimizes the need for peripheral equipment and electronic components, which in any case are difficult to integrate into a microfluidic chip. Apt application of this method can realize many desirable micropump [75], micromixer [76], and micro-platform [77] functions such as those illustrated in Fig. 7.

Given the need to drive fluid inside chips, a micropump is a necessary component of all microfluidic chips, regardless of specific applications. Various types of pumps have been designed and fabricated using different mechanisms, e.g., the air pump is actuated and controlled by gas air-pressure [9, 47], and the piezoelectric transducer actuator pump [78] utilizes electricomechanical energy conversion. The GERF-actuated microfluidic pump (Fig. 7a–c) is digitally programmable, and exhibits good performance at high pumping frequencies along with uniform liquid flow characteristics [75]. In this design, a five-layer structure is embedded inside the PDMS chip, the bottom layer channeling GERF that can affect the flow of the circulating fluid on the top layer via the pull–push movement of the diaphragms

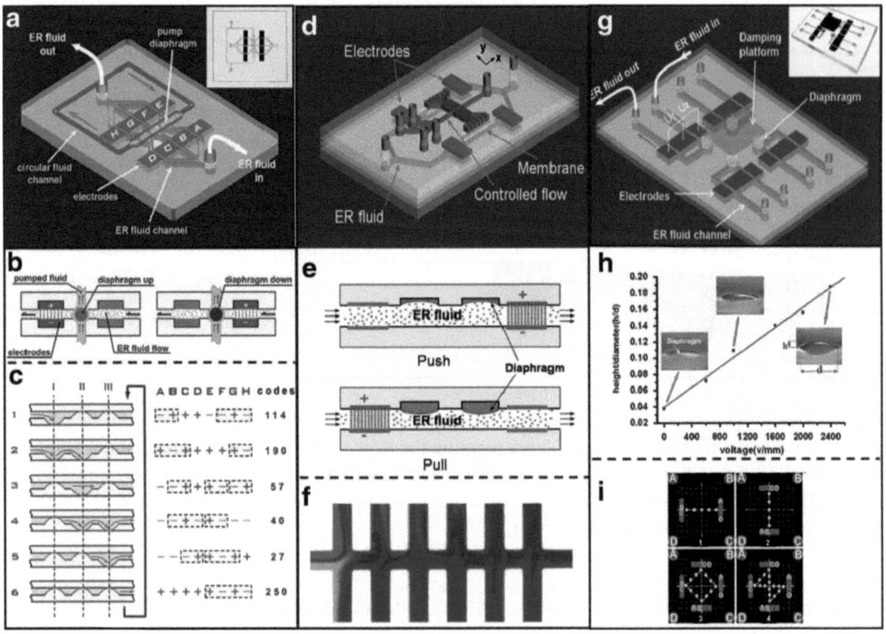

Fig. 7 Microfluidic devices realized by integration of GERF-based microvalve. (**a–c**) Micropump by integration of three microvalves [75]. (**a**) Micropump of 3D structure. The GERF-actuated chip controls the fluid circulation in the upper layer. (**b**) Single diaphragm valve via pumped GERF. (**c**) The diaphragm's pumping sequences and their corresponding signals. (**d–f**) GERF-actuated mixer design [76]. (**d**) Scheme of mixer construction. (**e**) Side view of push-and-pull GER valve. (**f**) Sinusoidal cross-stream flows in the six pairs of side channels. (**g–i**) Microfluidic platform [77]. (**g**) Flexible platform of microfluidic chip. (**h**) Displacement of diaphragms plotted as function of applied electric field across GERF channels. (**i**) Time traces of laser spot reflected from platform are shown on screen with coordinates. Digitally programmed electrical signals to the four ER valves generate the complex leveling modes of the platform to direct the laser spot

between two pairs of electrodes (Fig. 7b). When a control DC signal is applied to the electrode pairs sequentially, pumped flux in the upper layer varies as a function of the intensity and frequency of the electric field. The direction of the fluid flow and the pumped flux also can be controlled through a programmed signal sequence applied to four-electrode pairs (electrodes A–H in Fig. 7b, c).

Mixing two or more streams of fluids is an important issue in various microfluidic devices. The mixing process is not trivial on the microscale, owing to the dominance of the viscous effect and, hence, laminar flow phenomenon. Passive mixers are designed to induce 3D helical fluid motions from patterned structural asymmetries that can fold the streams into highly nested thin slices, so as to facilitate local molecular interdiffusion. Another approach is that of active mixers employing dynamic control to help achieve chaotic mixing. A PDMS-based GERF-driven cross-stream active mixer was reported by Wen et al.

Figure 7d shows a schematic depiction of the active mixer chip design, consisting of a main flow channel and six pairs of orthogonal side channels [76]. Operation of the mixer chip relies on the perturbation of the main x-directional channel flow by y-directional cross-stream side-channel flows. The side-channel flows are perturbated by pressure changes through thin membranes affected by a GERF microvalve (Fig. 7e). Square-wave electrical voltage signals (0–800 V) are applied between the electrodes to control the microvalve and, in turn, the perturbation, leading to pulsating sinusoidal cross-stream flows in the six pairs of side channels, as shown in Fig. 8f.

The microplatform (sketched in Fig. 7g–i) also works by way of the assistance of microvalves, integrated into the four corners of the chip [77]. The programmed push-and-pull of each microvalve will deform the diaphragms (Fig. 7h) of each valve sequentially, thereby realizing the desired platform function, as shown in Fig. 7i.

3.4 "Smart" Droplet Control by GERF

Despite the successful applications of multilayer-structured valves, researchers have derived a new methodology for on-chip fluid control utilizing the soft/droplet valve. This approach avoids the complexity of the multilayer architecture while maintaining the robustness of GERF-based devices. In these new designs, GERF and target fluid are operated in the same layer (or the same channel), with integrated electrodes on the channel sides. Two protocols have been tested: (1) GERF as carrier flow for target droplets, and (2) GERF droplets as an in-channel soft valve for target fluid or droplets. Thus, new research ground has been broken: GERF-modulated droplet microfluidics, also known as digital microfluidics, in which many digital and logic functions were and have been demonstrated.

As shown in Fig. 8a–f, fluids that are immiscible with GERF can be generated in GER carrier flow [56, 79]. The droplet generation schemes (flow focusing and T junction) are illustrated separately in Fig. 8a, b. As GERF (the carrier flow) is electrically controlled by electrodes placed on the sides of the GER inlet channel and near the fluid junction area, the droplets generated and dispersed (i.e., water, oil, and gas) are no longer uniform in size but can be determined by the controlling electric signal, as shown in Fig. 8c, d. The desired droplet length and separation speed, in other words, can be achieved by electrically controlling the pressure of the carrier flow.

An even more useful functionality of this configuration is active modulation of the relative positions of droplets [56]. Because an electric field can be applied to chip-integrated electrodes, GERF can "solidify" between them and the flow in the relevant channel will be "frozen," as shown in Fig. 8e, f. The relative droplet position can be adjusted in this way. Moreover, as the influence is achieved through medium fluid (GERF), direct application of electric field to target droplets is

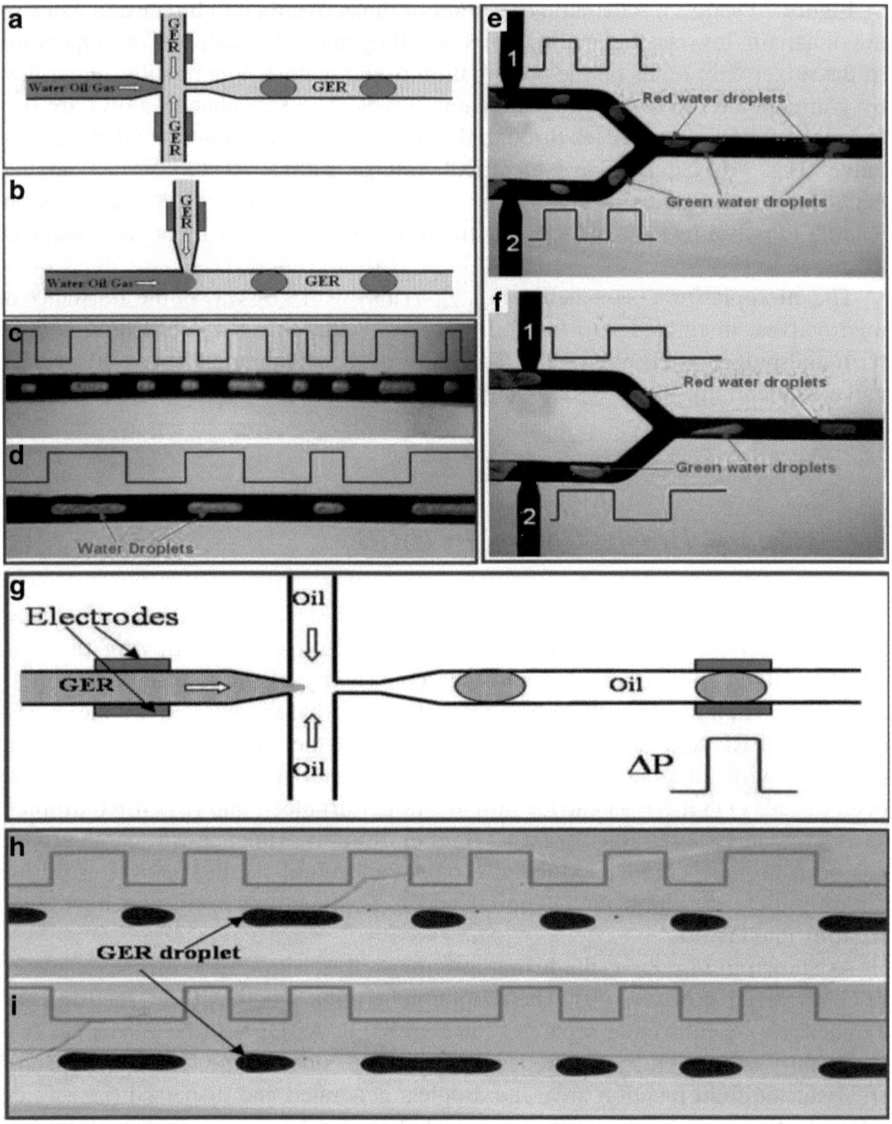

Fig. 8 (**a–d**) GERF-assisted droplet generation: (**a**) flow-focusing approach; (**b**) T junction. (**c**, **d**) Electric-field-controlled generation of droplets in a microchannel. *Lines* indicate the electric control signal applied to the electrodes embedded on both sides of the GERF inlets. (**e**, **f**) Digitalized controlled droplets distance [56]. (**g–i**) Digital GERF droplet generation: (**g**) schematic view; (**h**,**i**) different droplet patterns under different electric pulse trains [79]

avoided, preventing electrolysis or electrical cell lyses in target droplets. A reverse application of this technique is to put GERF as the dispersed phase, i.e., with fluid as the control, as shown in Fig. 8g. Examples of active digital control of GERF droplets are shown in Fig. 8h, i. The length of GERF droplets responds exactly to the input electric signals.

These approaches are significant not only for digitalizing in-channel GERF control but also for in-channel soft-valve schemes for more advanced applications, such as droplet display and droplet position modulation. In these applications, GERF (or its droplet) exhibits the capability of an in-channel soft valve, redirecting itself (Fig. 9a, b), guiding other fluid or droplets (Fig. 9c), and even reversing droplet order in channels (Fig. 9d).

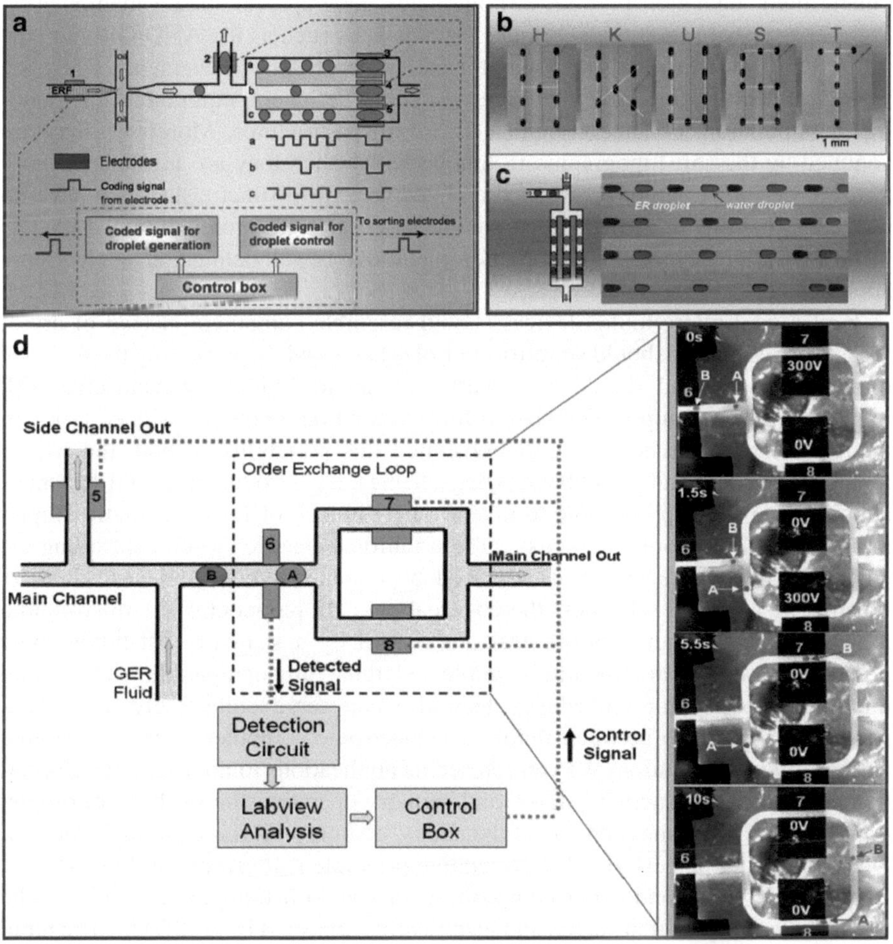

Fig. 9 (**a–d**) GERF-assisted smart droplet manipulation. (**a**) Flow chart and control circuit for the generation of a smart droplet display. (**b**) Optical images of the smart droplet display. (**c**) *Left*: chip component showing the orthogonal channels to form the water droplet "packages." *Right*: optical images of the generated packages formed with different numbers of water droplets sandwiched between two smart droplets [79]. (**d**) Flow chart and control circuit for droplet order exchange. *Right*: optical images taken at different times during the exchange process [56]

3.5 Droplet Logic

All the latest popular electronic devices, including the iPad, have evolved from the first vacuum tube, i.e., the first electronic logic gate. Now there are more than twenty million logic gates functioning in the CPU of any personal computer. Scientists have endeavored to reinvent this near-legendary component in other systems. Some binary logic functions have been successfully mimicked by fluidic diodes, microelectrochemical logic [80], and conducting-polymer-coated micro-electrode arrays [81]. In microfluidic domains, researchers have scrutinized both kinetic fluid regulation [47, 82, 83] and static geographical stream manipulation [84–87] as possible solutions. Simple logic devices such as the AND/OR gate, the static fluid transistor, and the oscillator are some of the achievements. They are limited, however, in that they entail either bulky peripheral equipment for round-trip manipulation, or have complicated 3D microstructures. Moreover, they are confined by the soft-lithographic technique with which they are formed. Because they are designed within pre-shaped architectures for distinct tasks, they have no reprogrammability or cascadability. Another solution is the aforementioned EWOD method, in which every single step of droplet move is well defined, in an electronic approach [10, 11, 88, 89].

Real digital microfluidic devices should resemble computers instead of preset "music boxes". They should simplify control schemes while preserving the delicacy of microdevices, and should be "smart" enough to "think" by themselves [7], indicating that the outputs that should fully depend on inputs in assigned tasks. In microfluidics, researchers have demonstrated this possibility in both the stream regulation method and in bubble/droplet schemes. By introducing the GERF smart colloid, Wen's group was able to invent a new branch of fully automatic droplet logic control: the droplet logic gate. The traditional electric switch controlling the GER ON/OFF phase change is replaced by conductive/high dielectric droplets flowing in a nearby channel, thereby realizing droplet-controlled microfluidic logic (on-chip droplet control) [90]. Fabricated by a standard soft-lithographic process [63], its planar structure is simple and, thus, fully compatible with existing microfluidics. Its chip-embedded electrodes can serve collectively as a data-exchanging interface between fluidic and electronic computer systems, enabling their seamless integration. We can foresee its applications to microfluidic information processors, transacting control, and memory operations on the basis of droplet trains in which nonlinear chemical dynamics, complex neuron communication, or DNA computing is performed. Its operative principle is illustrated in Fig. 10.

Thanks to soft conducting composites, through soft-lithography, microfluidic channels can be planar arranged and electrically connected by AgPDMS. The input of this device, droplets between electrodes as switches, can be modeled as impedance in circuits. That is, alterable impedance in fluidic form can be used to adjust the voltage applied to GERF or as the ON/OFF switch of the GERF phase change. A sufficiently large constant voltage is supplied across the two electrodes to solidify the GERF under the desired conditions. Preferably, the signal fluid is an ionized,

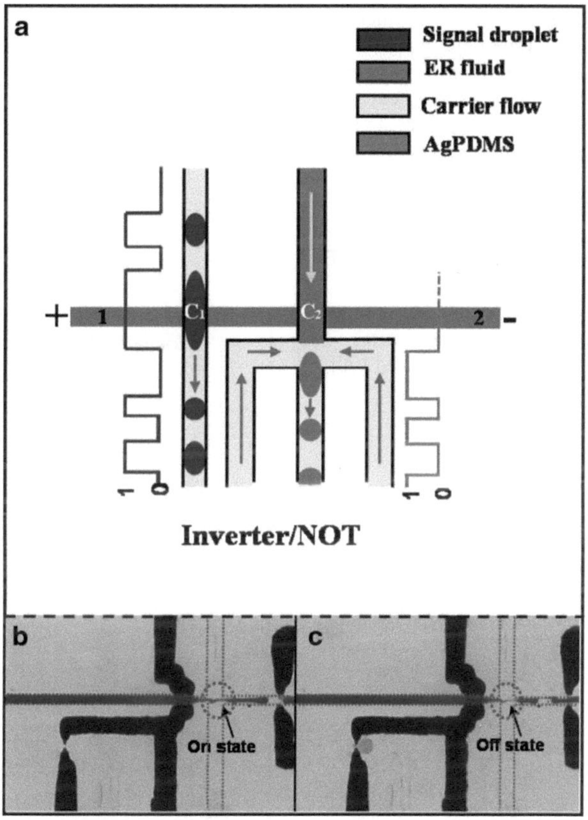

Fig. 10 Basic working principle of logic gate, illustrated by microfluidic inverter. (**a**) Basic working principle. The presence of a signal droplet between the signal electrodes will solidify GERF, whereas the presence of carrier flow will release GERF. (**b**) When the carrier flow is flowing between the signal electrodes, GERF flows continuously. (**c**) When the signal droplet passes by the signal electrodes, GERF solidifies between those electrodes, and is cut into droplets [90]

high-conductivity solution (which can be modeled as a resistor or conductor) or a high-dielectric-constant fluid (which can be modeled as a capacitor), and the carrier flow is an insulating fluid. When the carrier flow presents between the signal electrodes (input 0), the circuit is in an open state, and the GERF continues flowing (defined output: binary 1). When the signal droplet presents between the signal electrodes (input 1), the circuit is in a closed state, and the electric field generated stops the flow of the GERF (defined output: binary 0). When the signal fluid is a dielectric fluid, the fluid in the signal channel can be modeled as a capacitor and denoted as $C_1(x)$, where $x = 0$ if the input is carrier flow or $x = 1$ if the input is signal droplets. The capacitance generated by the GERF is denoted as C_2. In this case, if the applied voltage is V and the voltage share of the GERF is $V_{\mathrm{ER}}(x)$, we obtain:

$$V_{\mathrm{ER}}(x) = \frac{C_2}{C_1(x) + C_2} V. \tag{1}$$

Further, we can adjust input voltage V to ensure that $V_{\mathrm{ER}}(1)$ is larger than the GERF solidification voltage and that $V_{\mathrm{ER}}(0)$ is smaller than the crucial value. By

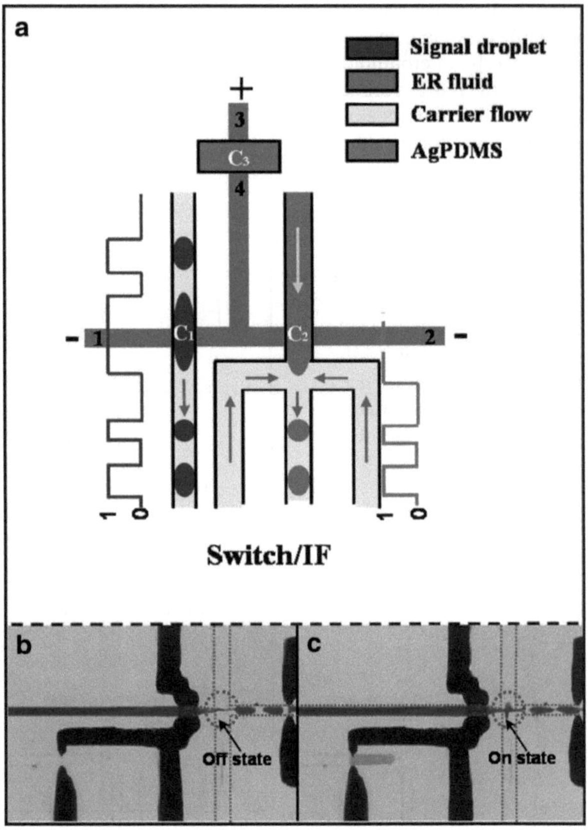

Fig. 11 Working principle of microfluidic switch. (**a**) Microfluidic logic switch is built by adding a capacitor (in the present case, another microfluidic channel filled with steady GERF) to the logic inverter. (**b**) When carrier flow presents between the signal electrodes, GERF solidifies. (**c**) When signal droplets present between the signal electrodes, GERF liquefies and flows out [90]

application of this principle, a fluidic inverter (NOT gate) can be realized. As shown in the experimental results, when there is signal fluid (water droplet) between the signal electrodes, the GERF will stop, and a GER droplet is formed (Fig. 10c); otherwise, the GERF will flow continuously (Fig. 10b).

A fluidic switch is designed according to the principle of a logic inverter, specifically by electrically connecting a capacitor to reverse the logic state of the inverter, as shown in Fig. 11a. The capacitor should have a capacitance comparable to that of the GERF, in order to effectively share voltage with the GERF output channel; when the signal droplet is present between the signal electrodes, the voltage share of the GERF is smaller than the crucial value for its phase change. A simple design methodology is to set the voltage input on electrode 1 to be the electrical conjugation of the voltage input on electrode 2, as shown in Fig. 11a. When the signal droplet is dielectric fluid, we can derive a simplified capacitance model: the input voltage from electrodes 1, 2, 3 is V_1, V_2 and V_3, and the capacitance in the signal channel is $C_1(x)$, where $x = 0$ when the carrier flow is between the signal electrodes and $x = 1$ when the signal droplet passes by the signal electrodes. C_2 is the capacitance of the GERF between the output electrodes, and C_3 is the capacitance of the added capacitor. $V_{ER}(x)$ is the voltage share of the GERF under a different input situation, the value of $V_{ER}(x)$ being:

$$V_{ER}(x) = \frac{(V_2 - V_3)C_3 + (V_1 - V_2)C_1(x)}{C_3 - C_1(x) - C_2}, \tag{2}$$

where V_1, V_2, and V_3 can be tuned for a desired voltage arrangement, making the GERF solidify when a signal droplet comes by or, conversely, change back to the fluid state when the droplet passes the signal electrodes. To rule out variation of the dielectric constant among different batches of GER powder, we used an additional steady GERF channel (with embedded electrodes) of identical dimensions with GERF output channel to compose an additional fluidic capacitor (shown in Fig. 11a). The case of ionized buffer signal droplets is straightforward to understand, and can be illustrated by our experimental results (see Fig. 11b, c).

The advantage of our microfluidic logic switch and inverter is that two counterpart functions are realized in one single-chip structure. Alternation between the two functions is achieved by selective voltage input, or by connection or disconnection of electrode pads. It is always desirable to have fewer units working to more purposes, for simplicity of logic device architecture and better reprogrammability [91]. Because a logic switch and inverter are the basic units of logic operations, a system comprising these two units can have additional functions simply by rearranging the voltage input, obviating any need for restructuring.

4 Family Tree and Development of GERF-Based Microdevices

By way of conclusion, we have sketched a family tree (Fig. 12) of ERF and the related techniques for realization of microfluidic control in microfluidic chips. Through improvements made to ER effects and the development of soft conducting composites, researchers have been able to integrate those techniques with microfluidics in order to digitalize droplets of nano- to picoliter size and achieve droplet logic, storage, and display modules.

The flexibilities in this family of microfluidic techniques are all functions of the distinctly smart material employed: GERF. This treble-function medium can be compared with an electric current: the fluidic output is the response, the dielectric information is the electric medium, and the control of fluid is the mechanical yield source. Sharing compatible working principles, the demonstrated GERF-actuated microfluidic mixer, storage, display, and droplet phase modulator functionalities are all compatible with each other. Liu et al. demonstrated a highly integrated DNA-amplifying microfluidic chip by employing technology in this family tree [92]. In the near future, simple combinations of IF/NOT microdroplet logics could lead to microfluidic processors, analogs to microelectronic computers. To take it

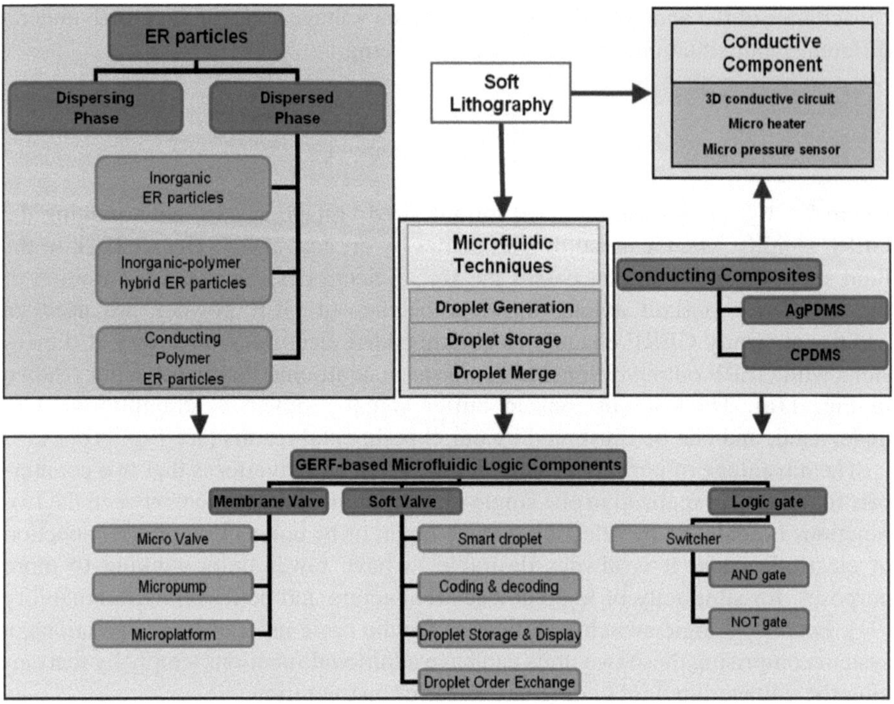

Fig. 12 Family tree of ERF-based microfluidic technologies

one step further, integration of all of the techniques described above might lead to a GERF-based microfluidic computing system. Moreover, the components of this suggested system all have chip-embedded electrodes, which can serve as information interfaces with electronic devices, promising a highly integrated system comprised of a computer and microfluidic processors.

Inevitably, the processing capability of this logic device (GERF response is 10 ms) will be compared with that of computers (electronic processes take nanoseconds). The delay can be meliorated by adjusting the flow speed and flow-focusing geometry, but not eliminated. It needs to be borne in mind that microfluidics and electronics deal with different issues: microfluidics is not expected to become a mainframe computing system but is earmarked for exploratory, LOC research and POC applications (e.g., portable diagnosis kits), areas in which conventional computers have their own intrinsic shortcomings. The future of microfluidics lies not in computing but in multidimensional information processing. Microfluidics in any case retains its inherent promise: an extension of the fluidic information realm beyond "binary 0/1" to the spatial, chromatic, or physiological dimension [69, 71, 88, 93]. Preloaded chemical or biological information can be well preserved in droplet form. Droplet PCR, for example, can easily store and recreate genetic information [94]. Microfluidics provides a unique tool for handling and processing biological, chemical, environmental, genetic, and chromatic information. Considering the contribution of DNA logic to fuzzy computing [95, 96], which indeed can be elaborated in picoliter droplets, it is really difficult to foresee a bound future of microfluidics if tools like DNA computing are incorporated.

We can imagine a microfluidic processor, performing important control and memory operations on the basis of droplet trains. Nonlinear chemical dynamics, complex neuron communication, or DNA computing might be carried out in every single droplet of this processor, and these droplets could couple together for more complex tasks. Electromagnetic technology extended the human sensory system and enabled us to sense the world through a portable device. Through microfluidic technology, we are extending a part of ourselves (blood, tissue, cell, or DNA) to microchips, and beyond. The GERF-assisted microfluidic technology can combine the extended "human body" and "human sensory system" on a piece of microfluidic chip, in a fully automatic sense. It will be interesting to see the outcomes of such a coupled system.

Acknowledgements This publication is based on work supported by Award No. SA-C0040/UK-C0016, made by King Abdullah University of Science and Technology (KAUST) and Hong Kong RGC grants HKUST 603608.

References

1. McDonald JC, Duffy DC, Anderson JR et al (2000) Electrophoresis 21:27
2. Quake SR, Scherer A (2000) Science 290:1536
3. Dudek MM, Lindahl TL, Killard AJ (2010) Anal Chem 82:2029
4. Srinivasan V, Pamula VK, Fair RB (2004) Lab Chip 4:310
5. Lin BC, Gao Y, Qin JH (2009) J Chin Chem Soc 56:1
6. Malic L, Brassard D, Veres T et al (2010) Lab Chip 10:418
7. Epstein IR (2007) Science 315:775
8. Beer NR, Rose KA, Kennedy IM (2009) Lab Chip 9:841
9. Unger MA, Chou H, Thorsen T et al (2000) Science 288:113
10. Pollack MG, Fair RB, Shenderov AD (2000) Appl Phys Lett 77:1725
11. Fair RB (2007) Microfluid Nanofluid 3:245
12. Su F, Chakrabarty K, Fair RB (2006) IEEE Trans Comput Aided Des Integr Circ Syst 25:211
13. Wen W, Huang X, Yang S et al (2003) Nat Mater 2:727
14. Tao R, Sun JM (1991) Phys Rev Lett 67:398
15. Halsey TC (1992) Science 258:761
16. Ma H, Wen W, Tam WY et al (2003) Adv Phys 52:343
17. Papadopoulos CA (1998) Mechatronics 8:719
18. Choi W, Tuteja A, McKinley GH (2009) Adv Mater 21:2190
19. Hao T (2001) Adv Mater 13:1847
20. Winslow WM (1949) J Appl Phys 20:1137
21. Li Y, Chen Y, Conrad H (1995) ASME 235:29
22. Conrad H, Li Y, Chen Y (1995) J Rheol 39:1041
23. Wu CW, Conrad H (1996) J Phys D 29:3147
24. Lu KQ, Shen R, Wang XZ et al (2005) Int J Mod Phys B 19:1065
25. Lu KQ, Shen R, Wang XZ et al (2006) Chin Phys 15:2476

26. Cheng Y, Wu K, Liu F et al (2010) ACS Appl Mater Interfaces 2:621
27. Shen R, Wang X, Lu Y et al (2009) Adv Mater 21:4631
28. Shen R, Wang XZ, Wen WJ et al (2005) Int J Mod Phys B 19:1104
29. Yin JB, Zhao XP (2004) Chem Mater 16:321
30. Zhao XP, Yin JB (2002) Chem Mater 14:2258
31. Yin JB, Zhao XP (2006) J Phys Chem B 110:12916
32. Yin J, Zhao X, Xiang L et al (2009) Soft Matter 5:4687
33. Shen C, Wen W, Yang S et al (2006) J Appl Phys 99:106104
34. Wen WJ, Huang XX, Sheng P (2004) Appl Phys Lett 85:299
35. Li J, Gong X, Chen S et al (2010) J Appl Phys 107:093507
36. Parmar KPS, Meheust Y, Schjelderupsen B et al (2008) Langmuir 24:1814
37. Yoshimoto S (2005) Macromol Rapid Commun 26:857
38. Kim JW, Liu F, Choi HJ et al (2003) Polymer 44:289
39. Cho MS, Choi HJ, Ahn WS (2004) Langmuir 20:202
40. Cho MS, Choi HJ, Kim KY et al (2002) Macromol Rapid Commun 23:713
41. Park SJ, Cho MS, Lim ST et al (2005) Macromol Rapid Commun 26:1563
42. Cho MS, Cho YH, Choi HJ et al (2003) Langmuir 19:5875
43. Gong X, Wu J, Huang X et al (2008) Nanotechnology 19:165602
44. Zeng S, Li B, Su X et al (2009) Lab Chip 9:1340
45. Hosokawa K, Maeda R (2000) Micromech Microeng 10:415
46. Hosokawa K, Fujii T, Endo I (1999) Anal Chem 71:4781
47. Groisman A, Enzelberger M, Quake SR (2003) Science 300:955
48. Yu Q, Bauer JM, Moore JS et al (2001) Appl Phys Lett 78:2589
49. Beebe DJ, Moore JS, Bauer JM et al (2000) Nature 404:588
50. Pal R, Yang M, Johnson BN et al (2004) Anal Chem 76:3740
51. Elizabeth Hulme S, Shevkoplyas SS, Whitesides GM (2009) Lab Chip 9:79
52. Zheng Y, Dai W, Wu H (2009) Lab Chip 9:469
53. Weibel DB, Kruithof M, Potenta S et al (2005) Anal Chem 77:4726
54. Weibel DB, Siegel AC, Lee A et al (2007) Lab Chip 7:1832
55. Yoshida K, Kikuchi M, Park JH et al (2002) Sens Actuators A Phys 95:227
56. Zhang M, Wu J, Niu X et al (2008) Phys Rev E 78:066305
57. Eddings MA, Johnson MA, Gale BK (2008) J Micromech Microeng 18:067001
58. Leclerc E, Sakai Y, Fujii T (2003) Biomed Microdevices 5:109
59. Moreira NH, Almeida AL, Piazzeta MH et al (2009) Lab Chip 9:115
60. Ng JMK, Gitlin I, Stroock AD et al (2002) Electrophoresis 23:3461
61. Sia SK, Whitesides GM (2003) Electrophoresis 24:3563
62. Niu X, Peng S, Liu L et al (2007) Adv Mater 19:2682
63. Niu X, Zhang M, Peng S et al (2007) Biomicrofluidics 1:044101
64. Liu L, Peng S, Niu X et al (2006) Appl Phys Lett 89:223521
65. Valero A, Post JN, van Nieuwkasteele JW et al (2008) Lab Chip 8:62
66. Fox MB, Esveld DC, Valero A et al (2006) Anal Bio Chem 385:474
67. Khine M, Lau A, Ionescu-Zanetti C et al (2005) Lab Chip 5:38
68. Young H, Boris R (2003) Sens Actuators A Phys 104:205
69. Derveaux S, Stubbe BG, Roelant C et al (2008) Anal Chem 80:85
70. DeMello AJ (2006) Nature 442:394
71. Kutter JP (2000) Trac Trends Anal Chem 19:352
72. Wang L, Zhang M, Yang M et al (2009) Biomicrofluidics 3:034105
73. Nakano M, Katou T, Satou A et al (2002) J Intell Mater Syst Struct 13:503
74. Niu X, Wen W, Lee YK (2005) Appl Phys Lett 87:243501
75. Liu L, Chen X, Niu X et al (2006) Appl Phys Lett 89:083505
76. Niu X, Liu L, Wen W et al (2006) Appl Phys Lett 88:153508
77. Liu L, Niu X, Wen W et al (2006) Appl Phys Lett 88:173505
78. Nguyen NT, Truong TQ (2004) Sens Actuators B 97:137–143

79. Niu X, Zhang M, Wu J et al (2009) Soft Matter 5:576–581
80. Zhan W, Crooks R (2003) J Am Chem Soc 125:9934
81. Wang X, Zhou J, Tam TK et al (2009) Bioelectrochemistry 77:69
82. Thorsen T, Maerkl SJ, Quake SR (2002) Science 298:580
83. Rhee M, Burns MA (2009) Lab Chip 9:3131
84. Weaver JA, Melin J, Stark D et al (2010) Nat Phys 6:218
85. Prakash M, Gershenfeld N (2007) Science 315:832
86. Mosadegh B, Kuo C, Tung Y et al (2010) Nat Phys 6:433
87. Cheow LF, Yobas L, Kwong D (2007) Appl Phys Lett 90:054107
88. Srinivasan V, Pamula VK, Fair RB (2004) Anal Chim Acta 507:145
89. Pamula VK, Srinivasan V, Chakrapani H et al (2005) Proc IEEE Int Conf Micro Electro Mech Syst MEMS :722
90. Wang L, Zhang M, Li J et al (2010) Lab Chip 10:2869
91. Rhee M, Burns MA (2008) Lab Chip 8:1365
92. Liu L, Cao W, Wu J et al (2008) Biomicrofluidics 2:034103
93. Wheeler AR, Throndset WR, Whelan RJ et al (2003) Anal Chem 75:3581
94. Beer NR, Hindson BJ, Wheeler EK et al (2007) Anal Chem 79:8471
95. Adleman LM (1994) Science 266:1021
96. Benenson Y, Gil B, Ben-Dor U, Adar R et al (2004) Nature 429:423

Top Curr Chem (2011) 304: 117–152
DOI: 10.1007/128_2011_143

Published online: 21 April 2011

Biosensors in Microfluidic Chips

Jongmin Noh, Hee Chan Kim, and Taek Dong Chung

Abstract A biosensor is a sensing device that incorporates a biological sensing element and a transducer to produce electrochemical, optical, mass, or other signals in proportion to quantitative information about the analytes in the given samples. The microfluidic chip is an attractive miniaturized platform with valuable advantages, e.g., low cost analysis requiring low reagent consumption, reduced sample volume, and shortened processing time. Combination of biosensors and microfluidic chips enhances analytical capability so as to widen the scope of possible applications. This review provides an overview of recent research activities in the field of biosensors integrated on microfluidic chips, focusing on the working principles, characteristics, and applicability of the biosensors. Theoretical background and applications in chemical, biological, and clinical analysis are summarized and discussed.

Keywords Biosensor · Electrochemical · Mass · Microfluidic chip · Optical · Transducer

Contents

1 Introduction 118
2 Detection Principles 120
 2.1 Optical Methods 120
 2.2 Electrochemical and Electronic Methods 125

J. Noh
Interdisciplinary Program, Biomedical Engineering Major, Graduate School, Seoul National University, Seoul, Korea

H.C. Kim
Department of Biomedical Engineering, College of Medicine and Institute of Medical and Biological Engineering, Medical Research Center, Seoul National University, Seoul, Korea

T.D. Chung (✉)
Department of Chemistry, Seoul National University, Seoul, Korea
e-mail: tdchung@snu.ac.kr

2.3 Mass-Sensitive Methods 127
2.4 Others 129
3 Analytes 129
3.1 Small Molecules 129
3.2 Proteins 130
3.3 Electrolytes and Gases 131
4 Functional Receptors Integrated in Microfluidic Systems 131
4.1 Enzyme Electrodes 131
4.2 Antibodies 133
4.3 Molecular Receptors 135
4.4 Aptamers 135
4.5 Miscellaneous 136
5 Microfluidic Components 136
5.1 Pretreatment 136
5.2 Surface Modifications 137
5.3 Beads 138
5.4 Droplets 140
5.5 Pumping and Valves 140
5.6 Peripherals 142
6 Perspectives 143
References 144

Abbreviations

PSA Prostate-specific antigen
TFT Thin film transistor

1 Introduction

Microfluidic technology for chemical or bioanalytical purposes concerns the precise control of fluids in a limited space, which may be intentionally patterned on chips, because a number of valuable benefits are expected from such systems [1, 2]. As many articles and reviews have pointed out, the alleged advantages include reduced reagent consumption, short analysis time, a small-sized scale, low cost, and high sensitivity. Over the last two decades, there has been an explosive development of miniaturized analytical systems and related techniques based on microfluidics for chemical analysis, bioanalysis, clinical diagnostics, and other applications [3–15].

In spite of the impressive opportunities, the use of microfluidics for chemical and biological analysis involves considerable challenges as well. The analytes in the fluids of microsystems are exposed to unusual physical conditions, such as high surface tension and high surface-to-volume ratio, which could be the reason why analytical information from the microfluidic systems might be significantly different from that predicted from conventional methods [16–19]. In many cases, pretreatment and handling of the samples for analysis in microfluidic systems are not merely a matter of scale-down. Analysis using microfluidic analyzers often faces troubles in adsorption, injection, mixing, and so on, which are not significant in conventional analysis in a large scale. Potential solutions for the problems that may occur in the microfluidic systems have been suggested in numerous papers, but many of them are still inadequate.

Another approach is omission or simplification of pretreatment and handling processes. This may be a preferable strategy for more effective analysis. However, better sensing parts are required to possibly share the burden by playing more roles than a simple detector, without any or negligible loss of the analytical quality. Biological functional units such as antibodies or enzymes can endow a detector with selective recognition ability so that the corresponding microfluidic analyzer should work without complicated and strict pretreatment prior to detection.

Traditionally, a biosensor is created by combining a biological component with a physiochemical detector [20–23]. Biosensors exploit biochemical molecular recognition properties for a selective analysis. The general structure of biosensors consists of analyte recognition, signal transduction, and readout, as shown in Fig. 1. The representative example of commercial biosensors is the blood glucose sensor. Since the first classical type from Clark and Lyons was introduced in 1962 for detecting glucose levels in blood, it has been followed by many different biosensors [24]. However, only a limited number of the biosensors fulfill the stringent requirements for commercialization, e.g., practical demand in terms of market size, and functional reliability in the real environments in which they are expected to operate. As informational technology becomes more widespread and more convenient applications are pursued, biosensors are increasingly demanded in modern life. On-site analysis at low cost and within a short time is undoubtedly important, not only for ordinary people but also for researchers working in laboratories.

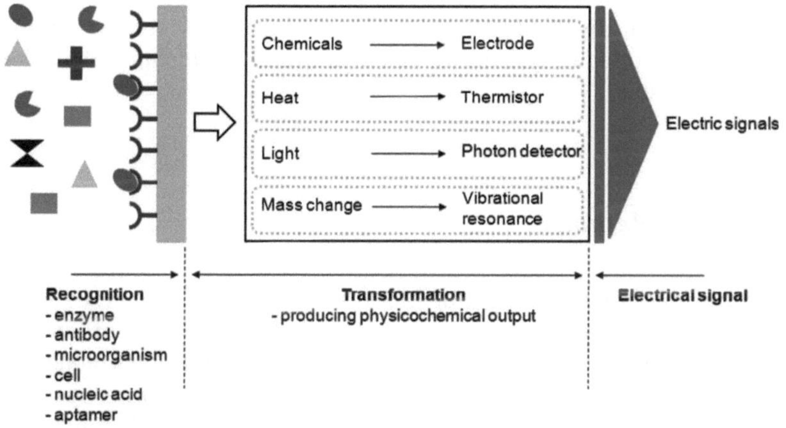

Fig. 1 Configuration of a biosensor

Generally, biosensors are designed to extract analytical information about certain analytes in given samples without pretreatment. Nonetheless, combination with microfluidic systems could provide conventional biosensors with opportunities for improvement in terms of analytical power or in applicability for a wider range of applications. For example, bioassays could benefit from automatic sequential separation and chemical synthesis that are linked to the detection part of a miniaturized system [25, 26]. Drug discovery, for instance, needs innovative bioassay technology that uses as small a volume as possible to minimize expensive, difficult, and inefficient processes. The lab-on-a-chip is a big target in the fields of medicine, biotechnology, and pharmacology and is one of the key motivations for integrating micro- or submicroscale biosensors into microfluidic systems.

Microfluidic-based biosensors have advanced greatly in various fields over the last few decades [8, 27]. The fundamental concept underlying the microfluidic biosensors that have been reported is to integrate the analytical functions necessary for biochemical analysis onto a single chip, including sample preparation, pretreatment, detection, and sometimes molecular separation or sorting.

This review summarizes the research achievements concerning integration of biosensors into microfluidic chips, which have been reported mostly since 2005. We discuss the latest developments by categorizing them on the basis of four different aspects; detection principles, analytes, functional receptors, and microfluidic components. Based on a balanced view of recent trends, the review is concluded by looking at the future perspectives in this field.

2 Detection Principles

The detection principles for biosensors integrated on microfluidic chips are classified into several types, including optical, electrochemical, and mass-sensitive methods. The trend in the development of detectors has been to constantly pursue two key virtues: sensitivity and selectivity. In this regard, the research issues on the detectors for microfluidic systems are not significantly different to those for conventional biosensors.

2.1 *Optical Methods*

Most optical detection methods for biosensors are based on ultra-violet (UV) absorption spectrometry, emission spectroscopic measurement of fluorescence and luminescence, and Raman spectroscopy. However, surface plasmon resonance (SPR) has quickly been widely adopted as a nonlabeling technique that provides attractive advantages. Fueled by numerous new nanomaterials, their unique, SPR-based or related detection techniques are increasingly being investigated [28–31].

The conventional methods for optical detection need various filters, lenses, light sources, and photomultiplier tubes (PMTs), which are essential for microfluidic systems as well. However, the functional components for optical detection in microfluidic analytical systems are seldom miniaturized, and necessarily sophisticated or expensive in many cases. Considering that microfluidic chips are basically intended to conduct handy, cheap, user-friendly, and on-site analysis, it is obvious that the crucial issue in the optical setup for microfluidic systems is how to overcome such a problem. To cope with this fundamental challenge, many researchers have made efforts to miniaturize waveguides, filters, and lenses so that they are tiny and simple enough to be integrated on a chip. Such optoelectronic components include laser diodes, light-emitting diodes (LEDs), charge coupled devices (CCDs), and complementary metal oxide semiconductors (CMOSs), which replace the conventional optical elements [32, 33]. Frank et al. provided a broad overview of current research and development in point-of-care diagnostic devices utilizing optical detectors in microfluidic platforms [13]. Various optical detections of biological samples covering absorbance, fluorescence, luminescence, and SPR are summarized in Table 1.

Absorption spectrometry is a classic detection method for chemical analysis. It requires a long path length of incident light through a sample at lower concentration. This is critical for absorbance measurements in microfluidic chips because the narrow microfluidic channels are obviously problematic for sensitive absorbance measurement. That is why there are not many cases employing absorbance detection. A few peptides have been separated by on-chip electrochromatography and detected by UV absorbance [34]. Optical fibers are normally used to enhance efficient absorbance in the microfluidic chips. For instance, the incident light from a deuterium–tungsten light source follows a optical fiber to reach the sample and then goes to a CCD array detector through the bottom fiber [13]. Incident light can be focused by integrating a microlens in a complex three-layer microfluidic chip [54]. This system has slit channels filled with "black ink" to absorb any scattered light, and also a cylindrical microlens in the polydimethylsiloxane (PDMS) layer on the chip. This creative design enabled detection of the peptide separation with enhanced sensitivity. Llobera et al. reported an optical air mirror that utilized the refractive index difference between PDMS and air [49]. Ro et al. showed efficient and sensitive absorbance detection using a light collimating system on a PDMS microchip for capillary electrophoresis (CE) [35]. This collimating setup consisted of various optical components features and gave significant reduction of stray light. Steigert et al. proposed an interesting concept of a centrifugal compact disk (CD)-based system that allowed a fast colorimetric assay of alcohol in a single droplet of whole blood [55]. In this optical system, the laser was illuminated perpendicular to the flat surface of the cyclic olefin copolymer disk, and the optical path length increased by one order of magnitude in comparison to the conventional direct incidence. Caglar et al. designed a glass–PDMS microfluidic chip for determination of calcium ions in samples by introducing an optical fiber and a ball lens [36].

Fluorescence detection involves molecular light absorption, which triggers the emission with longer wavelengths than for excitation. This method is often

Table 1 Optical detection

Method	Sensing parts	Analytes	LOD	References
Absorbance	Optical fiber, CCD detector	Thiourea	167 μM	[34]
	Optical fiber, microlens, slits, PMT detector	Fluorescenin	S/N = 3	[35]
	Optical fiber, a ball lens, CCD detector	Ca^{2+} ions	0.027 mM	[36]
	Optical fiber, PMT detector	Ammonia	2 ppm	[37]
Fluorescence	Microscope-based PMT detector	Protein	12 μg/mL	[38]
	Optical path length, CCD detector	Flavin	1 mM	[39]
	Optical fiber, pinhole filter, PMT detector	FITC	10^{-7} M	[40]
	Spherical lens, filter, PMT detector	Fluorescence	0.01 nM	[41]
	LED, SSPM	Fluorescence	N/A	[42]
Chemiluminescence	VersaDoc	CRP	0.1 mg/L	[43]
	Multianode-photomutiplier array	SEB *E. coli* M13	0.1 ng/L 10^4 cfu/mL 10^5 pfu/mL	[44]
	TFT photosensor	DNA, horseradish peroxidase	0.5 nM	[45]
	PMT detector	Cytochrome c Myoglobin Horseradish peroxidase	~250 nM ~160 nM ~240 nM	[46]
Bioluminescence	Microscope, PMT detector	ATP	~μM	[47]
Electrochemiluminescence	ITO electrode, PMT detector	Proline	1.2 μM	[48]
Internal reflection spectroscopy	LED, optical fiber, air mirror	Phosphate buffer	41 nM	[49]
Surface plasmon resonance	Direct capture	Cowpea mosaic virus	12.5 μg/mL	[50]
	Nucleotide hybridization	*Fusarium culmorum*	0.06 pg	[51]
	Substractive inhibition	*Puccinia striiformis*	10^5 sp/mL	[52]
	Substractive inhibition	*Phyththora infestans*	10^6 sp/mL	[53]

CCD charge-coupled device, *CRP* C-reactive protein, *FITC* Fluorescein-5-isothiocyanate, *ITO* indium tin oxide, *LED* light-emitting diode, *LOD* limit of detection, *PMT* photomultiplier tubes, *SEB* Staphylococcal enterotoxin B, *S/N* signal-to-noise ratio, *SSPM* solid-state photomultiplier, *TFT* thin film transistor employed for sensing in microfluidic chips because of its superior sensitivity and selectivity for a variety of samples, in addition to its compatibility with microdevices [56, 57]. However, the fluorescence of microfluidic systems has some drawbacks, e.g., autofluorescence and nonspecific emission from biomolecules in the sample. Hofmann et al. showed that sensitivity was improved by monolithically integrating an optical long-pass filter on a PDMS microfluidic chip [58]. Schmidt et al. developed a chip-sized spectrometer by combining a linear variable band-pass filter with a CMOS camera [59]. A liquid–liquid waveguide was introduced to the fluorescence detection system on a microfluidic chip by Vesenov et al. [60]. Balslev et al. made a monolithic optoelectronic hybrid platform including an on-chip optically pumped liquid dye laser, waveguides, and microchannels with passive diffusive mixers [32]. Pais et al. reported a disposable lab-on-a chip system equipped with a thin-film organic LED excitation source and an organic photodiode detector for on-chip fluorescence analysis [33]. On the other hand, a portable fluorescence biosensor with quantum dots (QDs) was developed for quantitative analysis and rapid screening of trace amounts of proteins [61]. Schulze et al. realized native fluorescence detection of the proteins separated by microchip electrophoresis using a deep UV neodymium-doped yttrium aluminum garnet (Nd:YAG) laser at 266 nm and a PMT detector [38]. Quin et al. utilized a pulsed nitrogen laser, a fluorescence emission guide, and a CCD detector for flavin metabolites detection [39]. Li et al. proposed a microfluidic chip into which the one end of an optical fiber was introduced, and the other terminal was coupled to a blue diode-pumped laser [40]. Renzi et al. demonstrated a hand-held microchip-based analytical instrument for detection of proteins [41]. Recently, a portable microfluidic flow cytometer with simultaneously detection of fluorescence and impedance was reported for cell analysis [42]. This system exploited an LED for excitation and detected fluorescent emission with a solid-state photomultiplier (SSPM).

Luminescence detection involves light emission as a result of chemical reaction, which may be from a living organism or artificial system like a luminol-labeled biosensor. Since chemiluminescence needs no light source, the system becomes even simpler than fluorescence methods while still producing sensitive responses. In many cases, detection based on luminescence is a very attractive choice for sensitive analysis because it is conducted in the complete absence of background light. QDs are a conspicuous example that show how powerful luminescence detection is. They emit strong luminescence and thus are increasingly being used for bioanalytical applications. Huang et al. reported that QDs have much higher photoluminescence quantum efficiency than their bulk materials [62]. In their study, a urea-sensing system composed of water-soluble mercaptosuccinic acid (MSA)–QDs and urease was prepared by a simple procedure and provided sensitive responses to urea. Hatakeyama et al. employed a DNA-arrayed thin film transistor (TFT) as a DNA chip platform and light detector [45]. The limit of detection (LOD) of their system was 0.5 nM, which was a much lower concentration than detected by a conventional device. A DNA chip-based protocol for clinical diagnostics was realized on a plastic card including an optoelectronic DNA chip and pyrotechnic microvalves for electrical fluid handling [63]. A plastic microfluidic immunosensor for C-reactive protein (CRP), a cardiac and inflammation marker, is another example of a sensing system that uses luminescence detection strategy [43]. This sensor gave an even better LOD than conventional enzyme-linked immunosorbent assay (ELISA) analysis. Multiplexing is believed to be the general trend of modern analytical techniques and not an exception for immunoassay based on luminescence. A multiplexed chemiluminescent capillary enzyme immunoassay was shown to be capable of simultaneously detecting toxins, bacteria, and viruses [44]. The LOD was 0.1 ng/mL for the toxin Staphylococcal enterotoxin B (SEB), 10^4 cfu/mL for *Escherichia coli* O157: H7, and 5×10^5 pfu/mL for the bacteriophage M13. Bioluminescence (BL) is the light emission from a living organism. Liu et al. integrated CE and BL detection into a chip for ATP analysis [47]. Electrochemiluminescence (ECL) is a kind of luminescence produced during electrochemical reactions. Qiu et al. presented a $Ru(bpy)_3{}^{2+}$ ECL detector for CE on a microfluidic chip [48].

Internal reflection spectroscopy (IRS) is another promising optical detection for immunoassay [64]. Llobera et al. reported PDMS-based multiple internal reflection systems comprised of self-alignment systems, lenses, microfluidic channels, and mirrors. It is noteworthy that the LOD was lowered down to 41 nM using an "air mirror" [49].

Raman spectroscopy has great potential in terms of analytical applications because it gives characteristic spectra in aqueous phase. Its critical problem is weak signals, and there have been ceaseless attempts to overcome this by exploiting surface-enhanced Raman scattering (SERS) [65]. The enhancement of Raman scattering is a phenomenon that is sensitive to the substrate surface. The enhanced signals by SERS come from the Raman-active bonds of the molecules adsorbed on rough surfaces [66, 67], nanoparticles [68–70], nanoshells [71], or extremely narrow gaps ("hot spots") [72, 73] of some metals such as Ag, Au, Pt, or Cu [74]. SERS-based detection is also advantageous because it is nondestructive [75]. Recently, Lee et al. reported a SERS decoding using thin microgold shells modified with Raman tags within a microfluidic chip [76].

SPR is a representative physical phenomenon that is widely utilized for label-free characterization of molecules on thin metal films. The basic principle and operation of SPR has been described in more detail in several review articles [77, 78]. The reports on SPR-based immune sensors have steeply increased for detection of analytes with low molecular weights in recent years. SPR detection in microfluidic systems can provide various advantages. Immunoreactions are completed within a short time due to small sample volumes down to the nanolitre scale. Kim et al. developed a simple and versatile miniaturized SPR immunosensor enabling parallel analyses of multiple analytes [79]. Their SPR sensor was claimed to exhibit good stability and reusability for 40 cycles and more than 35 days. Feltis et al. demonstrated a low-cost handheld SPR-based immunosensor for the toxin Ricin [80]. Springer et al. reported a dispersion-free microfluidic system with a four-channel SPR sensor platform, which considerably improved the response time and sensitivity [81]. The sensor was able to detect short sequences of nucleic acids down to a femtomole level for 4 min. Waswa et al. demonstrated the immunological detection of *E. coli* O157:H7 in milk, apple juice, and meat juice extracted from ground beef [82]. Wei et al. detected *Campylobacter jejuni* in poultry meat [83]. A palm-sized SPR biosensor has been developed in which the light source is modulated by a rotating mirror [84]. This system introduced a folded light path and a diode laser. The same group devised another briefcase-style sensor that is capable of simultaneously detecting small molecules, proteins, viruses, bacteria, and spores [80].

Nanostructures provide new opportunities for SPR sensing. It was shown that both sensitivity and selectivity can be improved by introducing an integrated array of nanoholes serving as substrate for SPR detection [85]. Huang et al. exploited the localized SPR properties of gold nanoparticles and successfully applied them for label-free monitoring of biomolecular interactions in real time [86]. They also discussed the unique localized surface plasmon resonance (LSPR) property of gold nanoparticles, which provides a basis for measuring the molecular adsorption onto the surface of metal nanoparticles [87]. LSPR is a phenomenon that is observed in noble metallic nanoparticles (silver, gold) and is caused by the collective oscillations of conductive electrons [88, 89]. Recently, the numerical simulation of LSPR microfluidic chips with grooved optical fibers was proposed to optimize the sensor [90]. Luo et al. fabricated a PDMS microfluidic device containing an array of gold spots onto which antigens or antibodies of interest could be attached [91]. The gold spots of this system are suitable for carrying out immunoreactions with SPR detection. Reportedly, the refractive index change caused by the molecules bound to the surfaces of the gold spots allows SPR imaging through real-time monitoring of immune reactions [92]. Pang et al. studied surface plasmon polariton dispersion related to a 2D nanohole array [93]. In this work, the sensitivity of the nanohole array sensor was attributed to the periodicity of the array and the excited surface plasmon polariton effect. Plant pathogen biosensors based on SPR were also reported [50–53].

2.2 Electrochemical and Electronic Methods

Electrochemical detection has been regarded as particularly appropriate strategy for microfluidic chip systems. Electrochemical biosensors in microfluidic chips enable high sensitivity, low detection limits, reusability, and long-term stability. And the detection mechanism and instrumentation for realization are simple and cost-effective. These valuable features have made electrochemical devices receive considerable attention [20, 94, 95]. The electrochemical detectors are commercially available for a variety of analyses [96]. The review written by Wang summarized the principles of electrochemical biosensors, important issues, and the state-of-the-art [97]. Lad et al. described recent developments in detecting creatinine by using electrochemical techniques [98].

In general, electroanalytical detection principles can be divided into three: potentiometry, amperometry, and conductometry (or impedometry). Potentiostats used for electrochemical biosensors are mostly equipped with amperometric and potentiometric modules. Potentiometry is used to measure the cell potential difference at virtually zero current (open circuit voltage). The representative potentiometric sensors are ion-selective electrodes (ISEs), which include the conventional glass electrodes widely used for pH determination. Bobacka et al. gave an overview of electrochemical theory related to potentiometric principles and nonequilibrium models for ISE [99]. Feridbod et al. reviewed potentiometric sensors that employ conducting polymers [100]. In amperometry, the current associated with the reduction or oxidation at working electrode is measured under a constant potential difference versus the reference electrode. Conductometry is used to measure the electrical conductance in solutions, in which the charge carriers are cations and anions. Applying an AC electrical field between two electrodes, the conductometer or impedance analyzer acquires the output data of changes in amplitude and/or phase angle. Amplitude change provides serial resistance value in the equivalent circuit, whereas phase angle shift carries capacitive and inductive components that are mostly involved in the electrochemical interface between the electrode surface and the solution. DC input signals give pure resistance and require even simpler instruments, which are favorable for miniaturized systems like microfluidic chips. Durand et al. used ionic conductance in a nanochannel for label-free determination of protein surface interaction kinetics [101]. However, DC input in conductometry out of microchannels or capillaries is normally avoided because the applied potential difference is concentrated on the electric double layer on the electrode surface rather than the solution.

With regard to biosensors and analytical chip systems, challenging problems of electrochemical detection strategy are deterioration in selectivity and stability of biological functional objects like enzymes and chronic passivation of the underlying electrodes. Insufficient selectivity or specificity is an intrinsic limitation of electrochemical detection, which has been addressed by combining chemically or biologically specific receptors.

A variety of electrode materials have recently been tried for electrochemical analysis, including nanostructured metals, carbon nanotubes, nanoparticles, and many more. Park et al. showed that the nanoporous platinum electrode offers the unprecedented advantage of very low polarizability due to the high exchange current density (i_0) of the electrode surface [102]. This technique also demonstrated near-Nernstian behavior with ignorable hysteresis and very short response time. Henry et al. suggested pretreated gold electrodes in a microfluidic cell to detect cancer markers [103]. Wang et al. analyzed amino acids on a PDMS device coated with titanium dioxide nanoparticles by indirect amperometry [95]. An interdigitated ultramicroelectrode array (IDUA) was introduced to enable highly sensitive detection of nucleic acid, giving 1 fmol of LOD and dynamic range of 1–50 fmol [104]. Boehm et al. identified and quantified bacteria in a microfluidic chip by simple and rapid impedometry (conductometry) [105]. By coupling an impedance-based detection system with monoclonal antibodies immobilized onto the surface in a microfluidic chip, they detected bacteria in a sample within a few minutes. Swensen et al. performed continuous, direct, and real-time detection of a small molecule analyte through combination between aptamer-based sensing and microfluidic sample handling technology [106]. A number of researchers have sought new ways to multiplex assay by using aptamer-based electrochemical biosensors that exhibit subpicomolar detection limits [107]. Javanmard et al. proposed the use of a microchannel with electrodes, on which functional receptors to target biomarkers are immobilized [108]. This attempt accomplished the detection of the anti-hCG antibody at a concentration of 1 ng/mL in a dynamic range of three orders of magnitude in less than 1 h.

Electronic or electric detection is hardly distinguished from electrochemical detection because it is linked to chemical or biological interaction in the vicinity of a conductor or semiconductor in which an electric current flows. In the sense that it extracts information from predominantly electric signals without directly involving an electrochemical reaction, it may be discussed separately from electrochemical detection. The working principle underlying an ion-sensitive field-effect transistor (ISFET) is similar to that of a metal oxide semiconductor field effect transistor (MOSFET). In ISFET, change in the ion concentration leads to the corresponding response in current between source and drain. A pH-sensitive field effect transistor (pH-FET) is a typical ISFET working sensitively and reliably. It provides a versatile basis for various chemical sensors, e.g., polymer membrane-coated ISFETs, enzyme-immobilized (ENFETs), and gas-sensing FETs coupled with gas-permeable membranes, because a lot of reactions are accompanied by stoichiometric generation or consumption of protons. The underlying electrode surface and the enzyme immobilization method are important factors that determine the sensitivity, stability, and selectivity. Masadome et al. devised a cationic surfactant ISFET using polystyrene plates and stainless wires as a template for fabricating the channel [109]. Truman et al. reported a single silicon thin film FET, which was designed to monitor transport and chemical properties of liquids in microfluidic systems [110]. A single polypyrrole nanowire-based FET sensor for real-time pH monitoring is another example [111]. In this system, polypyrrole nanowires were fabricated by electrochemical deposition inside the pores of an anodized aluminum oxide template, providing higher sensitivity and selective performance toward pH variation. Kim et al. demonstrated an extended gate FET (EGFET)-based biosensor combined with a silicon microfluidic channel for the electronic detection of streptavidin–biotin protein complexes [112]. The EGFET device was also used for the electronic detection of nucleic acids [113].

2.3 Mass-Sensitive Methods

Mass-sensitive sensors involve piezoelectric effects and surface acoustic waves. The piezoelectric effect was discovered in 1880. Piezoelectricity is the ability of some materials, mostly crystals and ceramics, to generate an electric potential in response to mechanical stress. The piezoelectric effect was mainly utilized for immunosensors and nucleic acid sensors because antigen–antibody association and DNA hybridization cause relatively large changes in mass. Mass-sensitive sensors may exploit surface acoustic waves. There have been reported a number of sensors based on quartz resonators: electrochemical quartz crystal microbalance (EQCM), surface acoustic wave (SAW), thickness shear mode (TSM), flexural plate wave (FPW), and shear-horizontal acoustic plate mode (SH-APM). In EQCM, both sides of a quartz disk are covered with thin film electrodes. SAW-based biosensors in microfluidic platforms have been applied to immunoassays and the detection of DNA, bacteria, and small molecules [114]. QCM sensors consist of a thin quartz disk and two electrodes. When an oscillating electric field is applied across the disk, an acoustic wave is induced at certain resonance frequency. Lee et al. showed that antibodies immobilized on the self-assembled monolayer of thiosalicylic acid on QCM allowed detection of *Bacillus cereus* and *Listeria monocytogenes* at 104 cells/mL and 450 spores, respectively [115]. Cooper et al. prepared QCM immunosensors that are potentially suited for detection of pathogens [116]. Actually, mass-sensitive sensors commonly suffer from difficulty in discriminating between mass changes arising from the binding of authentic target molecules and from the nonspecific adsorption of contaminants. A TSM sensor is a piezoelectric-based sensor in which perturbation in mechanical shear strain induces change in AC voltage and vice versa. Ergezen et al. monitored adhesion and aggregation of platelets by a TSM sensor in real time basis [117]. FPW sensors attracted interest due to their high sensitivity, high accuracy, and low operating frequency in clinical, environmental, and biological fields. A representative example is the allergy sensor based on a very high mass-sensitivity FPW detector [118]. SH-APM sensors involve leaky waves, where the wave is only partially confined to the surface. Rocha–Gaso reviewed the SH-APM sensor and other biosensors that exploit surface-generated acoustic waves for the detection of pathogens [119].

An important tool for mass detection is atomic force microscopy (AFM), which utilizes cantilevers for surface characterization. Adsorption onto the sensing electrode, composed of two chemically different materials, produces a differential stress between the surfaces, resulting in bending. Molecules adsorbed on a cantilever also bring about vibration frequency changes. Microfabricated cantilever functionalized with antibodies has been sought as a new detection device for biosensors. The sensors based on this physical phenomenon are summarized in Table 2. Philip et al. reviewed the technologies and recent developments in micro-/nanoscale cantilevers, which have been applied to the detection of relatively small analytes [128]. Recently, nanomechanical motion induced by protein–protein interactions on a microfabricated cantilever was reported to allow detection of prostate-specific antigen (PSA) [120, 121]. Cherian et al. investigated the adsorption characteristics of calcium ion on gold and silicon surfaces of a cantilever through variation in resonance frequency and bending [122]. Watari et al. confirmed the capability of multiple cantilever arrays to monitor bending forces induced by the ionization reactions in aqueous samples [123]. A cantilever was also used to measure the surface stress on micromechanical structures produced by DNA hybridization or biomolecular reaction [124]. Arntz et al. showed continuous label-free detection of two cardiac markers, creatine kinase and myoglobin, using an array of microfabricated cantilevers on which anticreatine kinases or antimyoglobin

Table 2 Mass-sensitive methods based on microcantilvers

Analyte	Receptors or sensing bodies	LOD	Typical stress (mN/m)	References
Prostate-specific antigen	SiN_x with Au coating	0.2 ng/mL	–	[120]
Prostate-specific antigen	Ta/Pt/PZT/Pt/Si O_2	10 pg/mL	–	[121]
Ca^{2+} ions	Bare silicon/Au cantilever	1 mM	1–450	[122]
pH	Thiol-modified cantilever	pH 4.5–9	1–30	[123]
DNA	Oligonucleotide	100 pM	1–30	[124]
Protein	Antibody	20 μg/mL	1–6	[125]
Protein	Oligonucleotide aptamers	100 pg/mL	1–10	[126]
Peptide	Antibody fragments	20 ng/mL	1–10	[127]

antibodies were immobilized [125]. DNA aptamer and single-chain antibody were also employed as receptor molecules [126, 127].

2.4 Others

Temperature is a useful variable that reflects analytical information by associating with biological functional elements. An enzyme thermistor utilizing a commercialized thermistor is a kind of resistor that detects resistance changes as a function of the ambient temperature. In this system, the molar enthalpies of enzyme-catalyzed reactions are sensed by the thermal detector.

3 Analytes

3.1 Small Molecules

Small molecules in this review are defined as organic compounds with low molecular weights. Small molecules can have a variety of biological functions. For instance, cell signaling, which may be triggered by small molecules, has a great importance for drug discovery, environmental analysis, mass production processes, and so on. Another example is bacteria modalities, which are becoming increasingly crucial. Boehm et al. demonstrated an on-chip microfluidic biosensor for impedometric bacterial detection and identification in the presence of monoclonal antibodies immobilized in the microfluidic system [105].

Label-free detection based on silicon nanowire FET devices was used to identify small molecule inhibitors [129]. High-throughput screening of small molecules is undoubtedly one of the hot issues in pharmaceutical drug discovery. Chan et al. quantified small molecule aggregation in a high throughput fashion without using any label, but a photonic crystal [130]. Wang et al. exploited optical extinction for probing interactions between nucleic acids and small molecules [131].

3.2 Proteins

Proteomics on a large scale have emerged as one of the major themes in modern biomedical research. Recently, proteomics based on the concept of protein analysis on biochip was reviewed [132]. Jonkheijm et al. summarized and discussed chemical, biological, and nanotechnological strategies that involve protein biochips as a new kind of tool [133]. Concerning detection of proteins and polypeptides, we should note the meaning and potential of multiplex analysis. Diercks et al. proposed a multiplexed protein detection strategy, which was expected to eliminate the difficulties in directly functionalizing PDMS of microfluidic devices by using prefunctionalized microparticles [134]. Osterfeld et al. accomplished multiplex protein detection of potential cancer markers at subpicomolar concentration levels with a dynamic range over four decades by introducing magnetic nanotags [135]. Fluorescence resonance energy transfer (FRET) was proved to be a powerful method for detection of protein–protein interactions, enzyme activities, and the presence of small molecules in the intercellular milieu [136]. Li et al. discussed the theoretical basis of FRET focusing on the key parameters responsible for the sensitivity of FRET biosensors [137]. In addition, Gales et al. used a bioluminescence resonance energy transfer (BRET) to directly monitor interactions between receptor and G protein in living mammalian cells [138]. A protein biosensor array based on SPR differential phase imaging was reported that worked by detecting a time-modulated differential phase [139].

Label-free protein biosensors witnessed great advance in recent years. An impedometric label-free detection was proposed to conduct immunoassay with high sensitivity and specificity [140]. Aptamers that were immobilized on solid substrates received appreciable attention because they have high permanent charge density, which is possibly exploited for label-free detection such electrochemical impedance spectroscopy (EIS) [141]. There are many cases of aptamers-based sensors, which are separately described in Sect. 4.4.

Siwy et al. demonstrated the utility of a single conically shaped gold nanotube that was embedded in a mechanically and chemically robust polymeric membrane [142]. They reported biofunctionalized conical Au nanotubes, which are potentially useful for obtaining highly sensitive and selective protein biosensors. So et al. introduced a single walled carbon nanotube field effect transistor (SWNT-FET) combined with aptamers as an alternative to the corresponding antibody [143].

Overall, current microfluidics is mature enough to provide the technology for protein analysis, including pretreatment and separation. Rather, it seems that biosensors with biological functional elements need significant progress to make contribution to practical protein research and practical applications.

3.3 *Electrolytes and Gases*

Most electrolyte sensors are ISEs that involve functional membranes containing appropriate ionopores and measure the concentration of ions in chemical, biological, or clinical samples. Especially for clinical uses, rapid and cost-effective analysis of sodium, potassium, chloride, calcium, and magnesium ions in physiological fluids including human blood are essential. ISEs are strongly advantageous in these aspects and thus widely used for quantitative analysis of electrolytes in chemical and clinical laboratories. Faridbod et al. reviewed applications of conducting polymers for potentiometric sensing [100]. For further improvement, much effort has been made for miniaturization. Fueled by the development of MEMS technology, many fabrication methods were introduced to address analytical concerns like LOD, biocompatibility, and sensor stability [144]. Wang et al. reviewed electrochemical biosensors for point-of-care testing of electrolytes and gases, and for immunoassays [95, 145].

4 Functional Receptors Integrated in Microfluidic Systems

4.1 *Enzyme Electrodes*

Enzymes integrated in a biosensor system catalyze the conversion of metabolite molecules to consume or produce detectable species. The change in concentration of the species resulting from enzyme reaction is detected by a corresponding signal transducer. Thus, the analytical performance of these biosensors should critically depend on the activity and stability of the immobilized enzymes. In many cases, the enzyme immobilization is the most important step that determines whether or not it is successful to develop reliable biosensors. In this respect, it is no wonder that a number of new immobilization schemes and materials have been proposed to improve the analytical capabilities of biosensors.

The targets of enzyme-based biosensors that have been reported are mostly small molecules that are involved in human metabolism or environmental cycles, e.g., glucose, urea, lactate, and many others [95, 97, 98, 146, 147]. The most popular strategy is the amperometric glucose sensor, in which glucose-specific enzymes like glucose oxidase are immobilized onto electrodes [148]. Besides electrochemical detection, optical signal transformation is also widely used with the enzyme-based biosensors. Whatever the signal conversion technique applied for enzymatic biosensors, the most valuable benefit obtained by using enzymes is excellent selectivity. The biological specificity of enzymes provides artificial systems with invaluable opportunities.

However, most enzymes may lose their activity when they are immobilized on substrates like electrodes or optical fibers. This is the core issue of enzymatic biosensors. Continuous glucose monitoring in the subcutaneous layer for diabetes mellitus is the example that shows how critical this issue is. Once a glucose sensor is implanted, it is not easy to calibrate its sensitivity. As the clotting and immune reactions proceed under the skin, the environment to which the implanted sensor is exposed keeps changing and possibly affects enzymatic activity in unexpected manner. Since the real-time information from such a continuous glucose sensor is more crucial to the patient than that of a disposable strip, variance in enzyme activity is of great concern. As a result, a variety of immobilization methods, including adsorption, capsulation, entrapment, crosslinking, and covalent bonding, have been developed to formulate an enzyme layer with minimal damage in enzyme activity and long term stability.

In addition to the enzyme activity, charge transfer rate in electrochemical types is another key issue in enzymatic biosensors. Charge transfer rate from the active site to the electrode is closely related to the overpotential that is applied. Lower overpotential reduces the probability of interference by redox active species. Facilitated charge transfer may also help maintaining the condition of diffusion limiting even when significant loss of enzyme activity deteriorates sensitivity.

The bananatrode reported by Sidwell et al. is interesting in that a banana slice was combined with gas-permeable dialysis membrane and acted as a membrane to detect dopamine in a biological sample [149]. Polymers still provide a lot of opportunities for better protocols of enzyme immobilization. Rauf et al. exploited cellulose acetate and poly(methyl methacrylate) (PMMA) for reducing enzyme leakage and enhancing the stability by 94% during 1 month of shelf life [150]. An array of hydrogel-entrapped enzymes in microfluidic channels was employed for simultaneous quantitative analysis of biological samples [151]. The hydrogel microarray was allegedly easy to fabricate in the microfluidic channels, which eliminated the potential cross-talk among the trapped enzymes. As for lactate biosensors, most of them are based on lactate dehydrogenase (LDH) or lactate oxidase [152, 153]. Lactate in serum was analyzed without pretreatment using screen printed carbon on which an LDH layer was coated [154]. Romero et al. reported an amperometric lactate sensor with a protective Nafion membrane, linearly responding to lactate concentrations between 2 and 1,000 μM [153]. Elevated levels of urea indicate possible problems in kidney and liver function. Nikoleli et al. used a urea sensor based on an air-stable lipid film that contained urease and was stabilized on a glass filter by photopolymerization [155]. Kuralay et al. immobilized urease in the poly(vinylferrocenium) (PVF^+) matrix [147] and Gutierrez et al. showed a potentiometric bioelectric tongue for the analysis of urea that employed PVC and carboxylate PVC [156]. In this work, the calibration process based on an artificial neural network and partial least squares was introduced to determine the concentration of urea without compensating endogenous ammonia or chemically rejecting the alkaline interferences.

Nanomaterials are a new class of research material that have been tested for enzyme immobilization. Sawicka et al. measured urea concentration by using nanocomposite fibers of urease and polyvinylpyrrolidone (PVP) [146]. Biocomposite nanofibers were prepared by electrospinning a solution in which urease and PVP were dissolved, leading to improvement in response time and sensitivity.

Barhoumi et al. discussed a ZnAl-based enzyme nanohybrid system for a urea biosensor, in which ureases were entrapped within layered double hydrides [157]. Tsai et al. employed a sol–gel method to fabricate an enzymatic optical biosensor array for the analysis of multiple samples [152].

4.2 Antibodies

Immunoglobulins (Igs) are gamma-globulin proteins that are present in blood and play important roles in identifying and neutralizing foreign objects such as bacteria and viruses in the immune system. An antibody is a kind of Ig composed of heavy (above 150 kDa) globular plasma proteins such as IgA, IgD, IgE, IgG, and IgM. An antigen is a molecule or pathogen that is capable of eliciting an immune response. Immunoassays make use of the sensitivity and specificity of the antibody–antigen interaction. There are direct and indirect methods that have been adopted for immunosensors. Direct methods acquire electrochemical, optical, or electrical signals resulting from the event of immunochemical complex formation without any labeling or secondary reactions, as shown Fig. 2 [158, 159].

Indirect methods utilize labeling. Typical labels for immunoassays are enzymes, fluorescent or radioactive molecules, nanoparticles, chemiluminescent probes, metal tags, and so on [158, 159]. By labeling, the immune reaction can be detected more sensitively through amplifying the corresponding electrochemical, optical, or other physical responses. That is why most conventional immunoassay protocols adopt labeling methods. Sandwich-type immunoassays are the most widely used

Fig. 2 Schematic of direct immunoassay

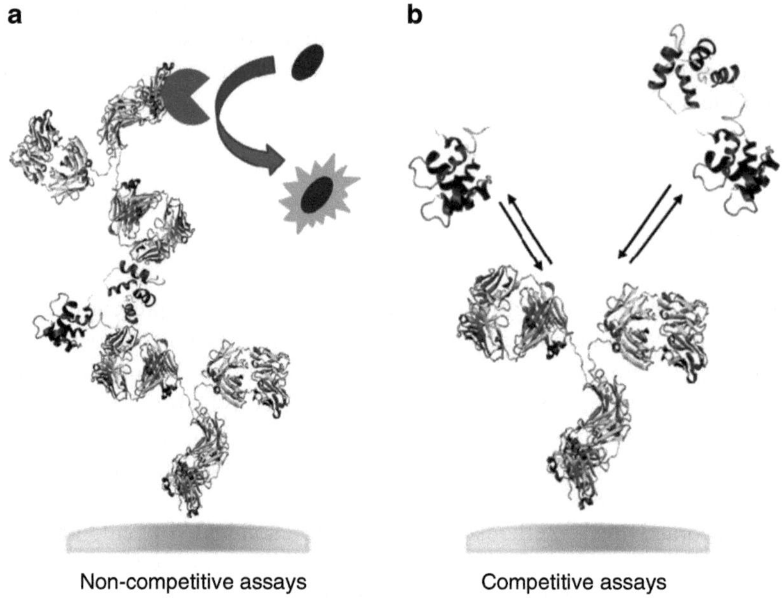

Fig. 3 Schematic of indirect immunoassay

method, which utilize secondary antibody with label. The general scheme of sandwich-type immunoassays is illustrated in Fig. 3a. A representative example is the ELISA, also known as an enzyme immunoassay (EIA), which is a fundamental technique in bioanalysis.

However, we should take into account the fact that labeling processes may increase cost and elongate the time of analysis because a few more steps in the analytical procedure are required. More importantly, the sandwich-type immune assay is not a good choice for small molecule analysis. If a target analyte is too small to be bound with another primary antibody, it is difficult to expect additional association events. In such cases, competitive immunoassays are often considered [160]. Unlabeled antigens in the sample are allowed to compete with labeled ones that used to be bound to antibody, as illustrated in Fig. 3b. The antigens in the sample replace the labeled antigens, which are released to the solution and measured. Higher concentration of unlabeled antigens (analytes) result in smaller amount of the labeled antigens that remain bound to the antibodies.

To avoid labeling hassle, label-free methods have been suggested. Some examples of label-free approaches are back-scattering interferometry [161], impedometric immunosensor for the stroke marker detection [162], and impedometry coupled with magnetic nanoparticle immunoseparation for direct detection of pathogenic bacterial cells in food samples [163]. Immunoassays in microfluidic devices employing SPR have also been reported. Luo et al. developed a PDMS microfluidic device with an array of gold spots onto which antigens or antibodies of interest were attached [91]. Recently, a promising strategy is to immobilize antibodies on mobile substrates instead of planar substrates. Such mobile platforms could be polymer-based microspheres such as latex beads or nanoparticles, which provide fundamental basis for multiplex immunoassay. Highly carboxylated polystyrene/polyacrylic acid latex microspheres coated with QDs is a good example [164]. The LOD was successfully enhanced as low as 50 ng/mL on a microfluidic chip. The immobilization of biological receptors like antibodies onto the solid supports is one of the core issues because it is closely related to how to maximize the efficiency of the recognizing ability with minimal damage to their activities. This stimulated intensive studies on the orientation-controlled immobilization of intact proteins [133, 165, 166].

4.3 Molecular Receptors

Valdes et al. suggested the utility of various ligands as molecular receptors for biosensors [167]. Molecular receptors have been used as substitutes for antibodies for specific binding and detection of pathogens or toxins, including carbohydrate derivatives, proteins, bacteriophages, and lysozymes. Koo et al. showed an efficient capture of target analytes for a specific detection of pathogenic microorganisms [168]. Jen et al. harnessed fiber-optic localized plasma resonance (FO-LPR) and the characteristic optical properties of self-assembled gold nanoparticles on the surface modified with a receptor to increase the sensitivity and shorten the response time on microfluidic chips [169].

4.4 Aptamers

Aptamers are artificial nucleic acids that specifically bind to target molecules. The aptamers with optimal sequences are obtained through a combinational selection process such as systematic evolution of ligands by exponential enrichment (SELEX) from random RNA or DNA libraries [170, 171]. Recently, aptamers biosensors based on fluorescence, electrochemistry, QCM, and infrared spectroscopy measurements have been actively studied [106, 107, 172–179]. Swensen et al. improved LOD and response time [106], and Xiao et al. constructed an electronic aptamer-based sensor by covalently attaching a methylene blue (MB)-labeled thrombin-binding DNA aptamer to a gold electrode. [176]. Gokulrangan et al. addressed a rapid aptamer-based bioanalysis for the sensitive detection of IgE using fluorescence polarization [172]. Hansen et al. presented a multianalyte electrochemical aptamer biosensors that exhibited subpicomolar (attomole) level LOD employing QD semiconductor nanocrystals [107]. This is an impressive achievement in regard to highly sensitive and selective simultaneous bioelectronic detection of several protein targets. Bang et al. used beacon-type aptamers based on intercalation of MB as an electrochemical marker and achieved an LOD of 11 nM thrombine [173]. Liao et al. took advantage of Fourier transform infrared attenuated total reflection (FTIR-ATR) and surface modification techniques [175]. Aptamer-modified carbon nanotube FETs were utilized for the label-free detection of IgE [178]. Centi et al. simplified the target capturing step by using aptamer-functionalized magnetic beads for sandwich electrochemical assay of thrombin [179]. Non-Faradic impedance spectroscopy was shown to possibly monitor the capacitance change induced by aptamer–protein association, with the aim of in vivo cytokine sensor development for chronic neural prostheses [180]. Wang et al. reviewed nanoparticle-based electrochemical bioassays by using aptamer-based immunoassays and nanoparticles [145]. Aptamer applications for the detection of a variety of samples such as thrombin, cytokines, and protein were also discussed [181, 182].

4.5 *Miscellaneous*

Beside enzyme, antibody, molecular ligands, and aptamers, significant research achievements have been made with the molecular imprinting method. The strategy of molecular imprinting is that the sites remember the local environment to selectively bind the analytes, even after the previously bound analytes are released [183, 184]. In recent years, the molecular imprinting method is widely studied because of its cost-effectiveness and high specificity. Alexander et al. reviewed the development of molecular imprinting science and technology [185]. Janiak et al. summarized molecularly imprinted polymers that had been synthesized by crosslinking of functional monomers in the presence of a template molecule. This review dealt with the molecular imprinting of peptides and proteins in aqueous media [186]. Eroz et al. exploited QCM with molecular imprinted layers to detect a variety of analytes, including glucose [187]. Xie et al. presented the molecular imprinting on walls of silica nanotubes for 2,4,6-trinitrotoluene (TNT) [188]. Hansen reviewed the molecular imprinting of proteins [189].

5 Microfluidic Components

5.1 *Pretreatment*

With regard to microfluidic-based integrated biochip technology, a number of laboratories have pursued devices that are capable of handling and analyzing real biological samples. Human whole blood is a typical example that has a variety of ingredients including blood cells, platelets, proteins, and many molecular species such as glucose, mineral ions, hormones, carbon dioxide, etc. For this reason, sample pretreatment is indispensable in many cases and thereby has been a critical issue in clinical, biological, and biochemical analysis on microfluidic chips. Traditional protocols of separating blood cells from whole blood involve labor-, cost-, and time-consuming processes that normally involve centrifuging steps. Therefore, separation technology for sample preparation is an urgent demand, to which rapid and efficient microfluidic technology could make the biggest contribution. Various technologies have been suggested as alternatives to centrifugal separation. For instance, plasma separation from whole blood was conducted by many kinds of membrane filters, which include a planar microfilter, bent channel structure, T-shaped channel, and micropost array [190]. Jaggi et al. observed the depletion of red blood cells (RBCs) from whole blood at high volume flow rates [191]. In this study, a high-aspect ratio channel bifurcation was used to separate RBCs. Moreover, particle density in a suspension was able to be tuned and the samples were filtered through microstructures such as micropillar arrays and membranes with circular, hexagonal, and rectangular holes. Chun et al. realized rapid and selective RBC lysis by quickly mixing the whole blood sample with a lysing reagent between a pair of salt bridges that were integrated on a microfluidic chip, and applying an AC electric field at low frequency [192].

5.2 Surface Modifications

Surface modification is crucial for improving the sensitivity of biosensors, which should minimize nonspecific binding. There have been a number of ways to modify the surfaces, e.g., wet chemical, organosilanization, ionized gas treatments, and UV irradiation [193]. Surface modification methods that have been published to date can be categorized into chemical and physical methods. Chemical modification immobilizes functional molecules to create surface properties desired so that the surfaces become passivated or activated with chemicals. On the other hand, physical modification may change the surface roughness, grain size, and grain boundary by exposure to lasers, plasmas, heat, and polishing. To achieve better sensitivity and reliable performance, a sufficient amount of analytes and functional molecules need to be immobilized onto the surface of interest in the microfluidic biosensors. A good example is the surface modification of protein-modified microcantilevers that were prepared from multiple surface conjugations [194]. Diao et al. tried a direct silanization with aminopropyltriethoxysilane and activation with glutaraldehyde [195]. Biofouling of the surfaces has been one of the major problems in biological analysis. Several variables like wettability, biocompatibility, and nonspecific adsorption were addressed to tackle the biofouling problem. Poly(ethylene glycol) (PEG) is well known for its protein-resistant property [196]. Bi et al. reported a simple method for the deposition of PEG brush on PMMA microfluidic channel surfaces [197]. The fabrication of PEG-coated microfluidic chips would provide a biocompatible surface for complex biological analysis. Prakash et al. showed covalent attachment of various polymers fused with silica or silica-coated silicon in glass [198]. The PDMS layer was adhered to the surface of another PDMS by gaseous plasma [199]. Lee et al. generated a protein micropattern on the PEG hydrogels by surface graft polymerization and photolithography [200]. It was reported that PEG layers were able to reduce nonspecific interactions [201].

5.3 Beads

Rapid and reliable quantitative analysis of cells is increasingly important. Although utility of beads are not limited to biological applications, cellular measurements such as cell counting, antibody-mediated agglutination, and diagnostics using cultured cells attract exceptional interest as they could benefit immensely from beads. Bead-based analysis for cell research and development can be discussed from two points of views: molecular/immune assay and flow cytometry. Since the multiplex bead array assay (MBAA) was introduced by Horan et al. in 1977 [202], polymeric microparticles have been considered and tried as solid supports for molecular/immunoassay. Such beads can reduce the volumes of reagents as well as samples by providing a high surface-to-volume ratio that is significantly enlarged in comparison to conventional microarray. Saunders et al. reported a competitive binding assay using beads on which antibodies were anchored and associated with nonlabeled antigens prior to exposure [203]. Holmes et al. presented a microfluidic rapid analysis of polymer beads labeled with fluorescent dye in small volumes [204]. This research showed the possibility that bead-based immunoassay could induce a substantial advance in sensitivity, precision, and accuracy by employing microfluidic chips. Gao et al. enlarged the reaction chambers for CD4-positive T lymphocyte capture to raise cell capture efficiency compared with conventional methods based on CD4 separation [205]. More importantly, it was demonstrated that bead-based immunoassay is a promising alternative to ELISA based on plates, and offers several valuable advantages [206]. Concerning the MBAA, there are two key issues. One is functionalization of the bead surfaces as desired and the other is recognition of individual beads that are suspended in the solution. The former issue has been intensively addressed by research and development in industry as well as academia. Bead-based multiplexing systems are commercially available from Luminex, Illumina, and BioArray Solutions, as summarized in Table 3. Beads are dispersed in the solution to have sufficient chances to interact with potential analytes and then need to be collected before rinsing or other subsequent processes. To successfully enable active dispersion and recollection, magnetic nanoparticles were impregnated in polymeric microbeads for immunoassay [207].

The bead surface critically affects nonspecific binding, which is particularly serious in the presence of physiological fluids, e.g., human whole blood. Conventional polymeric surfaces need chemical coupling processes to immobilize functional molecules and are vulnerable to contamination originating from nonspecific adsorption of proteins that are not analytes, like albumins in blood. Polymers including ethylene glycol oligomers have been tried in an effort to

Table 3 Bead-based multiplex technologies and companies

Company	Model	Analysis targets	Related methods
Luminex	xMAP	Cardiac markers	Gene expression profiling
		Cancer markers	Nuclear receptors
		Metabolic markers	HLA testing
		Neurobiology	Infectious disease
		Cytokines	–
		Chemokines	Cystic fibrosis
		Growth factor	–
Illumina	Sentrix	SNP genotyping	Gene expression
BioArray Solutions	BeadChip	Human erythrocyte antigen genotyping	Ashkenazi genetic disease genotyping

HLA Human leukocyte antigen, *SNP* Single nucleotide polymorphism

reduce nonspecific adsorption [201, 208], but it is still challenging. A thin gold shell on a polymeric microsphere was proposed as a new platform to tackle a couple of issues that were involved in bead surfaces [76, 209]. This approach is still at an early stage, but seems to deserve watching to see what will happen in the future.

As for identification of the suspended beads, it is necessary to achieve not only cost-effective and reliable encoding technology but also high throughput and a facile decoding strategy. Unlike microarrays, the positions of sensing sites are not fixed in the suspension array system. So, beads dispersed in solution should be recognized individually to reveal what kind of receptors is immobilized on their surfaces. Microfluidic devices are expected to provide important opportunities of higher throughput and better reliability to encoding as well as decoding technologies, eventually making multiplex analysis possible. The encoding–decoding methods that have been reported up to now are based on fluorescence, size, or shape [210–212]. The most popular way to encode microbeads is to incorporate fluorescent dyes and QDs into the beads [213, 214]. To read out the barcodes of the beads, flow cytometry is widely used due to its multifunctional capabilities of counting, characterizing, and sorting. It can count and characterize a variety of small particles, including cells, beads, and microparticles. The fluorescence-activated cell sorter (FACS) is a sophisticated readout device that allows fast decoding and sorting [215, 216]. However, the FACS system is normally a large and expensive instrument so that it deters rapid expansion of applicability to multiplex suspension arrays at low cost. Recently, research effort has been directed to microfluidic chip-based flow cytometers, which require small sample volume, less reagent consumption, and precise control of particles [217]. A miniaturized flow cytometer was proposed that used a unique polyelectrolytic salt bridge for accurate discriminative detection of cells as a function of their size and fluorescence [42, 218]. Fan et al. reported a magnetic-bead-based chemiluminescent immunoassay in microfluidic chips [219]. Antibody-conjugated magnetic beads were shown to be useful for detecting viruses, with the assistance of sample pretreatment devices for purification and preconcentration [220].

5.4 Droplets

Droplet-based microfluidics is expected to solve the problems of mixing and reagent diffusion in laminar flow-based microfluidics. Micro- and nanoscale emulsions of immiscible liquid droplets were applied in various applications such as DNA encapsulation [221] and drug delivery systems [222]. Microfluidic droplets have a few parameters of droplet size, frequency, composition, and so on. Properties of droplets are very sensitive to their physical size and its distribution. The conventional way to fabricate the droplets is to use macroscale instruments such as high-pressure homogenizers, high-power generators, and ultrasonic generators. However, it is very difficult to control the size of droplets using traditional methods. For these reasons, microfluidic techniques have been sought for producing droplets. Liu et al. chose dichloroethane as the continuous organic phase to create droplets inside a hydrodynamic focusing chip [223]. By this method, uniform droplets were made and their size was under control by varying the flow rates of the organic and water phases.

Rapid and reliable characterization of droplets in microfluidic environments has been widely addressed. Electrochemical methods can offer a simpler route to that goal than traditional optical imaging. Monpichar et al. reported a droplet-based microfluidic platform to study streptavidin–biotin binding kinetics with millisecond time resolution [224]. In this study, a droplet microfluidic platform was integrated with a confocal fluorescence detection system. Several microfluidic chips were reported to make droplets with better emulsion quality [225–227]. Recently, separation and collection of emulsion microdroplets have been explored. For example, a passive and active satellite droplet filtration system was demonstrated [227]. There were a few reports on a microfluidic chip that was capable of generating uniform and tunable sizes of emulsion droplets by using microchoppers [228].

5.5 Pumping and Valves

Microfluidic analytical systems require active or passive micropumps to control the reagent and sample delivery. A large number of micropumps have been devised since the microfluidic system was first introduced. Several mechanisms have been suggested for transporting the fluids in microfluidic systems, and can be discussed in two categories: displacement and dynamic pumping. Displacement pumps exert pressure forces on the fluid through one or more moving boundaries. Micropumps are based on periodic displacement involve reciprocating or rotary actuations [229–232], while periodic type micropumps have piezoelectric [233–238], peristaltic [239, 240], thermopneumatic [241, 242], electrostatic [243, 244], pneumatic [245], and electromagnetic moving units [246]. A number of reports have been published on membrane- or diaphragm-based displacement micropumps. Peristaltic pumps are reciprocating displacement micropumps that have check or active valves [247]. Cho et al. reported a fully integrated pumping system using centrifugal microfluidics on a polymer-based CD [248].

Dynamic pumps infuse energy to the fluid in a manner that increases either its momentum (centrifugal pumps) or its pressure (electroosmotic and electrohydrodynamic pumps) as shown in Table 4. They involve centrifugal or hydrodynamic actuations and, more specifically, are driven by electro-/magneto-hydrodynamic [257], electroosmotic [254–256], electrokinetic [251–253], electrowetting [258], and acoustic forces [259]. Centrifugal pumps are typically less effective for fluids with low Reynolds numbers and have limitation in miniaturization.

Sometimes electromagnetic field and acoustic-wave are regarded as suitable means for creating pressure to make flow. The field-free micropump driven by electroosmotic flow is an interesting example of swift electromagnetic manipulation of field-sensitive biological objects [260]. The design of the proposed system was a "Y"-shaped channel network with two arms, which have positively and negatively charged surfaces on the inner walls by depositing polyelectrolytes on the respective surfaces. The fluid on this chip is under quick and precise control by a programmed external electric field, while the designated region, in which biological elements are being handled, is free from electromagnetic field. On other hand, microelectromechanical systems technology was introduced to a micropump that needs no external power source [261]. This work showed a new actuation mechanism for the self-generated peristaltic motion of cascaded actuators that employ the fluidic circuit of an elastic tube. A self-priming low-power micropump based on a polypyrrole actuator can be fabricated without microfabrication equipment [262].

Microvalves are sometimes regarded as a part of micropumps in many reviews [263]. In recent years, new methods for the valves have been reported, as summarized in Table 5. A magnetic hydrogel nanocomposite valve was utilized for remote

Table 4 Classification of micropumps

Micropump type	Operating principle	Key factors	References
Mechanical	Diaphragm	Piezoelectric	[233–238]
		Thermopneumatic	[241, 242]
		Electrostatic	[243, 244]
		Pneumatic	[245]
		Peristaltic	[239, 240]
Displacement	Rotary	Rotating gear	[229, 230]
		Viscous force	[231, 232]
	Fluid	Ferrofluid	[249]
		Phase change	[250]
	Electromagnetic	Magnet	[246]
Dynamic	Centrifugal	Momentum	[248]
	Electrokinetic	Induction	[251, 252]
		Injection	[253]
	Electroosmotic	DC	[254, 255]
		AC	[256]
	Electro-/magneto-hydrodynamic	Lorentz force	[257]
	Electrowetting	Electric	[258]
	Acoustic	Waves	[259]

Table 5 Classification of microvalves

Microvalve type	Operating principle	Operating principle	Key factors	References
Active	Mechanical	Magnetic	External magnetic fields	[264]
			Integrated magnetic inductors	[265]
		Electric	Electrostatic	[266]
			Electrokinetic	[246]
		Piezoelectric	Frequency	[267]
		Thermal	Bimetallic, thermopneumatic	[268]
	Nonmechanical	Electrochemical	–	[269]
		Phase change	Hydrogel, sol–gel, paraffin	[270–272]
		Rheological	Electro-rheological, ferrofluids	[273]
	External	Modular	Built-in, rotary	[274]
		Pneumatic	Membrane, in-line	[275, 276]
Passive	Mechanical	Check valve	Flap	[277]
			Membrane	[278]
			Ball	[249]
			In-line mobile structure	[279]
	Nonmechanical	Capillary	Hydrophobic valve, abrupt	[280]

control of the flow in a microfluidic device by applying alternating magnetic fields [281]. This approach chose temperature control to regulate the collapse of a nanocomposite, leading to opening of the valves. Yoo et al. suggested a paraffin-actuated microvalve in which phase transition of the paraffin from solid to liquid is triggered by a thermal cue and results in volume change [282]. This system is practical for controlling the mass transport of reagents and samples in a microfluidic chip precisely and efficiently. Many valving principles have been introduced to steer the flow by a centrifugation protocol. Honda et al. reported a rotary disk driven by centrifugal force for highly integrated immunoassay [283]. They demonstrated that the CD based on centrifugal force successfully handled small volumes of samples and reagents with sufficiently good precision. With respect to flow control in more complicated microfluidic networks, Hisamoto et al. harnessed the surface wettability of the inner walls of microchannels, which were coated with a thermo-responsive polymer [284].

5.6 Peripherals

The micromixer is an essential microfluidic component for specific applications. Compared with conventional macroscale mixers, micromixers need a somewhat different approach to working principle, specification, and design. First of all, it is hard to make turbulence in such a limited volume [285, 286]. And, more than two solutions in a microfluidic system are necessarily mixed as rapidly as possible for many applications, e.g., studies on the kinetics of enzyme reactions, fast liquid reactions, discriminative cell lysis with minimal side effects, etc.

There are two types of micromixers: passive and active micromixers. Passive micromixers operate without an external driving force so that the mixing process depends on diffusion and chaotic advection. In order to increase the contact interface between the liquids, various attempts have been made, including multi-layer lamination [287], fluid injection [288], and 2D or 3D structures [289]. Most passive mixers rely on the chaotic advection principle to enhance mass transport by incorporating complex structure [290, 291], grooves in the channel well [292, 293], or zig-zag-shaped channels [294]. Hardt et al. discussed passive micromixers for applications in microfluidics [295]. Active micromixers involve the disturbance caused by external forces for the mixing process. External forces that have been suggested for active micromixers are pressure [296], electrodynamic [297], dielectrophoretic [298], electrokinetic [299], and magneto hydrodisturbance [300]. Chun et al. used polyelectrolytic salt bridges to obtain an efficient active micromixer operating by low voltage and straight/smooth surfaces [192]. In this system, a pair of positively charged polyelectrolytic gel electrodes, which face to each other, play the key role of inducing ion depletion/enrichment phenomenon. Chang et al. reviewed the recent progress in micromixing phenomenon based on electrowetting-on-dielectric, dielectrophoresis, and electroosmosis [301]. Hessel et al. reviewed the mixing principles of passive and active micromixers [302].

6 Perspectives

Biosensors and microfluidic analysis systems possess an intrinsic feature that is common to both. This is the fact that chemical and bioanalyses involving biosensors and microfluidic systems could be conducted by conventional methods in the laboratory as well, or even better. The reason why these systems are needed is not to enable chemical or bioanalysis of target analytes that conventional methods have been unable to analyze. They were developed to make the analytical system better for specific purposes through size reduction, and thereby widen their range of use. It is no wonder that tremendous research activities in this area have been concentrated on finding solutions to the problems that come from miniaturization or integration on a chip, and not on seeking new analytical principles or methodologies.

In this regard, the future direction of biosensors and microfluidic analytical systems is predicted to be towards better efficiency, convenience, cost-effectiveness, compatibility, and reliability. Label-free technology is a particularly good example that deserves special attention. It would provide valuable advantages in efficiency, convenience, and cost-effectiveness by simplifying the procedure, requiring less reagents, and cutting the cost and time of analysis. However, label-free detection still has its own challenges, and subsequent research efforts must be made to accomplish the ultimate goal, as argued by Fan et al. [303].

Another example is multiplex suspension analysis. As more and more knowledge about nature becomes available, the demand for chemical and bioanalyses increases correspondingly. The revolutionary advance in biotechnology is accelerating the growth of analytical needs. New analytes that need to be analyzed are ceaselessly appearing and some of them may be extremely important in terms of clinical, economical, industrial, or environmental concerns. Moreover, analytes with a very low population, such as circulating tumor cells, need at least a few milliliters of sample, otherwise analysis is impossible. According to the conventional concept of chemical and bioanalysis, an analyte requires a sample for its own analysis. This leads to research efforts to improve analytical methods so that they consume as small a sample as possible. However, minimizing the sample volume cannot ultimately solve this fundamental problem of conventional methodology. This situation unequivocally indicates the request of multiplex analysis techniques. 2D microarrays or biosensor arrays integrated on a chip or a strip are examples of ideas on the ways to cope with these new analytical demands. Noting that the entire surface of the 2D arrays should be covered by the sample solution, we should consider the microsuspension array as a promising alternative to conventional arrays on flat substrates. Microbeads are expected to offer a useful platform for multiplex suspension array. A large number of microbeads would provide sufficient amount of data to ensure reliability. The current trend in modern bioanalysis strongly indicates that efficient barcoding technology and a handy readout device for rapid decoding/detecting the beads will make a breakthrough towards ubiquitous chemical and bioanalysis using small but powerful analytical equipment that is compatible with home appliances and mobile smart phones.

Acknowledgments The authors acknowledge Mid-career Researcher Program through the NRF (National Research Foundation) grant funded by the MEST (Ministry of Education, Science and Technology) of Korea (No. 2010-0011751), the Public welfare & Safety research program through the National Research Foundation of Korea (NRF) funded by the Ministry of Education, Science and Technology (No. 2010-0020772), and the Converging Research Center Program through the Ministry of Education, Science and Technology (2010K001297).

References

1. Gravesen P, Branebjerg J, Jensen S (1993) J Micromech Microeng 3:168–182
2. Bao M, Wang W (1996) Sens Actuators B 56:135–141
3. Namasivayam V, Lin R, Johnson B, Brahmasandra S, Razzacki Z, Burke DT, Burns MA (2004) J Micromech Microeng 14:81–90
4. Jakeway SC, de Mello AJ, Russell EL (2000) Fresenius J Anal Chem 366:525–539
5. Tudos AJ, Besselink GAJ, Schasfoort RBM (2001) Lab Chip 1:83–95
6. Brivio M, Fokkens RH, Verboom W, Reinhoudt DN (2002) Anal Chem 74:3972–3976
7. Schulte TH, Bardell RL, Weigl BH (2002) Clin Chim Acta 321:1–10
8. Srinivasan V, Pamula VK, Fair RB (2004) Lab Chip 4:310–315
9. Roman GT, Kennedy RT (2007) J Chromatogr A 1168:170–188
10. Briss CL, McMullin JN, Backhouse CJ (2007) Lab Chip 7:1280–1287

11. Herr AE, Hatch AV, Throckmorton DJ, Tran HM, Brennan JS, Giannobile WV, Singh AK (2007) Proc Natl Acad Sci USA 104:5268–5273
12. Boden R, Lehto M, Margell J, Hjort K, Schweitz J-A (2008) J Micromech Microeng 18:1–7
13. Myers FB, Lee LP (2008) Lab Chip 8:2015–2031
14. Fang X, Liu Y, Kong J, Jiang X (2010) Anal Chem 82:3002–3006
15. Arora A, Simone G, Salieb-Beugelaar GB, Kim JT, Manz A (2010) Anal Chem 82:4830–4837
16. Whitesides GM (2006) Nature 442:368–373
17. Janasek D, Franzke J, Manz A (2006) Nature 442:374–380
18. Craigheed H (2006) Nature 442:387–393
19. Yager P, Edwards T, Fu E, Heltor K, Melson K, Tam M, Weigl B (2006) Nature 442:412–418
20. Turner AD, Karube I, Wilson G (1986) Biosensors: Fundamentals and Applications, Oxford Science Publications, Oxford
21. Frieder S, Florian S (1992) Biosensors, Elsevier, Newyork
22. Cooper J, Cass T (1990) Biosensors, Oxford University Press
23. Zhang X, Ju H, Wang J (2008) Electrochemical sensors, biosensors and their biomedical applications, Elsevier
24. Clark LC, Lyons C (1962) Ann N Y Acad Sci 102:29–45
25. Dittrich PS, Manz A (2006) Nat Rev Drug Discov 5:210–218
26. El-Ali J, Sorger PK, Jensen KF (2006) Nature 442:403–411
27. Thorsen T, Maerkl S, Quake S (2002) Science 298:580–584
28. Shan X, Huang X, Foley KJ, Zhang P, Chen K, Wang S, Tao N (2010) Anal Chem 82:234–240
29. Wang S, Huang X, Shan X, Foley KJ, Tao N (2010) Anal Chem 82:935–941
30. Huang X, Wang S, Shan X, Chang X, Tao N (2010) J Electroanal Chem 649:37–41
31. Shan X, Patel U, Wang S, Iglesias R, Tao N (2010) Science 327:1363–1366
32. Balslev S, Jorgensen AM, Bilenberg B, Mogensen KB, Snakenborg D, Geschke O, Kutter JP, Kristensen A (2006) Lab Chip 6:213–217
33. Pais A, Banerjee A, Klotzkin D, Papautsky I (2008) Lab Chip 8:794–800
34. Jindal R, Cramer SM (2004) J Chromatogr A 1044:277–285
35. Ro KW, Lim K, Shim BC, Hahn JH (2005) Anal Chem 77:5160–5166
36. Caglar P, Tuncel SA, Malcik N, Landers JP, Ferrance JP (2006) Anal Bioanal Chem 386:1303–1312
37. Park J-S, Park K-B, Shin K-S, Park H-D, Kim M-C, Kim J-R, Park S-J, Song Y-H (2006) Sens Actuators B 117:516–622
38. Schulze P, Ludwig M, Kohler F, Belder D (2005) Anal Chem 77:1325–1329
39. Qin J, Fung Y, Zhu D, Lin B-C (2004) J Chromatogr A 1027:223
40. Li H-F, Cai Z, Lin J-M (2006) Anal Chim Acta 565:183–189
41. Renzi RF, Stamps J, Horn BA, Ferko S, VanderNoot VA, West JAA, Crocker R, Wiedenman B, Yee D, Fruetel JA (2005) Anal Chem 77:435–441
42. Joo S, Kim KH, Kim HC, Chung TD (2010) Biosens Bioelectron 25:1509–1515
43. Bhattacharyya A, Klapperich CM (2007) Biomed Microdevices 9:245–251
44. Yacoub-George E, Hell W, Meixner L, Wenninger F, Bock K, Lindner P, Wolf H, Kloth T, Feller KA (2007) Biosens Bioelectron 22:1368–1375
45. Hatakeyama K, Tanaka T, Sawaguchi M, Iwadate A, Mizutani Y, Sasaki K, Tateishi N, Matsunaga T (2009) Lab Chip 9:1052–1058
46. Huang X, Ren J (2005) Electrophoresis 26:3595–3601
47. Liu B-F, Ozaki M, Hisamoto H, Luo Q, Utsumi Y, Hattori T, Terabe S (2005) Anal Chem 77:573–578
48. Qiu H, Yan J, Sun X, Liu J, Cao W, Yang X, Wang E (2003) Anal Chem 75:5435–5440
49. Llobera A, Demming S, Wilke R, Buttgenbach S (2007) Lab Chip 7:1560–1566

50. Torrance L, Ziegler A, Pittman H, Pater M, Toth R, Eggleston I (2006) J Virol Methods 134:164–170
51. Zezza F, Pascale M, Mule G, Visconti A (2006) J Microbiol Methods 66:529–537
52. Skottrup P, Hearty S, Frokiaer H, Leonard P, Hejgaard J, O'Kennedy R, Nicolaisen M, Justesen AF (2007) Biosens Bioelectron 22:2724–2729
53. Skottrup P, Nicolaisen M, Justesen AF (2007) J Microbiol Methods 68:507–515
54. Ligler FS (2009) Anal Chem 81:519–526
55. Steigert J, Grumann M, Brenner T, Riegger L, Harter J, Zengerle R, Ducree J (2006) Lab Chip 6:1040–1044
56. Xu F, Datta P, Wang H, Gurung S, Hashimoto M, Wei S, Goettert J, McCarley RL, Soper SA (2007) Anal Chem 79:9007–9013
57. Liu P, Seo TS, Beyou N, Shin K-J, Scherer JR, Mathies RA (2007) Anal Chem 79:1881–1889
58. Hofmann O, Wang X, Cornwell A, Beecher S, Raja A, Bradly DDC (2006) Lab Chip 6:981–987
59. Schmidt O, Bassler M, Kiesel P, Knollenberg C, Johnson N (2007) Lab Chip 7:626–629
60. Vezenov DV, Mayers BT, Conroy RS, Whitesides GM, Snee PT, Chan Y, Nocera DG, Bawendi MG (2005) J Am Chem Soc 127:8952–8953
61. Li Z, Wang Y, Wang J, Tang Z, Pounds JG, Lin Y (2010) Anal Chem 82:7008–7014
62. Huang C-P, Li Y-K, Chen T-M (2007) Biosens Bioelectron 22:1835–1838
63. Marchand G, Broyer P, Lanet V, Delattre C, Foucault F, Menou L, Calvas B, Roller D, Ginot F, Campagnolo R, Mallard F (2008) Biomed Microdevices 10:35–45
64. Schwarz RA, Arifler D, Chang SK, Pavlova I, Hussain IA, Mack V, Knight B, Richards-Kortum R (2005) Optics Lett 30:1159–1161
65. Qian X, Peng X-H, Ansari DO, Yin-Goen Q, Chen GZ, Shin DM, Yang L, Young AN, Wang MD, Nie S (2008) Nat Biotechnol 26:83–90
66. Jeanmaire DL, Van Duyne RP (1977) J Electroanal Chem 84:1–20
67. Gao X, Zhang Y, Weaver MJ (1992) Langmuir 8:668–672
68. Hildebrandt P, Stockburger M (1984) J Phys Chem 88:5935–5944
69. Kneipp K, Wang Y, Kneipp H, Perelman LT, Itzkan I, Dasari RR, Feld MS (1997) Phys Rev Lett 78:1667–1670
70. Nie S, Emory SR (2010) Science 275:1102–1106
71. Oldenburg SJ, Westcott SL, Averitt RD, Halas NJ (1999) J Chem Phys 111:4729–4735
72. Wustholz KL, Henry A-I, McMahon JM, Freeman RG, Valley N, Piotti ME, Natan MJ, Schatz GC, Van Duyne RP (2010) J Am Chem Soc 132:10903–10910
73. Lim D-K, Jeon K-S, Kim HM, Nam J-M, Suh YD (2009) Nat Mater 9:60–67
74. Haynes CL, Duyne RP (2001) J Phys Chem B 105:5599–5611
75. Banholzer MJ, Millstone JE, Qin L, Mirkin CA (2008) Chem Rev 37:885–897
76. Lee S, Joo S, Park S, Kim S, Kim HC, Chung TD (2010) Electrophoresis 31:1–7
77. Cooper MA (2002) Nature 1:515–528
78. Homola J, Yee SS, Gauglitz G (1999) Sens Actuators B 54:3–15
79. Kim SJ, Gobi KV, Iwasaka H, Tanaka H, Miura N (2007) Biosens Bioelectron 23:701–707
80. Feltis B, Sexton B, Glenn F, Best M, Wilkins M, Davis T (2008) Biosens Bioelectron 23:1131–1136
81. Springer T, Piliarik M, Homola J (2010) Sens Actuators B 145:588–591
82. Waswa J, Irudayaraj J, DebRoy C (2007) Food Sci Technol 40:187–192
83. Wei D, Oyarzabal A, Huang T, Balasubramanian S, Sista S, Simonian AL (2007) J Microbiol Methods 69:78–85
84. Shin Y-B, Kim HM, Jung Y, Chung BH (2010) Sens Actuators B 150:1–6
85. Leebeeck AD, Kumar LKS, Lange VD, Sinton D, Gordon R, Brolo AG (2007) Anal Chem 79:4094–4100

86. Huang C, Bonroy K, Reekmans G, Laureyn W, Verhaegen K, Vlaminck ID, Lagae L, Borghs G (2009) Biomed Microdevices 11:893–901
87. Huang C, Bonroy K, Reekmans G, Verstreken K, Lagae L, Borghs G (2009) Miroelectron Eng 86:2437–2441
88. Frederix F, Friedt KH, Choi W, Laureyn W, Campitelli A, Mondelaers D, Maes G, Borghs G (2003) Anal Chem 75:6894–6900
89. Willets KA, Van Duyne RP (2007) Annu Rev Phys Chem 58:267–297
90. Huang C-T, Jen C-P, Chao T-C, Wu W-T, Li W-Y, Chau L-K (2009) Sensors 9:6456–6470
91. Luo Y, Yu F, Zare RN (2008) Lab Chip 8:694–700
92. Zordan MD, Grafton MMG, Acharya G, Reece LM, Cooper CL, Aronson AI, Park K, Leary JF (2008) Cytometry A 75A:155–162
93. Pang L, Hwang GM, Slutsky B, Fainman Y (2007) Appl Phys Lett 91:123112
94. Wang J, Polsky R, Tian B, Chatrathi MP (2000) Anal Chem 72:5285–5289
95. Wang J (2008) Chem Rev 108:814–825
96. Privett BJ, Shin JH, Schoenfisch MH (2010) Anal Chem 82:4723–4741
97. Wang J (2006) Biosens Bioelectron 21:1887–1892
98. Lad U, Khokhar S, Kale GM (2008) Anal Chem 80:7910–7917
99. Bobacka J, Ivaska A, Lewenstam A (2008) Chem Rev 108:329–351
100. Faridbod F, Ganjali MR, Dinarvand R, Norouzi P (2008) Sensors 8:2331–2412
101. Durand NFY, Renaud P (2008) Lab Chip 8:319–324
102. Park S, Boo H, Kim Y, Han J-H, Kim HC, Chung TD (2005) Anal Chem 77:7695–7701
103. Henry OY, Fragoso A, Beni V, Laboria N, Sanchez JLA, Latta D, Germar FV, Drese K, Katakis I, O'Sullivan CK (2009) Electrophoresis 30:3398–3405
104. Goral VN, Zaytseva NV, Baeumner AJ (2006) Lab Chip 6:414–421
105. Boehm DA, Gottlieb PA, Hua SZ (2007) Sens Actuators B 126:508–514
106. Swensen JS, Xiao Y, Ferguson BS, Lubin AA, Lai RY, Heeger AJ, Plaxco KW, Soh HT (2009) J Am Chem Soc 131:4262–4266
107. Ja H, Wang J, Kawde A-N, Xiang Y, Gothelf KV, Collins G (2006) J Am Chem Soc 128:2228–2229
108. Javanmard M, Talasaz AH, Nemat-Gorgani M, Pease F, Ronaghi M, Davis RW (2009) Lab Chip 9:1429–1434
109. Masadome T, Yada K, Wakida S-I (2006) Anal Sci 22:1065–1069
110. Truman P, Petra U, Manfred S (2006) Lab Chip 6:1220–1228
111. Shirale DJ, Bangar MA, Chen W, Myung NV, Mulchandani A (2010) J Phys Chem C 114:13375–13380
112. Kim D-S, Park J-E, Shin J-K, Kim PK, Lim G, Shoji S (2006) Sens Actuators B 117:488–494
113. Friitz J, Cooper EB, Gaudet S, Sorger PK, Manalis SR (2002) Proc Natl Acad Sci USA 99:14142–14146
114. Lange K, Rapp BE, Rapp M (2008) Anal Bioanal Chem 391:1509–1519
115. Lee SH, Stubbs DD, Carney J, Hunt WD (2005) IEEE Sens J 5:737–743
116. Cooper MA, Singleton VT (2007) J Mol Recognit 20:154–184
117. Ergezen E, Appel M, Shah P, Kresh JY, Lec RM, Wootton DM (2007) Biosens Bioelectron 23:575–582
118. Huang I-Y, Lee MC (2008) Sens Actuators B 132:340–348
119. Rocha-Gaso M-I, March-Iborra C, Montoya-Baides A, Arnau-Vives A (2009) Sensors 9:5740–5769
120. Wu G, Datar RH, Hansen KM, Thundat T, Cote RJ, Majumdar A (2001) Nat Biotechnol 19:856–860
121. Lee JH, Hwang KS, Park J, Yoon KH, Yoon DS, Kim TS (2005) Biosens Bioelectron 20:2157–3162
122. Cherian S, Mehta A, Thundat T (2002) Langmuir 18:6935–6939
123. Watari M, Galbraith J, Lang H-P, Sousa M, Hegner M, Gerber C, Horton MA, Mckendry RA (2006) J Am Chem Soc 129:601–609

124. Stachowiak JC, Yue M, Castelino K, Chakraborty A, Majumdar A (2006) Langmuir 22:263–268
125. Arntz Y, Seelig JD, Lang H-P, Zhang J, Hunziker P, Ramseyer JP, Meyer E, Hegner M, Gerber C (2003) Nanotechnology 14:86–90
126. Savran CA, Knudsen SM, Ellington AD, Manalis SR (2004) Anal Chem 76:3194–3198
127. Backmann N, Zahnd C, Huber F, Bietsch A, Pluckthun A, Lang H-P, Guntherodt H-J, Hegner M, Gerber C (2005) Proc Natl Acad Sci USA 102:14587–14592
128. Waggoner PS, Craighead HG (2007) Lab Chip 7:1238–1255
129. Wang WU, Chen C, K-h L, Fang Y, Lieber CM (2005) Proc Natl Acad Sci USA 102:3208–3212
130. Chan LL, Lidstone EA, Finch KE, Heeres JT, Hergenrother PJ (2009) J Assoc Lab Autom 14:348–359
131. Wang J, Coffey PD, Swann MJ, Yang F, Lu JR, Yang X (2010) Anal Chem 82:5455–5462
132. Phizicky E, Bastiaens PIH, Zhu H, Snyder M, Field S (2003) Nature 422:208–215
133. Jonkheijm P, Weinrich D, Schroder H, Niemeyer CM, Waldmann H (2008) Angew Chem 47:9618–9647
134. Diercks AH, Ozinsky A, Hansen CL, Spotts JM, Rodriguez DJ, Adrem A (2009) Anal Biochem 386:30–35
135. Osterfeld SJ, Yu H, Gaster RS, Caramuta S, Xu L, Han S-J, Hall DA, Wilson RJ, Sun S, White RL, Davis RW, Pourmand N, Wang SX (2008) Proc Natl Acad Sci USA 105: 20637–30640
136. Ai H-W, Hazelwood KL, Davidson MW, Campbell RE (2008) Nat Methods 5:401–403
137. Li IT, Pham E, Truong K (2006) Biotechnol Lett 28:1971–1982
138. Gales C, Rebois RV, Hogue M, Trieu P, Breit A, Hebert TE, Bouvier M (2005) Nat Methods 2:177–184
139. Wong CL, Ho HP, Suen YK, Kong SK, Chen QL, Yuan W, Wu SY (2008) Biosens Bioelectron 24:606–612
140. Li CM, Chen W, Yang X, Sun CQ, Gao C, Zheng ZX, Sawyer J (2005) Front Biosci 10:2518–2526
141. Xu D, Xu D, Yu X, Liu Z, He W, Ma Z (2005) Anal Chem 77:5107–5113
142. Siwy Z, Trofin L, Kohli P, Baker LA, Trautmann C, Martin CR (2005) J Am Chem Soc 127:5000–5001
143. So H-M, Won K, Kim YH, Kim B-K, Rhu BH, Na PS, Kim H, Lee J-O (2005) J Am Chem Soc 127:11906–11907
144. Bakker E, Pretsch E (2008) Trends Anal Chem 27:612–618
145. Wang J (2007) Electroanalysis 19:769–776
146. Sawicka K, Gouma P, Simon S (2005) Sens Actuators B 108:585–588
147. Kuralay F, Ozyoruk H, Yildiz A (2006) Sens Actuators B 114:500–506
148. Updike SJ, Hicks GP (1967) Nature 214:986–988
149. Sidwell JS, Rechnitz GA (1985) Biotechnol Lett 7:419–422
150. Rauf S, Ihsan A, Akhtar K, Ghauri MA, Rahman M, Anwar MA, Khalid AM (2006) J Biotechnol 121:351–360
151. Heo J, Crooks M (2005) Anal Chem 77:6843–6851
152. Tsai H-C, Doong R-A (2005) Biosens Bioelectron 20:1796–1804
153. Romero MR, Ahumada F, Garay F, Baruzzi AM (2010) Anal Chem 82:5568–5572
154. Piano M, Serban S, Pittonson R, Drago GA, Hart JP (2010) Talanta 82:34–37
155. Nikoleli G-P, Nikolelis DP, Methenitis C (2010) Anal Chim Acta 675:58–63
156. Gutierrez M, Alegret S, Valle MD (2007) Biosens Bioelectron 22:2171–2178
157. Barhoumi H, Maaref A, Rammah M, Martelet C, Jaffrezic N, Mousty C, Vial S, Forano C (2006) Mater Sci Eng C 26:328–333
158. Kuwana T, Bouchier-Hayes L, Chipuk JE, Bonzon C, Sullivan BA, Green DR, Newmeyer DD (2005) Mol Cell 17:525–535
159. Moreno-Bondi MC, Benito-Pena ME, Urraca JL, Orellana G (2010) Top Curr Chem 1–54

160. Murphy BM, He X, Dandy D, Henry C (2008) Anal Chem 80:444–450
161. Bornhop DJ, Latham JC, Kussrow A, Markov DA, Jones RD, Soresen HS (2007) Science 317:1732–1736
162. Barton AC, Davis F, Higson SPJ (2008) Anal Chem 80:6198–6205
163. Varshney M, Li Y, Shinivasan B, Tung S (2007) Sens Actuators B 128:99–107
164. Lucas LJ, Chesler JN, Yoon J-Y (2007) Biosens Bioelectron 23:675–681
165. Vallina-Garcia R, Garcia-Suarez MDM, Fernandez-Abedul MT, Mendez FJ, Costa-Garcia A (2007) Biosens Bioelectronics 23:210–217
166. Jung HJ, Hwang I, Kim BJ, Min H, Yu H, Lee TG, Chung TD (2010) Langmuir 26:15087–15091
167. Valdes JJ, Wall JG, Chambers JP, Eldefrawi ME (1998) Johns Hopkins Univ Appl Phys Lab Tech Dig 9:4–10
168. Koo OK, Liu Y, Shuaib S, Bhattacharya S, Ladisch MR, Bashir R, Bhunia AK (2009) Anal Chem 81:3094–3101
169. Jen C-P, Huang C-T, Lu Y-H (2009) Microelectron Eng 86:1505–1510
170. Tuerk C, Gold L (1990) Science 249:505–510
171. Ellington AD, Szostak J (1990) Nature 346:818–822
172. Gokulrangan G, Unruh JR, Holub DF, Ingram B, Johnson CK, Wilson GS (2005) Anal Chem 77:1963–1970
173. Bang GS, Cho S, Kim B-G (2005) Biosens Bioelectron 21:863–870
174. Liao W, Guo S, Zhao XS (2006) Front Biosci 11:186–197
175. Liao W, Wei F, Liu D, Qian MX, Yuan G, Zhao XS (2006) Sens Actuators B 114:445–450
176. Xiao Y, Lubin AA, Heeger AJ, Plaxco KW (2005) Angew Chem Int Ed 44:5456–5459
177. Hianik T, Ostatna V, Zajacova Z, Stoikova E, Evtugyn G (2005) Bioorg Med Chem Lett 15:291–295
178. Maehashi K, Katsura T, Kerman K, Takamura Y, Matsumoto K, Tamiya E (2007) Anal Chem 79:782–787
179. Centi S, Tombelli S, Minunni M, Mascini M (2007) Anal Chem 79:1466–1473
180. Liao W, Cui XT (2007) Biosens Bioelectron 23:218–224
181. Xu Y, Cheng G, He P, Fang Y (2009) Electroanalysis 21:1251–1259
182. Cho EJ, Lee J-W, Ellington AD (2009) Annu Rev Anal Chem:241–264
183. Turner NW, Jeans CW, Brain KR, Allender CJ, Hlady V, Britt DW (2006) Biotechnol Prog 22:1474–1489
184. Ye L, Mosbach K (2008) Chem Mater 20:859–868
185. Alexander C, Andersson HS, Andersson LI, Ansell RJ, Kirsch N, Nicholls IA, O'sMahony J, Whitecombe MJ (2006) J Mol Recognit 19:106–180
186. Janiak DS, Kofinas P (2007) Anal Bioanal Chem 389:399–404
187. Ersoz A, Denizli A, Ozcan A, Say R (2005) Biosens Bioelectron 20:2197–2202
188. Xie C, Liu B, Wang Z, Gao D, Guan G, Zhang Z (2008) Anal Chem 80:437–443
189. Hansen DE (2007) Biomaterials 28:4178–4191
190. Kim YC, Kim S-H, Kim D, Park S-J, Park J-K (2009) Sens Actuators B 19:861–868
191. Jaggi RD, Sandoz R, Effenhauser CS (2007) Microfluid Nanofluid 3:47–53
192. Chun H, Kim HC, Chung TD (2008) Lab Chip 8:764–771
193. Goddard JM, Hotchkiss JH (2007) Prog Polym Sci 32:698–725
194. Gao H, Buchapudi KR, Harms-Smyth A, Schulte MK, Xu X, Ji H-F (2008) Langmuir 24:345–349
195. Diao J, Ren D, Engstrom JR, Lee KH (2005) Anal Biochem 343:322–328
196. Lee S, Voros J (2005) Langmuir 21:11957–11962
197. Bi H, Meng S, Li Y, Guo K, Chen Y, Kong J, Yang P, Zhong W, Liu B (2006) Lab Chip 6:769–775
198. Prakash S, Long TM, Selby JC, Moore JS, Shannon MA (2007) Anal Chem 79:1661–1667
199. Patrito N, McCague C, Norton PR, Petersen N (2007) Langmuir 23:715–719
200. Lee W, Choi D, Lee Y, Kim D-N, Park J, Koh W-G (2008) Sens Actuators B 129:841–849

201. Kitagawa F, Kubora K, Sueyoshi K, Otsuka K (2010) J Pharm Biomed Anal 53:1272–1277
202. Horan PK, Wheeless LL (1977) Science 198:149–157
203. Saunders GC, Jett JH, Martin JC (1985) Clin Chem 31:2020–2023
204. Holmes D, She JK, Roach PL, Morgan H (2007) Lab Chip 7:1048–1056
205. Gao D, Li H-F, Guo G-S, Lin J-M (2010) Talanta 528–533
206. Sachdeva N, Asthana D (2007) Front Biosci 1:4682–4695
207. Yang S-Y, Lien K-Y, Huang K-J, Lei H-Y, Lee G-B (2008) Biosens Bioelectron 24:855–862
208. Sokolsky-Papkov M, Agashi K, Olaye A, Shakesheff K, Domb AJ (2007) Adv Drug Deliv Rev 59:187–206
209. Piao L, Park S, Lee HB, Kim K, Kim J, Chung TD (2010) Anal Chem 82:447–451
210. Pregibon DC, Toner M, Doyle PS (2007) Science 315:1393–1396
211. Lee H, Kim J, Kim H, Kim J, Kwon S (2010) Nat Mater 22:745–749
212. Qian X, Peng X-H, Ansari DO, Yin-Goen Q, Chen GZ, Shin DM, Yang L, Young AN, Wang MD, Nie S (2007) Nat Biotechnol 26:83–90
213. Zhu X, Duan D, Madsen S, Publicover NG (2010) Anal Bioanal Chem 396:1345–1353
214. Tennico YH, Hutanu D, Koesdjojo MT, Bartel CM, Roemcho VT (2010) Anal Chem 82:5591–5597
215. Wang MM, Tu E, Raymond DE, Yang JM, Zhang H, Hagen N, Dees B, Mercer EM, Forster AH, Kariv I, Marchand PJ, Butler WF (2005) Nat Biotechnol 23:83–87
216. Breslauer DN, Lee PJ, Lee LP (2006) Mol Biosyst 2:97–112
217. Chung TD, Kim HC (2007) Electrophoresis 28:4511–4520
218. Chun H, Chung TD, Kim HC (2005) Anal Chem 77:2490–2495
219. Fan A, Lau C, Lu J (2005) Anal Chem 77:3238–3242
220. Elshal MF, McCoy JP (2006) Methods 38:317–323
221. Korobko AV, Jesse W, van der Maarel JRC (2005) Langmuir 21:34–42
222. Moghimi SM, Hunter AC, Clifford MJ (2005) FASEB J 19:311–330
223. Liu S, Gu Y, Le Roux RB, Matthews SM, Bratton D, Yunus K, Fisher AC, Huck WTS (2008) Lab Chip 8:1937–1942
224. Monpichar S-A, Emily CD, Adrew JD, Edel JB (2008) Anal Chem 80:7063–7067
225. Tan Y-C, Lee AP (2005) Lab Chip 5:1178–1183
226. Willaime H, Barbier V, Kloul L, Maine S, Tabeling P (2006) Phys Rev Lett 96:054501
227. Chen C-T, Lee G-B (2006) J Microelectromech Syst 15:1492–1498
228. Lai C-W, Lin Y-H, Lee G-B (2008) Biomed Microdevices 10:749–756
229. Maruo S, Inoue H (2006) Appl Phys Lett 89:144101
230. Lei K, Law W, Suen Y-K, Li W, Yam Y, Ho H, Kong S-K (2007) Proc Inst Mech Eng 221:129–141
231. Blanchard D, Ligrani P, Gale B (2005) Sens Actuators A 122:149–158
232. Al-Halhouli A, Al-Salaymeh A, Kilani M, Buttgenbach S (2007) Microfluid Nanofluid 3:537–546
233. Feng G-H, Kim E (2005) J Microelectromech Syst 14:192–199
234. Kim Y, Kim J, Na K, Rhee K (2005) Proc Inst Mech Eng 219:1139–1145
235. Cui Q, Liu C, Zha X (2007) Microfluid Nanofluid 3:377–390
236. Izzo I, Accoto D, Menciassi A, Schmitt L, Dario P (2007) Sens Actuators B 133:128–140
237. Doll A, Heinrichs M, Goldschmidtboeing F, Schrag H, Hopt U, Woias P (2006) Sens Actuators A 130–131:445–453
238. Chen S, Cheng C, Lin Y (2007) Sens Actuators A 135:1–9
239. Goulpeau J, Trouchet D, Ajdari A, Trabeling P (2005) J Appl Phys 98:044914
240. Geipel A, Doll A, Jantscheff P, Esser N, Massing U, Woias P, Goldschmidtboeing F (2007) J Micromech Microeng 17:949–959
241. Kim J-H, Na K-H, Kang C, Kim Y-S (2005) Sens Actuators A 120:365–369
242. Boden R, Lehto M, Simu U, Thornell G, Hjort K, Schwietz J-A (2006) Sens Actuators A 127:88–93
243. Machauf A, Nemirovsky Y, Dinnar U (2005) J Micromech Microeng 15:2309–2316

244. Astle A, Kim H, Bernal L, Najafi K, Washabaugh P (2007) Sens Actuators A 134:245–256
245. Inman W, Domansky K, Serdy J, Owens B, Trumper D, Griffith L (2007) J Micromech Microeng 17:891–899
246. Yamahata C, Lacharme F, Burri Y, Gijs MAM (2005) Sens Actuators B 110:1–7
247. Jang L-S, Kan W-H (2007) Biomed Microdevices 9:619–626
248. Cho Y-K, Lee J-G, Park J-M, Lee B-S, Lee Y, Ko C (2007) Lab Chip 7:565–573
249. Yamahata C, Chastellain M, Parashar V, Petri A, Hoffmann H, Nam G (2005) J Microelectromech Syst 14:96–102
250. Yin Z, Prosperetti A (2005) J Micromech Microeng 15:643–651
251. Woias P (2005) Sens Actuators B 105:28–38
252. Singhal V, Garimella S (2007) Sens Actuators A 134:650–659
253. Go D, Garimella S, Fisher T, Mongia R (2007) J Appl Phys 102:053302
254. Chen Z, Wang P, Chang H-C (2005) Anal Bioanal Chem 382:817–824
255. Hu J, Chao C (2007) Int J Refrigeration 30:290–298
256. Wu J, Lian M, Yang K (2007) Appl Phys Lett 90:234103
257. Duwairi H, Abdullah M (2007) Microsyst Technol 13:33–39
258. Mugele F, Baret J-C (2005) J Phys Condens Matter 17:705–774
259. Nguyen N-T, Meng A, Black J, White R (2000) Sens Actuators A 79:115–121
260. Joo S, Chung TD, Kim HC (2007) Sens Actuators B 123:1161–1168
261. Jeong OC, Konishi S (2008) J Micromech Microeng 18:085017
262. Kim JH, Lau KT, Diamond D (2008) Fabrication of microfluidic pump using conducting polymer actuator. In: Proceedings IEEE international conference on sensor networks, ubiquitous, and trustworthy computing, 11–13 June 2008, Taichung, Taiwan, pp 457–463
263. Laser DJ, Santiago JG (2004) J Micromech Microeng 14:35–64
264. Goettsche T, Kohnle J, Willmann M, Ernst H, Spieth S, Tischler R, Messner S, Zengerle R, Sandmaier H (2005) Sens Actuators A 118:70–77
265. Oh K, Han A, Bhansali S, Ahn C (2002) J Micromech Microeng 12:187–191
266. Teymoori MM, Abbaspor-sani E (2005) Sens Actuator A 117:222–229
267. Peirs J, Reynaert D, Van Brussel H (2000) Sens Actuators A 85:409–417
268. Takao M, Miyamura K, Ebi H, Ashiki M, Sawada K, Ishida K (2005) Sens Actuators A 119:468–475
269. Suzuki H, Yoneyama R (2003) Sens Actuators B 96:38–45
270. Hu Z, Zhang X, Li Y (1995) Science 269:525–527
271. Liu Y, Rauch C, Stevens R, Lenigk R, Yang J, Rhine D, Grodzinski P (2002) Anal Chem 74:3063–3070
272. Pal R, Yang M, Johnson B, Burke D, Burns M (2004) Anal Chem 76:3740–3748
273. Yamahata C, Chastellain M, Parashar V, Petri A, Hoffmann H, Gijs MAM (2005) J Microelectromech Syst 14:96–102
274. Oh K, Park C, Namkoong K (2005) A world-to-chip microfluidic interconnection technology with dual functions of sample injection and sealing for a multichamber micro PCR chip. In: Proceedings MEMS 2005, 18th international conference on micro-electromechanical systems, 30 Jan–3 Feb 2005, Miami, FL, pp 714–717
275. Baek J, Park J, Ju J, Lee T, Lee S (2005) J Micromech Microeng 15:1015–1020
276. Unger M, Chou H-P, Thorsen T, Scherer A, Quake S (2000) Science 288:113–116
277. Feng G-H, Kim E (2004) J Micromech Microeng 14:429–435
278. Li B, Chen Q, Lee D-G, Woolman J, Carman G (2005) Sens Actuators A 117:325–330
279. Reichmuth D, Shepodd T, Kirby B (2004) Anal Chem 76:5063–5068
280. Futterer C, Minc N, Bormuth V, Codarbox J-H, Laval P, Rossier J, Viovy J-L (2004) Lab Chip 4:351–356
281. Satarkar NS, Zhang W, Eitel RE, Hilt JZ (2009) Lab Chip 9:1773–1779
282. Yoo J-C, La G-S, Kang CJ, Kim Y-S (2008) Curr Appl Phys 8:692–695
283. Honda N, Lindberg U, Anderson P, Hoffmann S, Takei H (2005) Clin Chem 51:1955–1961
284. Takei G, Nonogi M, Hibara A, Kitamori T, Kim H-B (2007) Lab Chip 7:596–602

285. Takahashi S, Yeh S-R, Das TK, Chan C-K, Gottfried DS, Rousseau DL (1997) Nat Struct Biol 4:44–50
286. Park HY, Qiu X, Rhoades E, Korlach J, Kwok LW, Zipfel WR, Webb WW, Pollack L (2006) Anal Chem 78:4465–4473
287. Wong S, Ward M, Wharton C (2004) Sens Actuators B 100:365–385
288. Voldman J, Gray M, Schmidt M (2000) J Microelectromech Syst 9:295–302
289. Haverkamp V (1999) J Anal Chem 364:617–624
290. He B (2001) Anal Chem 73:1942–1947
291. Chen H, Meiners J (2004) Appl Phys Lett 84:2193–2195
292. Stroock A (2002) Science 295:647–651
293. Stroock A, Whitesides GM (2003) Acc Chem Res 36:597–604
294. Mengeaud V, Josserand J, Girault H (2002) Anal Chem 74:4279–4286
295. Hardt S, Drese KS, Hessel V, Schonfeld F (2005) Microfluid Nanofluid 1:108–118
296. Niu X, Lee Y (2003) J Microelectromech Syst 13:454–462
297. El Moctar A, Aubry N, Batton J (2003) Lab Chip 3:273–280
298. Deval J, Tabeling P, Ho C (2002) A dielectrophoretic chaotic mixer. In: Proceedings MEMS2002, 15th IEEE international workshop on micro-electromechanical systems, 20–24 Jan 2002, Las Vegas, NV, pp 36–39
299. Tang Z (2002) J Micromech Microeng 12:870–877
300. Bau H, Zhong J, Yi M (2001) Sens Actuators B 79:207–215
301. Chang C-C, Yang R-J (2007) Microfluid Nanofluid 3:501–525
302. Hessel V, Lowe H, Schonfeld F (2005) Chem Eng Sci 60:2479–2501
303. Fan X, White IM, Shopova SI, Zhu H, Suter JD, Sun Y (2008) Anal Chim Acta 620:8–26

Top Curr Chem (2011) 304: 153–169
DOI: 10.1007/128_2011_142

Published online: 27 April 2011

A Nanomembrane-Based Nucleic Acid Sensing Platform for Portable Diagnostics

Satyajyoti Senapati, Sagnik Basuray, Zdenek Slouka, Li-Jing Cheng, and Hsueh-Chia Chang

Abstract In this perspective article, we introduce a potentially transformative DNA/RNA detection technology that promises to replace DNA microarray and real-time PCR for field applications. It represents a new microfluidic technology that fully exploits the small spatial dimensions of a biochip and some new phenomena unique to the micro- and nanoscales. More specifically, it satisfies all the requisites for portable on-field applications: fast, small, sensitive, selective, robust, label- and reagent-free, economical to produce, and possibly PCR-free. We discuss the mechanisms behind the technology and introduce some preliminary designs, test results, and prototypes.

Keywords Depletion · Dielectrophoresis · Ion-selective membranes · Limiting-current

Contents

1 Introduction 154
2 Membrane-Induced Deionzation, Debye Layer Extension, and Induced Vortex Molecular Concentration 157
3 On-Chip Membrane Synthesis and Functionalization 159
4 Dielectrophoretic and Electrokinetic Molecular Concentration 161
5 Polarization and Warburg Impedance Signals of Membrane Sensors: Label-Free and Non-Optical Detection 164
6 Selectivity Enhancement 165
7 Integrated Units 166
8 Conclusion and Commercialization Issues 168
References 168

S. Senapati, S. Basuray, Z. Slouka, L.-J. Cheng, and H.-C. Chang (✉)
Department of Chemical and Biomolecular Engineering, University of Notre Dame, 46556 Notre Dame, IN, USA
e-mail: hchang@nd.edu

Abbreviations

AC	Alternating current
CNT	Carbon nanotube
DC	Direct current
DEP	Dielectrophoresis
DNA	Deoxyribonucleic acid
FCW	Fluorescence correlation spectroscopy
FET	Field-effect transistor
I–V	Current–Voltage
kb	Kilobase
PCR	Polymerase chain reaction
pM	Picomolar
RNA	Ribonucleic acid
SNP	Single-nucleotide polymorphism
ssDNA	Single-stranded DNA

1 Introduction

A new molecular sensing platform promises to significantly advance existing electrochemical/capacitance/field-effect transistor (FET) sensing technology into a probe-functionalized, multitarget and smart (automated) electrode sensing platform, whose assay time (minutes), detection limit (picomolar concentrations), selectivity (single-base mismatch discrimination), dynamic range, and robustness are orders of magnitude better than the current state-of-the-art techniques. The platform involves no moving parts, no valves, no optical detection, and will be fully automated with regenerable probes for prolonged usage. Most importantly, the assay time is shorter than the hour-long degradation half-life of RNAs, enabling realization of a polymerase chain reaction (PCR)-free nucleic acid detection platform [1]. The new DNA/RNA sensing technology is based on several new on-chip ion-selective membrane and nanoslot technologies developed in our group and elsewhere [2–7].

Development of rapid and portable detection devices for point-of-care application is an important aspect of the modern diagnostics industry for effective detection of diseases in developing countries, from anti-terrorism and biowarfare applications to environmental monitoring, including the detection of harmful organisms on beaches. The most specific sensing platform is the genetic detection platform, which identifies a particular sequence of the target pathogen's genome. As a result of active research in this area, small pretreatment units are now available that can concentrate the pathogens with membranes and beads, lyse cells, and remove chromosomal DNA for amplification in an integrated PCR chip [8].

However, the key technological bottleneck remains the detection and quantification of the amplified DNAs.

Two gold standards for genetic detection have appeared in the last decade, both involving labeling of fluorophores or quenchers onto the target molecule during PCR amplification: DNA microarray and real-time PCR. DNA microarrays offer sensitivity and large library volume. However, the assay time is long due to diffusion limitations. It also requires periodic rinsing to avoid nonspecific binding. Finally, the fluorescent confocal detection instrumentation is still too bulky and costly for portable applications. Quantification of the number of target DNAs is also impossible. Real-time PCR sacrifices large library volume for rapid and quantifiable detection, higher sensitivity, and good selectivity. However, it still requires expensive and bulky fluorescent detection instrumentation. (Model ViiA^{TM}7 of Applied Biosystems is the size of a small refrigerator and costs US $200,000.) The main challenge for portable diagnostics is then a miniature label-free nucleic acid sensing platform without any sophisticated instruments and reagents. The elimination of the PCR step would also be advantageous, as it would remove the 30-min thermal cycling time and the need for a PCR unit. In many medical applications, over a million DNA and RNA copies are available in a typical sample volume of 100 μL. Consequently, a detection platform capable of sensing one million copies of DNA/RNA can be PCR-free. For bacterial pathogens, each cell produces a million copies of mRNA and only one copy of DNA. However, the tradeoff for this relative abundance of RNA is its short life-time (less than an hour) due to rapid degradation [1]. Hence, an RNA detection platform with an assay time of less than 1 h (and without reverse-transcription PCR) would be the first RNA detection platform of its kind.

Several label-free field-use DNA/RNA sensing technologies have been intensively studied in the last decade. The most viable field-use sensing technology to date is, in our opinion, electrochemical sensing. Electrochemical sensing with molecular probe functionalized electrode sensors can measure the change in electron-transfer rate upon docking of the target DNA/RNA molecules and redox reporter agents that can magnify this electrochemical current. Because many current carriers and inhibitors in the buffer can affect this electrochemical signal, even in the presence of surface-assembled monolayers, this sensing technology lacks robustness and is difficult to calibrate [9]. Capacitance, conductance, and FET electrode sensors have also attracted considerable interest recently. For such non-Faradaic sensors, excess charges brought to the surface by the docked DNA/RNA molecules and their associated potential can produce a local change in Debye double-layer conductance/capacitance and sub-surface current of the sensor. Conductance measurements are typically insensitive at practical ionic strengths because the presence of the DNA/RNA molecules in the high-conductivity Debye layer would not significantly affect the local conductance [10, 11]. Moreover, the same Debye layer is only a few nanometers thick for practical RNA samples, and only the lower fraction of the charges on the long (>10 kb) linear DNA/RNA is responsible for the capacitance signal, again resulting in low sensitivity [12, 13]. At its current state, conductance/capacitance/FET sensors have a detection limit higher than nanomolar, which translates into 10^8 copies of nucleic acid molecules for practical sample volumes [13], which is too high for field-based detection. Most importantly, the largest drawback of all electrode sensors is their long assay time. At the low target molecule concentrations (picomolar) of practical samples, the diffusion time of long (more than kilobase) nucleic acids to the electrode sensor often exceeds hours, thus rendering such a platform ineffective for rapidly degrading RNA.

Several techniques have been suggested for removing the slow transport of long nucleic acid molecules to the electrode sensor. One technique involves the activation of a high voltage at the electrode sensor to electrophoretically attract nearby DNAs [14]. However, this electrophoretic concentration technique is highly nonspecific and other like-charge molecules can also be attracted to the sensor. Moreover, for buffers of high ionic strength, the elevated voltage can produce undesirable Faradaic reactions that can produce false current or voltage signals. Internal vortices, generated on microelectrodes by various ingenious but unreliable mechanisms, have also been suggested as a means of concentrating the target molecules towards the sensor [15, 16]. Generation of internal vortices remains, however, an imperfect science. It would be more desirable for the sensor to generate such vortices automatically at a precise location and for the vortices to exhibit a strong electric signal such that they can be detected and automatically controlled; this new technology will be described in Sect. 2.

The missing technologies for portable DNA/RNA diagnostics are therefore a label-free electrode sensor that does not suffer from diffusion limitation (i.e., short assay time), is highly selective and sensitive, and yet is insensitive to buffer ionic strength and chemical composition. We propose here that the ion-selective membrane sensor technologies, with properly tuned electrokinetic features and dynamic feedback actuation, can meet these specifications. Our group has recently developed an on-chip sol–gel silica fabrication technique [17, 18] and a nanocolloid assembly technique for on-chip membrane synthesis [3]. We have also applied several photocuring polystyrene sulfonate or polyallylamine synthesis techniques to fabricate on-chip membranes [19]. Recently, we have developed the technology to fabricate nanoslots on chips [5], which behave like single-pore membranes, for application in diagnostic chips. The membranes are used for molecular detection and involve continuous pumping of the sample solution in a cross-flow (tangential to the membrane surface) format to minimize hydrodynamic resistance. On-chip electrodes control the ionic current and voltage drop across these membrane components to produce the desired phenomena for rapid molecular concentration, transport, and detection. A first-generation integrated chip is shown in Fig. 1 for rapid detection of kilobase DNA with probe-functionalized nanocolloid assemblies (membranes). These passive chips are not automated and do not involve feedback control because they are missing several sensors and activation components that our group has recently developed. We will discuss our recent attempts to add and integrate, via on-chip feedback control circuitry, these new components to the first-generation devices to produce a multitarget smart DNA/RNA sensor platform.

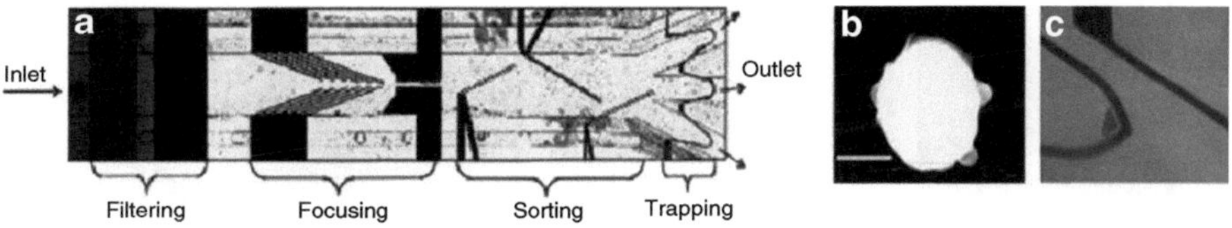

Fig. 1 (**a**) Open-flow DEP chip through which nanocolloids functionalized with complementary oligonucleotides are pumped. (**b**) SEM image of a larger colloid (500 nm) with a long oligonucleotide. *Scale bar*: 200 nm. These nanocolloids are focused, sorted and assembled passively at a microelectrode gate with symmetric and aligned top-down electrode pairs. (**c**) Magnification of a microelectrode gate; the *triangle* shows the trapping of nanocolloids within a micrometer-sized region. DNA solutions, ranging from picomolar to nanomolar concentrations, are then pumped over the nanocolloid assembly (membrane). Fluorescent imaging (see Fig. 6) is used to quantify the specificity and concentration factor, whereas label-free detection yields quantifiable electrical signals (see Figs. 4, 8 and 10)

2 Membrane-Induced Deionzation, Debye Layer Extension, and Induced Vortex Molecular Concentration

One solution to the robustness issue is to deplete the inhibitors and chemicals around the sensor such that close to deionized water conditions are always produced near the sensor, regardless of the buffer ionic strength and composition. Our laboratory has recently developed several of these depletion technologies based on fabricated ion-selective nanoslots [5–7] and on-chip nanoporous membranes [17, 18]. Significant counterion transport can rapidly deplete the counterions on one side of the membrane. To sustain electroneutrality, the co-ions also deplete rapidly to produce an ion-depleted zone. Sufficiently high DC fields (>100 V/cm) can deionize a 100 μm neighborhood (the depletion zone) near the membrane. The depletion layer with low interfacial ionic strength produces the maximum possible ion current without convection and exhibits a distinct limiting-current plateau in the polarization *I*–*V* or cyclic voltammetry spectrum (Fig. 2b). This nonlinear *I*–*V* polarization is not due to electron-transfer reactions but bulk-to-membrane ion flux across the extended and depleted interfacial double layer. Its sensitivity to the interfacial charge in the depleted double layer allows sensitive conduction/capacitance detection of hybridization with the same actuation on-chip electrodes that drive the ion current.

At another critical voltage, the limiting current gives way to a sharp increase in the current, the overlimiting current, which is a very sensitive signature of vortices driven by an extended polarized (Debye) layer at the membrane interface, as shown in Fig. 2b [4, 10, 20]. Nonequilibrium (counter)ion transport across the ion-selective membrane produces an extended polarized layer and nonequilibrium over-potential that is orders of magnitude thicker/higher than the Debye screening length and the equilibrium zeta-potential. As such field-induced polarization is curvature- and perturbation-sensitive, the induced electro-osmotic flow is not uniform and the resulting backpressure can drive microvortices of specific dimension, and linear velocity at

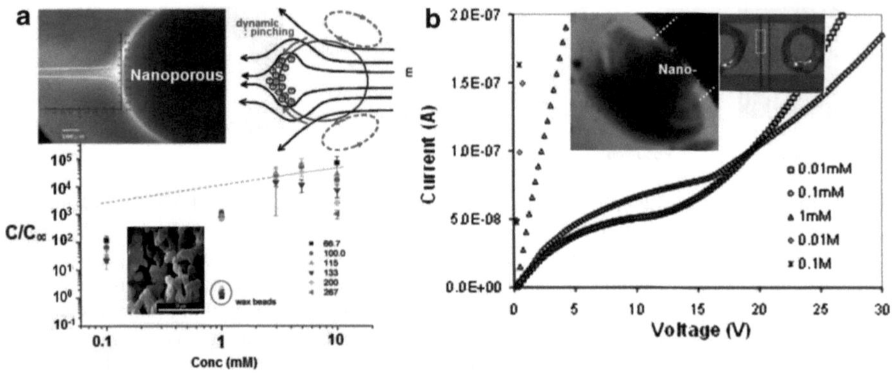

Fig. 2 (**a**) Enrichment and depletion across a nanoporous silica granule synthesized within a glass chip by sol–gel chemistry, producing a five orders of magnitude concentration of ions on one side of the granule and a comparable degree of ion depletion on the other side. *Top left*: High magnification SEM image of silica granule with superimposed plot showing ion concentration. *Top right*: Scheme illustrating counterion movement. *Bottom*: Concentration factor c/c_{∞} as a function of the ionic concentration (c_{∞}) of the fluorescent solution for different sizes of silica beads. *Inset*: SEM image of silica beads. (**b**) Depletion of charged fluorescent dye (*left image*) at one entrance of a 50 nm nanoslot between two circular microreservoirs (*right image*). The depletion has a very distinct polarization signature: the current plateaus at a limiting current value when depletion occurs. When vortices are observed in both the silica granule and the nanoslot beyond a critical voltage, the polarization (single-sweep cyclic voltammetry) curve shows a large overlimiting current beyond the limiting current plateau. The same overlimiting current is shown in Fig. 4 before and after hybridization. The plot shows the polarization characteristics of the nanoslot for different ionic concentrations of solution. Linear polarization curves missing the limiting region can be observed for concentrations above 0.1 mM. The disappearance of the limiting region is given by the loss of the ion-selective properties of the nanochannel as a result of decreasing Debye layer thickness inside the nanochannel

precise voltage windows. Such microvortices enhance the ion current through the membrane or nanoslot (hence the overlimiting current) and thus exhibit a sensitive polarization or single-sweep cyclic-voltammetry overlimiting signal as shown in the polarization curve in Fig. 2b. This strong conductance signature allows us to develop a smart platform that can generate such vortices on demand. Concentration of the charged dye by five orders of magnitude (shown in Fig. 2a) is mostly due to convective concentration of the molecules at the stagnation points of the vortices. Other than the distinctive conductance signals of the membrane depletion/vortex phenomena, their actuation and sensing time is also very rapid. With thin membranes and short nanoslots (Fig. 2b), the ion depletion and hydrodynamic timescales range from microseconds to seconds, allowing for rapid automation.

The ion current across an ion-selective medium can be very sensitive to the charge polarity and density on the surface outside the medium. Our previous work on alumina nanochannels demonstrates that with negatively charged SiO_2 entrance side-walls, the ion conductance across the positive-charged Al_2O_3 nanochannel is suppressed and shows a nonlinear *I*–*V* characteristic (Fig. 3a). The ion charge inversion induced by the heterogeneous entrance charge enhances ion depletion

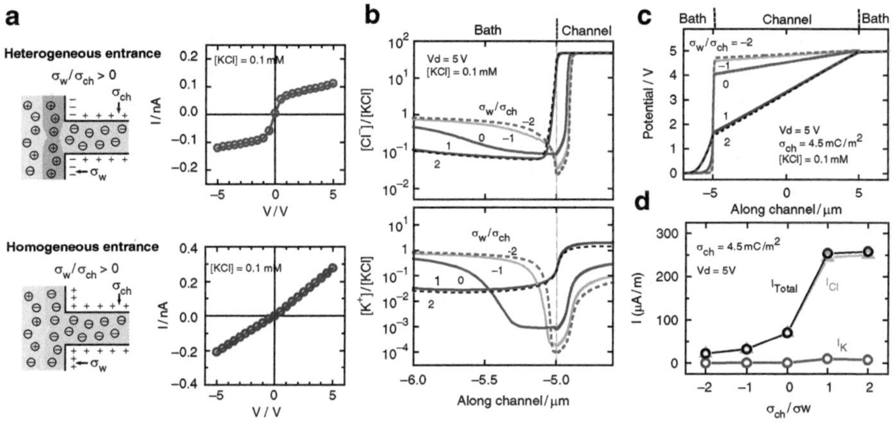

Fig. 3 Effect of entrance surface charge density and polarity on the ion transport in a 20 nm thick, 60 μm long, positively charged Al_2O_3 nanochannel. (**a**) Heterogeneous nanochannel entrance (the charge of entrance side-walls, σ_w and the charge of nanochannel σ_{ch} appear in opposite polarities, $\sigma_w/\sigma_{ch} < 0$) induces ion charge inversion at channel access. Experimental *I–V* characteristics of an Al_2O_3 nanochannel device with negatively charged silica entrance side-walls (*top*) and Al_2O_3 entrance side-walls (*bottom*) measured with 0.1 mM KCl. (**b**) Calculated Cl ion and K ion distributions near left channel entrance with values of σ_w/σ_{ch} varying from 2 to −2. (**c**) Calculated potential profile along the nanochannels with varied σ_w/σ_{ch} ($\sigma_{ch} = 4.5$ mC/m^2) under V_d (Voltage applied across the nanochannel) = 5 V. (**d**) Summarized theoretical ion current density (current per channel width) of the nanochannels with varied σ_w/σ_{ch}

(Fig. 3b) and hence creates a large voltage drop at the channel entrance (Fig. 3c). The heterogeneous entrance charge efficiently suppresses the flow of counterions through the nanoslot (anions in the case of the positively charged Al_2O_3 nanochannel). This effect is clearly seen in Fig. 3d and is reflected in the measured *I–V* curves depicted in Fig. 3a. The ion conductance is found to change significantly when the surface charge of the entrance side-walls converts its polarity and density. The shift of ion conductance induced by surface charge conversion will be utilized as a basis of DNA/RNA sensing. Hybridization of DNA or RNA on a positively charged anion-selective medium can be detected by measuring the nonlinear *I–V* characteristics.

3 On-Chip Membrane Synthesis and Functionalization

Another key step is the proper development of surface chemistry to attach addressable probes onto different membrane sensors. This can be achieved by patterning UV-curable acrylic-based polymers inside the microfluidic channel doped with different monomers containing charged or functional groups. Such polymers are ion-selective and provide reactive chemical groups on their surfaces for the attachment of DNA/RNA probes. The functionality of all the devices proposed here relies on the ion-selectivity of the polymeric material, which is less dependent on ionic strength than the nanofluidic counterparts. Briefly, using photolithographic techniques, cation- and anion-exchange membranes are defined in glass microfluidic channels by crosslinking positively charged diallyldimethylammonium (DADMA) and negatively charged 2-acrylamido-2-methyl-1-propanesulfonic acid (AMPSA) using a crosslinker (*N*, *N′*-methylene bisacrylamide) and photo-initiator. Each membrane has a defined width and length of few tens to hundreds of micrometers, bridging two microfluidic channels that are about 20 μm deep and 20–100 μm wide. The pore size of the nanoporous membrane can be controlled by varying the concentration of the monomers and crosslinker. To achieve surface functionalization of the oligo probes, the surface of an anion-exchange membrane is modified with amino groups by using allylamine as an additive in the prepolymer solution. The DNA or RNA probe (~27 bases) pre-attached with functional groups of choice can then be used to functionalize the probes onto membrane surface. Through examination by microscope and measurement of the ability to deplete ions, the polymerization time and the concentrations of crosslinker and photo-initiator have been optimized to produce reproducible, well-defined ion-selective membranes with functional chemical groups inside microchannels.

In Fig. 4, we show the first experimental evidence that the onset voltage and the onset of overlimiting current, key features of the nonlinear *I–V* curve of our sensor, are sensitive to nucleic acid hybridization onto oligo probes functionalized onto the surface of the ion-selective medium, as the resulting change in the surface charge can enhance or eliminate the extended Debye layer. The voltage differential is particularly large because of the nearly infinite differential resistance at the limiting current conditions. In contrast, the low-voltage linear ohmic region, where classical electrochemical sensors operate, registers an insignificant shift. Most conveniently, the depleted and extended double layer, which can be three orders of magnitude thicker than the Debye layer, also allows more charges on the target RNA to contribute to the effective surface charge. If the membrane is oppositely charged from the hybridized or functionalized molecules, the latter can even invert the charge on the membrane surface, eliminating the overlimiting current completely

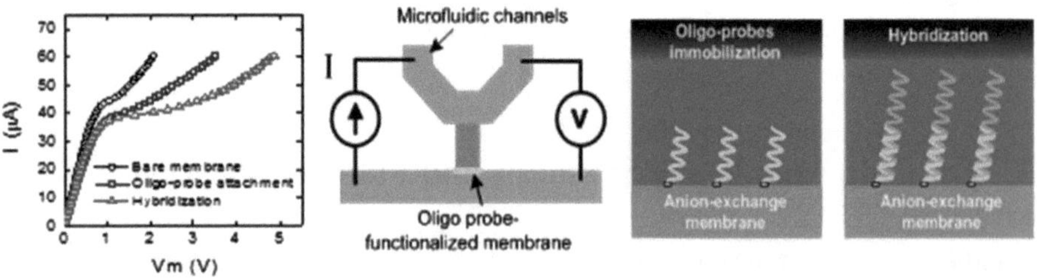

Fig. 4 Ion-selective membrane as a sensitive sensor for the detection of biomolecules. *Left*: Significant change of *I–V* characteristics in the overlimiting current regime is observed after RNA hybridization from a picomolar sample. The 50% change in conductance is compared to typical 5% changes of electrochemical electrode sensors at the same concentration (low-voltage region). *Center*: Diagram of sensor showing position of membrane. *Right*: Schematic presentation of nucleic acid hybridization onto immobilized oligoprobes

when the surface is effective electroneutral with exact compensation. The result is a very sensitive RNA sensor with picomolar sensitivity, compared to the nanomolar sensitivity of most electrode electrochemical sensors, as seen in Fig. 4.

4 Dielectrophoretic and Electrokinetic Molecular Concentration

Dielectrophoresis (DEP), a molecular force due to induced molecular dipoles, has been shown to be an effective means of concentrating large DNA/RNA molecules into the depleted region near the membrane surface (see Figs. 3, 4) where the probes are functionalized [9–12]. The electric field is focused by the nanopores in the membrane to produce a high field gradient at the membrane interface. A polarizable molecule in the bulk, with a large induced dipole, would then experience a net force towards the high-field region (the membrane surface). With the intense field amplification of nanopores, this DEP force on the molecules can overcome molecule–membrane repulsive interaction.

In a recent fluorescent correlation spectroscopy (FCS) experiment in collaboration with Yingxi Elaine Zhu (University of Notre Dame), our laboratory was able to confirm this domination of dielectrophoretic attraction over like-charge repulsion with floating probe-functionalized carbon nanotubes (CNTs) and the fluorescently labeled kilobase target single-stranded DNA (ssDNA). Because CNTs quench the fluorophores on hybridization of target DNA, reduction in the fluorescent intensity can be used to quantify the hybridization degree and the attraction of the molecules to the nanoelectrode. As seen in Fig. 5, dielectrophoretic attraction due to the

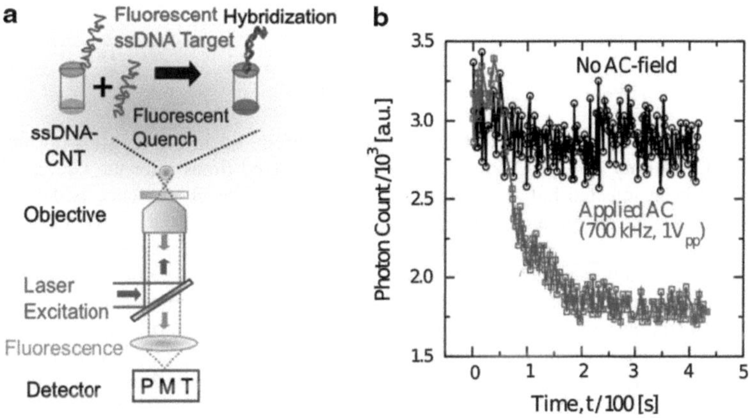

Fig. 5 (**a**) FCS detection of DNA hybridization from a picomolar solution. (**b**) Accelerated DNA fluorescence quench upon DNA docking with oligo-functionalized CNT probe under AC fields, in sharp contrast to the hour-long diffusive docking process without AC-fields

field-focusing CNTs allows hybridization in less than 2 min at picomolar concentrations. In contrast, the diffusion time for the long ssDNA at this concentration is hours.

Instead of floating nanoelectrodes (CNTs), our laboratory [3–7] is able to fabricate 50 nm nanoslots on glass (inset in Fig. 3b) and is able to show concentration of ssDNA to the nanoslot. The same DNA concentration is shown with nanoporous membranes in Fig. 3, with a concentration factor of up to five orders of magnitude. Alternatively, 100 nm nanocolloids can be assembled into a nanocolloid crystal (a membrane) at a top-down electrode pair by nanocolloid DEP (Figs. 1c, 6). The 10 nm spacing between the nanocolloids focuses the electric field of the electrode gate and can rapidly (order of seconds) trap and concentrate ssDNA

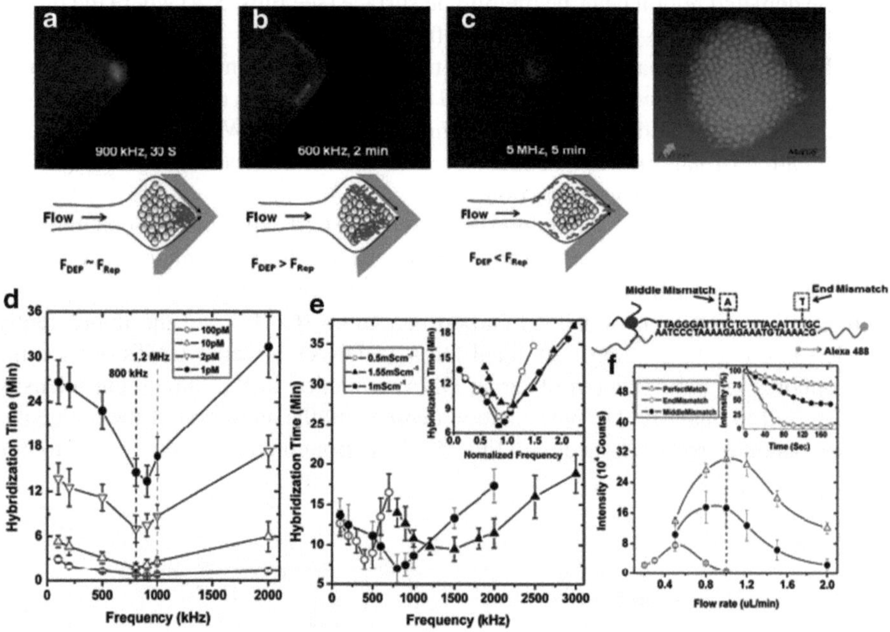

Fig. 6 (**a–c**) Fluorescence images of the trapping electrode tip in Fig. 2, showing the 100 nm nanocolloid assembly (see SEM image on the *right*) at fixed times after the Green Crab DNA solution had been injected but at different AC frequencies. The fluorescence is detectable only when the ssDNA is concentrated beyond the micromolar level from the undetectable concentrations (nanomolar to picomolar) of the injected solution. The DNA trapping location is further clarified in the schematics below the images. Trapping at the assembly is achieved at low frequencies, whereas none occurs at high frequencies. F_{DEP} dielectrophoretic force, F_{Rep} repulsive force. (**d**) Detection time increases with decreasing concentration. (**e**) There is an optimum frequency with a sharp minimum in detection time, which scales as D/λ^2 where λ is the Debye length for the given electrolyte strength. (**f**) Fluorescence intensity at 2 min from different flow rates of 100 pM of a 1 kb ssDNA target from a Green Crab species with a 26 base docking segment in the middle (*solid cricle*) and with a complementary 26 base oligo on the nanocolloid (*open triangle*) or with a single end mismatch (*open circle*). The flow rate window with single-mismatch discrimination is indicated by a *vertical dashed line*. The scheme above the plot shows the actual 26 base ssDNA docking sequence and the location of mismatched bases

molecules of a Green Crab species [3] from a picomolar solution onto the on-chip nanocolloid membrane by molecular DEP. A properly tuned DEP force can drive the DNAs towards the nanostructure against electrostatic repulsion from the like-charged structures, but they will not deposit onto the surface until they are convected to a sharp tip (~10 μm) at the nanostructure (Fig. 6a); intermolecular interaction can be adjusted to minimize nonspecific binding. A concentration factor exceeding 10^5 within minutes is observed from the fluorescent imaging in Fig. 6, thus rapidly and significantly enhancing the sensitivity of any sensor at the trapping location. The shear rate and AC frequency can be optimized so that the sensor can selectively discriminate against kilobase target molecules with a single mismatch in the 26 base pairing segment in the middle (Fig. 6f). This shear-enhanced selectivity eliminates the need for rinsing and washing steps.

Apart from dielectrophoretic concentration, which is not effective for small nucleic acids because the DEP force scales as the cubed power of the hydrodynamic radius of the molecule, our group has successfully demonstrated rapid analyte preconcentration based on ion depletion at an ion-selective membrane in microfluidic chips. As shown in Fig. 7a, a cation-exchange membrane UV-polymerized in a microslot bridging two microfluidic channels can induce deionization under voltage biases. The ion-depletion region functioning as an energy barrier traps the molecule passing across it in an electroosmotic flow tangential to the membrane on the side. The UV-curable ion-selective membrane offers superior concentration efficiency and proccessability compared to the microfabricated nanochannels reported previously [21, 22] or Nafion resins [23, 24]. Unlike the 100 nm thick nanochannels and surface-patterned Nafion thin films, the proposed membrane slot has the same depth as the microfluidic channel, yielding a large junction area. The large cross-section area provides greater ion current and better control of ion-depletion in the microchannels. Therefore, preconcentration can be achieved in few seconds. The fluorescence image in Fig. 7b shows the concentration of labeled molecules by several orders of magnitude in 10 s from a solution being pumped by

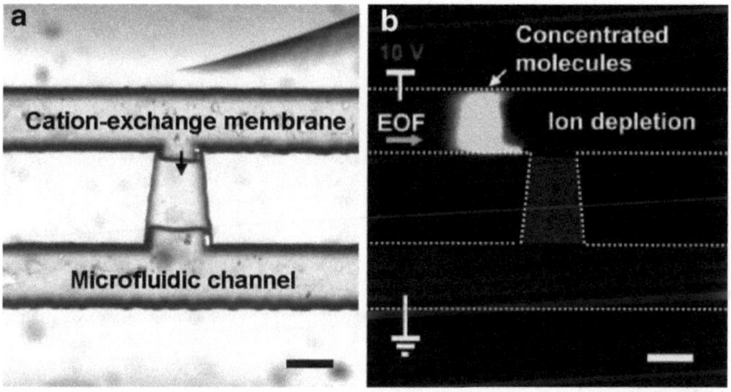

Fig. 7 (**a**) Optical microscopic image of a preconcentrator based on a charge-selective membrane. (**b**) Concentration of fluorescently labeled molecules taking place 10 s after applying a voltage bias of 10 V. Scale bars: 50 μm. *EOF* electroosmotic flow

electro-osmosis from the left to right in the top microfluidic channel, after 10 V is applied across the membrane. Moreover, the proposed membrane adheres to acryl-functionalized glass surfaces well; whereas Nafion has poor adhesion to most solid surfaces and the process is more operator-dependent.

5 Polarization and Warburg Impedance Signals of Membrane Sensors: Label-Free and Non-Optical Detection

The presence of the docked RNA/DNA and their mobile counterions produce a large conductivity change at the depleted region, which is where most of the voltage drop occurs. Moreover, the extended Debye (polarization) layer [4, 25] allows more of the charges on a long (>2 nm) DNA/RNA molecule to contribute to the charging capacitance and surface-charge compensation on the surface. As described earlier, the surface charge can sensitively alter the onset voltage for microvortices and the overlimiting currents that the vortices contribute to. These effects greatly enhance the capacitance, conductance and polarization signatures of the docked nucleic acids, resulting in sensitive nonlinear *I–V* polarization signatures, such as those due to the charge-inversion after hybridization in Fig. 4.

The dynamics of depletion layer formation with strong charging also exhibits a distinct capacitance signature in the AC impedance spectrum (Fig. 8a). This Warburg spectrum has a constant phase angle of $\pi/4$, whose modulus increases with decreasing frequency and is classically associated with diffusion controlled ion transport. In a recent paper [7], we have shown that, under an AC field, the depletion region next to a membrane sensor is created periodically during the

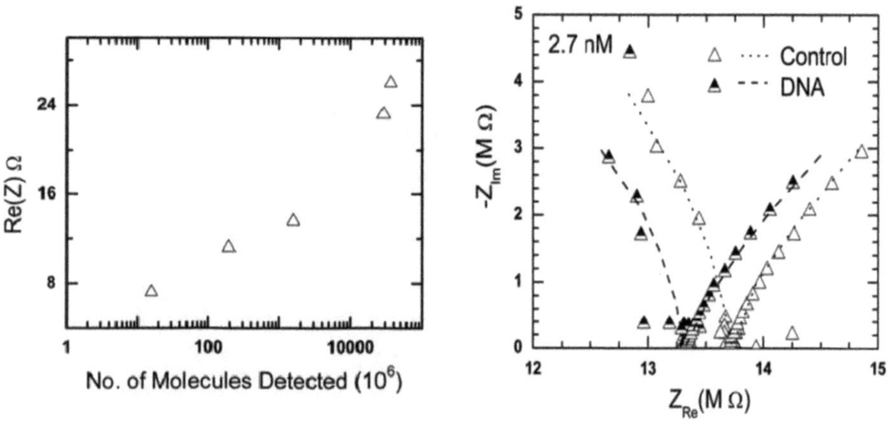

Fig. 8 (**a**) Warburg impedance spectrum of the nanoslot in Fig. 2b, showing a shift to lower resistance with a 2.7 nM 1 kb *E. coli* ssDNA solution relative to the control without DNA. (**b**) The shift in the intercept with the real axis allows precise quantification of the number of ssDNA molecules in the microreservoir down to 10^7 copies

half-cycle when the mobile counterions are driven into the nanoslot or on-chip membrane. The depletion layer dynamics was verified by high-speed confocal imaging to be a diffusive one such that its thickness grows in a self-similar manner as $\sqrt{Dt}$ [5] and was shown to exhibit the Warburg spectrum, with a constant phase of $\pi/4$ (Fig. 8a). The intercept of the Warburg spectrum with the real axis represents the limiting ion flux when the depletion layer is smallest in dimension – just above the critical voltage where the depletion phenomenon can be sustained. It hence offers an accurate estimate of the low conductivity in this small region, as most of the voltage drop occurs there. As mentioned earlier, the presence of a few macro-ions attracted to this small depleted region by DEP can significantly change its local conductance. In Fig. 8b, we indicate sensitive detection of *E. coli* ssDNA below nanomolar concentrations or 10^7 molecules. With a reduction of the nanoslot width, down to the micrometer size of the nanocolloid assembly in Fig. 6, the detection limit is expected to reach below picomolar concentrations or 10^5 copies of nucleic acid. The same Warburg signal can be captured with the field across the nanocolloid assembly of Fig. 6 to allow label-free quantification of the docked DNA/RNAs. This large-voltage AC impedance technique is quite distinct from the classical low-amplitude impedance spectroscopy for electron transfer rates because we induce nonequilibrium ion transport through the ion-selective nanoslot or membrane to produce extended polarized Debye layer and concentration depletion layers.

6 Selectivity Enhancement

The single mismatch (SNP) discrimination capability of the experiment shown in Fig. 6f is due to hydrodynamic shear. In a recent MD project [15], it has been shown that shear is most discriminating because it can meter small thermal-energy-level

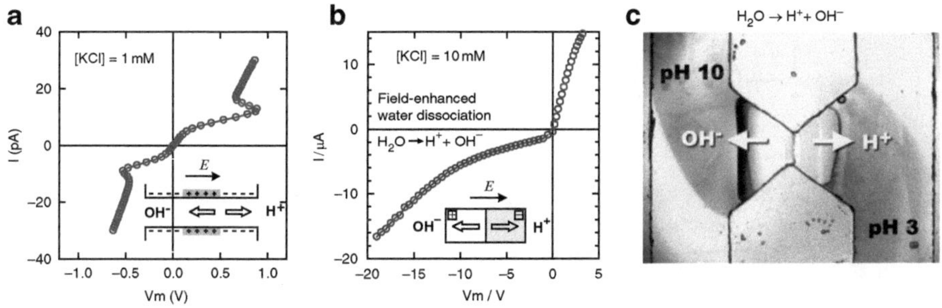

Fig. 9 Field-enhanced water dissociation increases ionic currents in (**a**) a reverse-biased 20 nm thick bipolar-junction nanofluidic channel containing positive and negative surface charges ($|V_m| > 0.6$ V), and (**b**) a UV-polymerized bipolar membrane ($V_m < -10$ V). (**c**) Hydroxide ions and protons are produced at the bipolar membrane junction and transport to opposite sides of the membrane. The pH change of the solution in the microchannels can be observed with a mixture of universal pH indicator. Left half of the bipolar membrane is positively charged whereas the right half is negatively charged

hydrogen bond energies to dehybridize the target molecules. One of the co-authors (S.Z.) and other groups have recently developed microscale bipolar membrane technologies that can be used to control the local pH in microfluidic chips to improve both the specificity and selectivity of the membrane sensor [26, 27]. These bipolar membranes/nanopores exhibit distinct hysteretic *I–V* polarization and cyclic voltammetry signatures due to local field-induced water-breaking reactions that generate more ions [25, 26, 28–30]. An image of the pH fronts generated by a UV-polymerized bipolar membrane composed of positively charged dimethylammonium and negatively charged sulfonic groups are shown in Fig. 9. It was found that the ion currents can drastically increase when reversely biased at a high voltage, forming a breakdown regime. In accordance with the second Wien effect, the ionic current breakdown results from the enhanced water dissociation into cation (H^+) and anion (OH^-) at the bipolar junction, in which a strong electric field greater than 10 MV/cm can build up at a reverse bias [29, 30]. These membrane actuation components can be used to control the local pH for our integrated devices, with feedback control based on the distinctive hysteretic polarization signals and *I–V* characteristics seen in Fig. 9.

7 Integrated Units

Other than the nanoslot chip of Fig. 2 and the DEP chip of Figs. 1, 6 for nanocolloid assemblies, our group has integrated several components into the first-generation of passive sensor chips [2–7]. One prototype is shown in Fig. 10. An assembly of oligo-functionalized CNTs (a CNT membrane) is used to effect the ion depletion and the Warburg quantification (Fig. 8) of hybridized ssDNA from a Green Crab invasive species. The detection limit of the Warburg impedance signal is picomolar concentrations or about the desired 10^5 copies, the detection time is about 15 min,

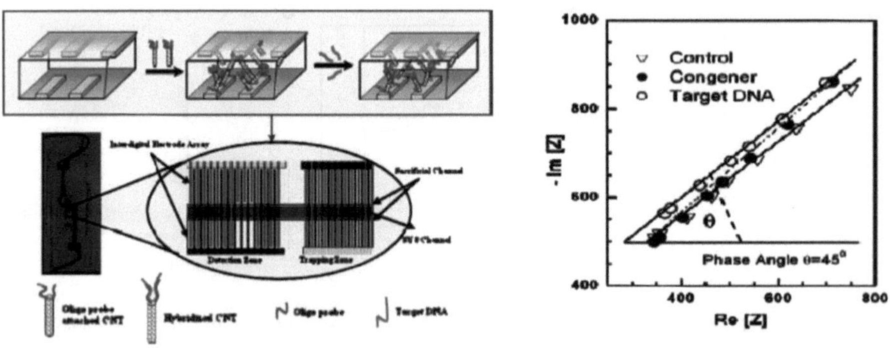

Fig. 10 *Left*: An integrated chip that uses an interdigitated electrode array to assemble oligo-functionalized CNTs into an ion-selective membrane. *Right*: The Warburg signal measured across the CNT assembly is able to detect picomolar concentrations (10^5 copies) of a kilobase-long DNA from a Green Crab species and differentiate a congener species with three mismatches over the 26 pair docking segment due to the hydrodynamic shear offered by the high through flow

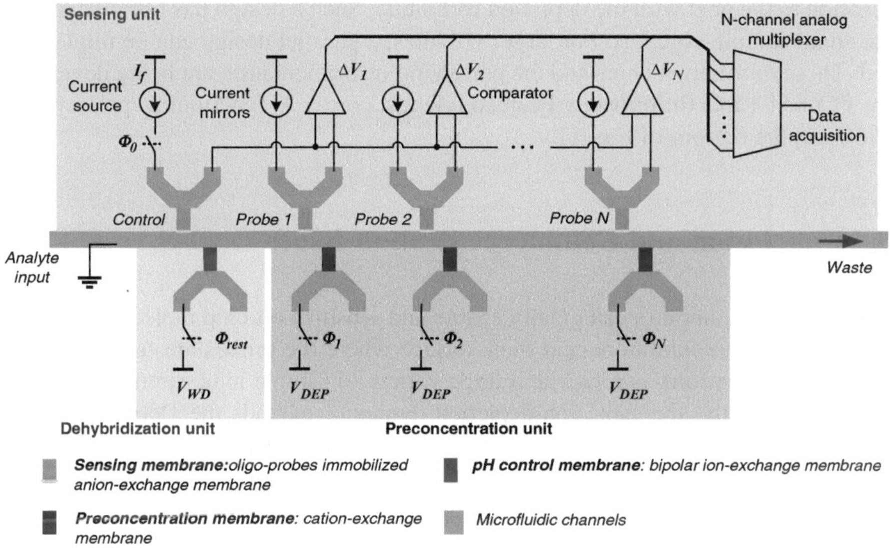

Fig. 11 *Top*: Integrated smart RNA hybridization sensor composed of sensing unit, preconcentration unit, and dehybridization unit. *Bottom*: A functioning prototype measuring 25 × 10 × 40 cm is shown in which a glass chip (like that shown in Fig. 10 but containing the multitarget design) is seen at the top of the instrumentation. A handheld prototype is expected in a year

and the selectivity is three mismatches in the 27 base pairing segments. The study showed that long kilobase target ssDNA produces a larger signal, consistent with the theory that the extended Debye layer allows more of the charges of a long molecule to contribute to the local charge capacitance and conductance.

A multitarget unit currently being developed is shown in Fig. 11. It offers sequential detection of different targets by moving the sample from one sensor location to the next with the depletion technique. Such a design has been developed for small sample volumes. For larger volumes, a parallel design can be implemented. These multitarget chips and the peripheral instrumentation are being developed by FCubed LLC (http://www.FCubed.iviehost.com). A functioning prototype is shown at the bottom of Fig. 11.

8 Conclusion and Commercialization Issues

Nanoporous membranes can greatly enable and sensitize on-chip molecular sensing. They can deplete inhibitors near their surface where the probes are functionalized, such that the platform is robust to a large variety of sample ionic strengths and pH. More importantly, the same ion-depletion dynamics extends the Debye layer and hence allows more sensitive conductance and capacitance detection of the hybridized molecules. The high field in the same depletion region can produce fast dielectrophoretic trapping of the larger target molecules. If the depletion region extends across the entire flow channel, it can also trap smaller molecules. Hence, by activating different membrane components on a chip, the molecules can be concentrated and transported to different sensors. The membrane's ability to invert its surface charge upon hybridization produces a large conductance signal for hybridization. A large capacitance signal is also produced, corresponding to the intercept of the Warburg spectrum with the real axis, when the depletion layer is formed periodically under an AC field such that the hybridized target molecules and their counterions are responsible for this asymptotic conductance when all other ions are depleted within the small depletion layer. These nanoporous membranes are fabricated on the chip and are situated on the side of the flowing channel without blocking the flow, such that a high throughput (>1 μL/min) can be achieved. Bipolar nanoporous membranes can also be used to split water and to exercise precise control of pH near the sensor, to enhance selectivity. This rapid and precise pH control can also allow multitarget sensing with the same probe if the probes are designed to be pH-sensitive.

Although the current membranes are synthesized on glass chips to allow easy inspection and testing, the same technology can be transferred to hard polymer chips, which should be cheaper to produce and easier to bond. This remains the final obstacle to commercialization.

Acknowledgment The authors are grateful to NSF, Great Lakes Protection Agency, Gates Foundation, NIH and ND-PDT for their generous support of this research.

References

1. Deana A, Celesnik H, Belasco JG (2008) Nature 451:355
2. Basuray S, Senapati S, Ajian A, Mahon AR, Chang H-C (2009) ACS Nano 3:1823
3. Cheng I-F, Senapati S, Cheng X, Basuray S, Chang H-C (2010) Lab Chip 10:828

4. Chang H-C, Yossifon G (2009) Biomicrofluidics 3:012001
5. Yossifon G, Chang H-C (2008) Phys Rev Lett 101:254501
6. Yossifon G, Chang Y-C, Chang H-C (2009) Phys Rev Lett 103:154502
7. Yossifon G, Mushenheim P, Chang Y-C, Chang H-C (2009) Phys Rev E 79:046305
8. Cady NC, Stelick S, Kunnavakkam M, Batt CA (2005) Sens Actuators B 107:332–341
9. Bakker E, Qin Y (2006) Anal Chem 78:3965
10. Umezawa Y, Aoki H (2004) Anal Chem 76:321
11. Basuray S, Chang H-C (2010) Biomicrofluidics 4:013205
12. Suni II (2008) Trends Anal Chem 27:604
13. Stern E, Wagner R, Sigworth FJ, Breaker R, Fahmy TM, Reed MA (2007) Nanolett 7:3405
14. Sosnowski R, Tu E, Butler WF, O'Conell JP, Heller MJ (1997) Proc Natl Acad Sci USA 94:1119
15. Kreft J, Chen YL, Chang H-C (2008) Phys Rev E 77:030801
16. Lin HT, Tsai LC, Chi PY, Chen CD (2005) Nanotechnology 16:2738
17. Wang P, Chen Z, Chang H-C (2006) Sens Actuators 113:500
18. Wang P, Chen Z, Chang H-C (2006) Electrophoresis 27:3964
19. Senapati S, Mahon SR, Gordon J, Nowak C, Sengupta S, Powell THW, Feder J, Lodge DM, Chang H-C (2009) Biomicrofluidics 3:022407
20. Ben Y, Chang H-C (2002) J Fluid Mech 461:229
21. Cheng L-J, Guo LJ (2010) Microfluid Nanofluid 9:1033
22. Wang Y-C, Stevens A, Han J (2005) Anal Chem 77:4293
23. Lee JH, Song Y-A, Han J (2008) Lab Chip 8:596
24. Liu V, Song YAK, Han J (2010) Lab Chip 10:1485
25. Cheng L-J, Guo LJ (2009) ACS Nano 3:575
26. Slouka Z, Pribyl M, Snita D, Postler T (2007) Phys Chem Chem Phys 9:5374
27. Zhang H, Mitrovski SM, Nuzzo RG (2007) Anal Chem 79:9014
28. Onsager L (1934) J Chem Phys 2:559
29. Ramirez P, Manzanares JA, Mafé S (1991) Ber Bunsenges Phys Chem 95:499
30. Mafé S, Manzanares JA, Ramirez P (1990) Phys Rev A 42:6245

Top Curr Chem (2011) 304: 171–201
DOI: 10.1007/128_2011_144

Published online: 21 April 2011

Optical Detection Systems on Microfluidic Chips

Hongwei Gai, Yongjun Li, and Edward S. Yeung

Abstract Optical detection continues to dominate detection methods in microfluidics due to its noninvasive nature, easy coupling, rapid response, and high sensitivity. In this review, we summarize two aspects of recent developments in optical detection methods on microfluidic chips. The first aspect is free-space (off-chip) detection on the microchip, in which the conventional absorption, fluorescence, chemiluminescence, surface plasmon resonance, and surface enhanced Raman spectroscopies are involved. The second aspect is the optofluidic (inside-chip) detection. Various miniaturized optical components integrated on the microfluidic chip, such as waveguide, microlens, laser, and detectors are outlined.

Keywords Lab-on-a-chip · Optofluidic · Point of care

H. Gai (✉)
School of Chemistry and Chemical Engineering, Xuzhou Normal University, Xuzhou 221116, China
and
State Key Laboratory of Chemo/Biosensing and Chemometrics, Hunan University, Changsha 410082, China
and
School of Biology, Hunan University, Changsha 410082, China
e-mail: gaihw@hnu.cn

Y. Li (✉)
State Key Laboratory of Chemo/Biosensing and Chemometrics, Hunan University, Changsha 410082, China
and
School of Chemistry and Chemical Engineering, Hunan University, Changsha 410082, China
e-mail: liyongjunef@gmail.com

E.S. Yeung
State Key Laboratory of Chemo/Biosensing and Chemometrics, Hunan University, Changsha 410082, China
and
School of Biology, Hunan University, Changsha 410082, China

Contents

1 Introduction 173
2 Conventional Free-Space Optical Detection 173
2.1 Absorbance Detection 173
2.2 Fluorescence Detection 176
2.3 Chemiluminescence 181
2.4 Surface Plasmon Resonance and Localized Surface Plasmon Resonance 182
2.5 Surface Enhanced Raman Spectroscopy/Imaging 184
3 Detection Inside the Chip 185
3.1 Planar Optical Waveguide 186
3.2 Microlens 189
3.3 Optofluidic Laser 193
3.4 Optical Detector 195
4 Conclusions 197
References 197

Abbreviations

μ-TAS	Micro total analysis system
ARROWs	Antiresonant reflecting optical waveguides
CCD	Charge-coupled device
CE	Capillary electrophoresis
CL	Chemiluminescence
CMOS	Complementary metal-oxide-semiconductor
DFB	Distributed feedback
HRP	Horseradish peroxidase
L^2 waveguide	Liquid-core/liquid-cladding waveguide
LCW	Liquid-core waveguide
LED	Light-emitting diode
L-GRIN	Liquid gradient refractive index
LIF	Laser-induced fluorescence
(L)SPR	(Localized) surface plasmon resonance
OFM	Optofluidic microscopy
OPD	Organic photodiode
PC	Photonic crystal
PDMS	Poly(dimethylsiloxane)
PMMA	Poly(methyl methacrylate)
PMTs	Photomultiplier tubes
QD	Quantum dot
SERS	Surface enhanced Raman spectroscopy
UV/Vis	Ultraviolet–visible

1 Introduction

Microfluidic chips have the potential to integrate all experimental processes in a research laboratory (sample preparation, reaction, separation, and detection) onto a microscale device. Ever since the micro total analysis system (μ-TAS) was firstly proposed by Manz et al. in the early 1990s [1], it has achieved great advances in many scientific research fields, such as medical diagnostics [2], drug screening [3], proteomics [4], and environmental and food analysis [5, 6]. To be able to read out the analytical results captured by a microchip, the detection system is the core part and is responsible for signal acquisition. To date, various detection modules including optical detection, electrochemical detection, mass spectrometry and others, have been assembled or partly assembled onto the microchip. The ultimate goal of having the detection system on the microfluidic chip is to incorporate it as an inseparable part of a portable microdevice with high sensitivity and fast response.

In this review, we mainly focus on the recent developments in optical systems assembled onto a chip because optical detection is easily implemented in chemical/biochemical laboratories, has acceptable sensitivity in regular analysis, and simplicity in coupling with other functional units. Such topics have been touched on in previous reviews. A general overview of earlier detection methods has been published elsewhere [7, 8]. The specialized subjects of optical systems such as detectors for separation and unconventional detection can be found in the literature [9–11]. The current optical detection systems on microfluidic chips can be summarized from two perspectives. The first is the traditional optics configuration strategy also called "off-chip" or "free space," in which light propagates in air before and after interaction with the sample (see Sect. 2). In this mode, the detection units (e.g., light source, mirrors, and detectors) are separated from the chip platform. We present the detection methods of absorption, luminescence and scattering in Sects. 2.1–2.5. The second strategy is the "on-chip" mode, in which the optics are integrated with or fabricated together with the fluidic functional units (see Sect. 3). When the optical property of the integrated optics can be tuned by fluidic technology, the microfluidic is also called "optofluidic." This mode, however, is in the development stage, and a highly integrated optofluidic is not fully realized yet. Some related optical components such as waveguides, microlens, detectors and microscopes will be discussed in Sects. 3.1–3.4. The nonoptical detection methods (e.g., electrochemical detection, mass spectrometers, and nuclear magnetic resonance) are not covered in this review but can be found in previous reviews [7, 8, 12].

2 Conventional Free-Space Optical Detection

2.1 Absorbance Detection

UV/Vis absorption spectroscopy is the most straightforward and popular detection method in separation and microseparation science due to its generality, sensitivity, and adequate detection limit. Absorption spectra are related to the structure and concentration of the analyte and are based on the capability of samples to attenuate the incident electromagnetic radiation at various wavelengths.

The depth of the channels in microchips is relatively shallow compared to their width due to the character of the fabrication method. If light is designed to pass through the channel perpendicular to the chip plane, the optical length through the sample is limited by the depth of the channel, which is shortened from millimeters in conventional analysis to tens of micrometers in microanalysis, adversely impacting the sensitivity. To address the issue of the low sensitivity caused by a short optical length, various approaches were developed to increase the optical path length.

With free-space optics, light first propagates in the air and then enters the detection cell. The optical length is enlarged simply by changing the geometry of the channel at the detection point, such as incorporating a short, wide bubble cell within the channel [13], or sandwiching an additional glass substrate with a through-hole at the detection point to increase the optical length in the vertical direction of the chip plane [14]. Despite that, these methods suffer from the limited improvement of optical length because the changes of channel geometry are detrimental to maintaining separation efficiency and to blocking stray light. Introducing the light along the channel axially is a practical alternative for enhancing absorbance. The light can be reflected between two mirrors by coating 200-μm aluminum films above and below the channel. Effective optical lengths ranging from 50 to 272 μm were achieved by adjusting the angle of laser incidence (Fig. 1) [15]. Billot et al. implemented a similar idea on a glass–PDMS–glass microchip by coating the mirrors on the inside surfaces of the channel and exciting the sample with a normal illumination configuration [16]. Besides making light reflect multiple times inside the channel, passing light along the channel axially in a straight line is another way of increasing the optical length. This idea is implemented by fabricating two parallel 45° mirrors at each end of the microchannel so that light was

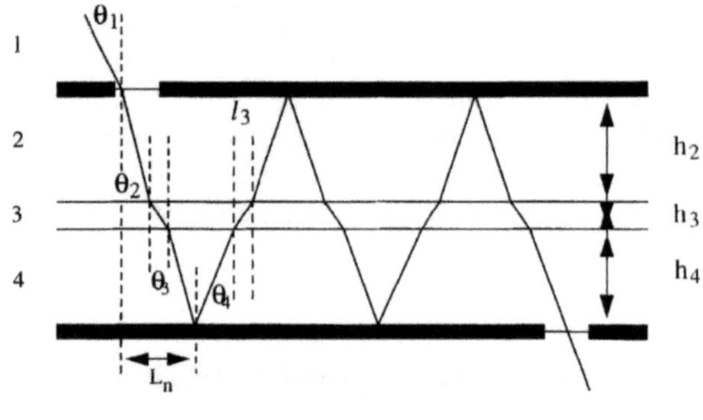

Fig. 1 Diagram showing a planar multireflection absorbance cell. Aluminum mirrors (*thick lines*) were coated above and below a channel. The narrowest region inside the cell is the fluid flow path (Reprinted from [15] with permission of Wiley-VCH). i (1,2,3,4) refers to different medium; θ_i is the angle within the medium i with respect to the surface normal; h_i is the thickness of the medium i; l_i is the longitudinal distance in each of medium and $l_i = h_i \tan(\theta_i)$; L_n is the total longitudinal travel for once reflection; n the total number of reflections

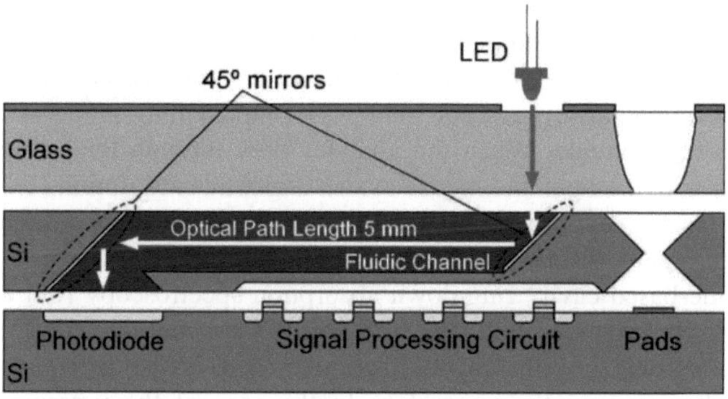

Fig. 2 Cross-sectional diagram of a microchip. Two parallel 45° mirrors were fabricated at each end of the microchannel to increase the optical length (Reprinted from [17] with permission of Elsevier)

reflected into the channel and out to the detector by the mirrors, making the 5-mm channel the optical path (Fig. 2) [17]. Within this category, other efforts were also made to improve the sensitivity of UV absorption and to maintain the high separation performance of microscale electrophoresis. For examples, Lin et al. fabricated a hybrid poly(dimethylsiloxane) (PDMS)/quartz chip with a 100-μm thick PDMS membrane on the detection window for UV absorbance detection [18]; Ou et al. developed a PDMS/SU-8/quartz hybrid chip incorporating UV absorbance-based whole-channel imaging detection [19].

Embedding waveguide optics into the microchannel is another way to enlarge the optical length. Integrated waveguides on-chip are similar to electrical circuits on electronic chips, carrying information about the analytes. This has good potential for design of a lab-on-a-chip because it can realize the practical miniaturization of the optical elements. Actually, one of the initial motivations for developing waveguides was to extend the optical length by coupling the light effectively into the axial direction of the channel. The channel length then becomes the light path length. Integrated optical waveguides on microchips offer the capability to deliver and capture light at nearly arbitrary positions and angles on the chip. With the help of an in-plane waveguide, various absorbance detection geometries have been developed, such as a 750-μm U-cell [20], a 1,000-μm U-cell on an electrochromatography chip [21] and a hollow Abbe prism on a PDMS chip [22]. However, the enhancement of optical losses with increasing optical path length was observed because of stray light and light dispersion, leading to a reduction in the signal-to-noise ratio [23, 24]. Accordingly, a light collimating system was integrated in the front of the extended optical path to block stray light and increase the sensitivity [25]. Llobera et al. integrated optical fibers, air mirrors, and collimation microlenses on a chip to realize a multireflection absorbance detection system in which the detection limit of 41 nM for fluorescein was achieved [26]. Fang et al. provided an example of miniaturized absorbance detection. They employed a U-shape bent capillary to couple light into the capillary and out to the photodetector with less light loss and less dead volume, resulting in a 15 mm effective optical path length, which significantly simplified the system for use as a hand-held photometer [27].

A technique based on absorbance of evanescent waves has been developed very recently, in which the evanescent field was coupled into a microchannel by a U-bend SU-8 waveguide. When the analytes flow through the evanescent field, they interact with the light and induce evanescent wave absorption. This technique is expected to also suppress the scattered light when light passes directly through the detection region, thereby enhancing the sensitivity [28].

Waveguide-based cavity ring-down absorption spectroscopy [29] can increase the optical path length by reflecting a pulse of light multiple times through the sample without changing the channel geometry, and is also a promising way for absorption detection on the microchip. In the case of the tartrazine detection, microfluidic-based fiber-loop ring-down absorption spectroscopy has shown a satisfactory detection limit of 5 μM [30, 31].

Despite the fact that absorbance is more limited in sensitivity than fluorescence, the potential of an inexpensive miniaturized optical setup still provides an advantage in certain applications, such as when changes in optical density or color are sufficient for clinical diagnosis.

2.2 Fluorescence Detection

Today's lab-on-a-chip devices take advantage of various optical detection strategies to achieve microvolume detection. Of these, fluorescence detection is still a preferred choice to be incorporated onto microdevices primarily due to its high sensitivity and high selectivity for cellular and molecular sensing. Fluorescence is the emission of light of fluorphores where the cycle of excitation and emission will continue until the excitation source is turned off or the fluorophore is photodecomposed. The fluorescence of biomacromolecules originates either from the native chromophore inside the molecules or from chemically linked fluorescent labels. Recently, the fluorescent labels for microanalysis have been extended from synthetic dyes, fluorescent proteins, and quantum dots (QDs) to nanoparticles. Development and exploitation of new types of fluorescent probes with high quantum yield and high specificity on lab-on-a-chip will be helpful to improve the detection sensitivity and minimize the size of optical components in the sensing system.

The complete optical setup of induced fluorescence involves an excitation part and an emission part. The intersection point of the two parts is the detection window in the microdevice. The excitation part starts from the light source and ends at the microdevice, while the emission part originates from the microdevice and stops at the detector. In free-space fluorescence detection, the common excitation sources include the laser, light-emitting diode (LED), and mercury or xenon arc lamp, and each light source has its distinctive spectroscopic property and practical benefits. Laser-induced fluorescence (LIF) is a highly sensitive optical detection method and is able to perform even single molecule detection. LIF has been introduced into the microchip at the early stage of lab-on-a-chip and remains to be a popular approach, especially for analyzing low-abundance molecules after combining with point detectors like photomultiplier tubes (PMTs). The high pressure vapor discharge lamp is the standard light source for modern fluorescence microscopy due to its broad spectrum, ranging from UV to visible regions, and because it is less expensive and more flexible by offering different wavelengths with the assistance of proper optical filter sets. The lamp-based fluorescence microscope is usually coupled with an imaging detector like a charge-coupled device (CCD) to analyze cellular fluorescence. An LED is a small semiconductor light source and can efficiently produce light. These merits indicate that LED is suitable for integration into microfluidic chips. Yao et al. succeeded in integrating a green organic LED and thin film interference filter into a microfluidic device through layer-by-layer stacking [32], and also designed a similar LED-induced fluorescence-based microdevice for the imaging of whole-column isoelectric focusing [33] and for high throughput capillary array electrophoresis on the chip [34].

Fluorophores emit photons with 4 pi steradians once they are excited. Therefore, fluorescence can be collected at any solid angle by an objective or a lens. One prevalent optical configuration for induced fluorescence detection is that the collection objective in the emission part is the same as the focus components in the excitation part. Fluorescence collected by the same objective passes through the dichroic and emission bandpass filter and reaches the detector. Such configuration is called epi-illumination in fluorescence microscopy and is also the basis of confocal LIF as long as a pinhole is placed in the conjugate plane to the sample plane to block the out-of-focus light and a laser diffraction-limited spot is used to excite samples. A typical confocal LIF mode on-chip is depicted in Fig. 3 and is almost the same as for the confocal microscope except that the microchip is used to replace the microscope stage. Additionally, the emission part of another setup for fluorescence detection does not share the same objective with the excitation part, and each part has its own focusing systems. In this configuration, the collection angle seriously affects the sensitivity and has been optimized to 45° [35] by Fang's group. Subsequently, Whiting et al. determined under- and oversampling of individual particle distributions in microfluidic electrophoresis with orthogonal LIF [36]. Using this orthogonal LIF detector, a detection limit of 1,761 molecules of fluorophore was obtained under a constant flow of fluorescein. Compared to the common LIF detection scheme used with microfluidic devices, this is a significant improvement.

From the viewpoint of instrumentation development, the aim of optical detection systems of lab-on-a-chip is miniaturization, but under the premise that the detection sensitivity is maintained at a practical level. Therefore, we shall review the fluorescence detection on-chip from the points of view of improving sensitivity and miniaturization.

LIF is appropriate to deal with the minute amounts of sample in the microchannel and to demonstrate the feasibility of the concept of μ-TAS in the early stage of development [37, 38]. Optimization of detection parameters to improve the sensitivity of confocal LIF on-chip was accomplished [39] and the authors detected 300 fM fluorescein with a signal-to-noise ratio of 6.1 in continuous-flow mode and

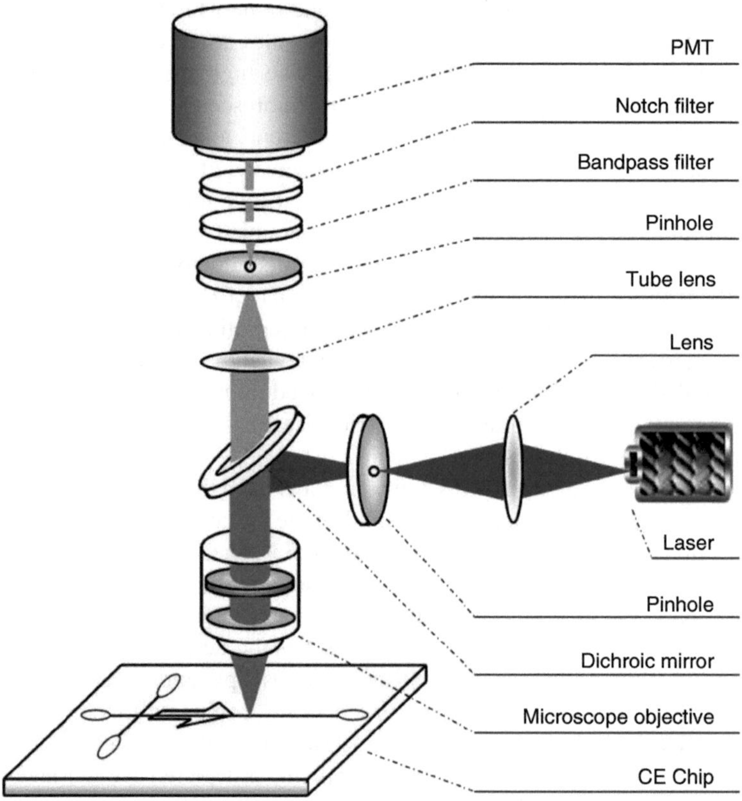

Fig. 3 Confocal fluorescence detection on a microchip. A laser was used as the excitation source. The laser light passed through a biconvex lens, was focused into an illumination pinhole, and was subsequently reflected by a dichroic mirror and focused in the channel on the capillary electrophoresis (CE) chip by a microscope objective. The fluorescence signal from the sample, collected by the microscope objective, was passed through the dichroic mirror, focused by the tube lens into a confocal pinhole, and then detected by a photomultiplier tube (PMT). To improve the signal-to-noise ratio, band-pass filter and notch filter were inserted above PMT for spectral filtering

1 pM fluorescein in capillary zone electrophoresis. Consequently, the confocal LIF was successfully used to detect single molecules of a 100–1,000 bp DNA ladder in a microchannel by electrodynamic focusing [40]. High-quality DNA sequencing separations have also been performed in microfabricated electrophoresis channels [41]. Early results of detection of single molecules labeled with organic dyes in microchannels were reviewed by Manz in 2005 [42]. The point detection mode suffers from low detection efficiency because the diffraction-limited focused laser spot is so small that the whole channel cannot be covered and most molecules do not pass through the laser beam. Chiu et al. designed a line-confocal detection geometry comprising an elliptically shaped probe volume to enlarge the fraction of molecules that flow through the laser beam without severely decreasing the

signal-to-noise ratio [43]. Another way to improve the single molecule detection efficiency in microchannels is to employ an imaging detector, either an intensified charged coupled device (ICCD) or an electron multiplied CCD (EMCCD), by imaging a larger detection area than covered by the point detector. Single moving lambda DNA molecules forced by pressure or electrokinetic flow in the microchannel are easily observed with an imaging detector [44]. When the total internal reflection excitation mode is coupled to the microchannel, the velocity of molecules near the channel wall can also be measured. Therefore, the velocity distribution in the microchannel can be profiled by adjusting the excitation depth of the evanescent wave [45–47]. In addition to organic dyes that are used as the fluorescent label, recently developed fluorescence nanoparticles are also used to label molecules. Our group has improved the fluorescence characteristic of core–shell QDs by adding thiol-group chemicals and labeled QDs to the oligonucleic acids and demonstrated the advantages of QDs for microchips [48–50].

The microdevice is also an ideal platform for single-cell analysis when the functional units of cell sampling, cell sorting, cell breakage, chemical separation, and detection are integrated onto a single chip. Meanwhile, single-cell chemical analysis on the microchip is a typical application of LIF detection on-chip. Yu developed a high-throughput single cell analysis method on PDMS microfluidic chips to simultaneously analyze glutathione and rhodamine 123 in single K562 cells. They could analyze about ten cells each minute in one channel [51]. For nucleic acid molecules, the labeling process can be implemented in the channel by intercalating dyes into the base pairs. Kleparnik [52] assessed the effects of doxorubicin, a cytostatic, on the apoptosis of single cardiomyocyte cells by separating DNA fragments labeled with ethidium bromide. For protein molecules, label-free LIF detection is a suitable method due to the UV fluorescence of the three amino acids tryptophan, tyrosine, and phenylalanine. Hellmich reported LIF detection of native protein lysed from single cells and excited with a 266-nm laser [53]. The authors further demonstrated the CE separation of two proteins from a single cell and improved the separation efficiency and detection sensitivity by shortening the electrical lysis time and improving the injection plug size [54].

LIF detection is the most sensitive optical method so far, but is hard to miniaturize in order to satisfy the ultimate goal of a microfluidic chip that assembles all analytical processes within one micrometerscale microstructure. Therefore, how to achieve the miniaturization of fluorescence detection on microdevices is becoming an active field for lab-on-a-chip research. Several examples demonstrate recent advances in miniaturized LIF detection on the microchip. In 2005, Renzi et al. designed a hand-held microchip-based analytical instrument that combines fluidic, optics, electrical power, and interface modules and integrates the functions of fluidics, microseparation, lasers, power supplies etc., into an $11.5 \times 11.5 \times 19.0$ cm system (Fig. 4a) [55]. This device allows picomolar (10^{-11} M) detection sensitivity for fluorescent dyes and nanomolar sensitivity (10^{-9} M) for fluorescamine-labeled proteins. Novak et al. built a low-cost mini-fluorescence detection system of $30 \times 30 \times 11$ mm (Fig. 4b), in which LED, photodiode, mirrors, lenses, and electronic circuit are housed together, with

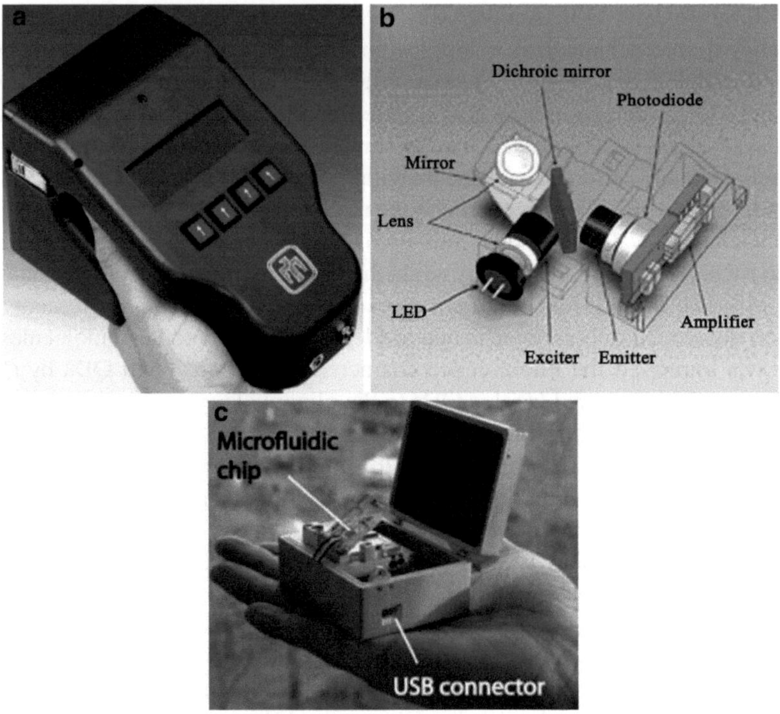

Fig. 4 Examples of portable microfluidic instruments. (**a**) The system integrates fluidics, microseparation chips, lasers, optics, high-voltage power supplies, electronic controls, data algorithms, and a user interface into a hand-portable instrument (Reprinted from [55] with permission of The American Chemistry Society). (**b**) A miniature LED-induced fluorescence microdevice (Reprinted from [56] with permission of The Royal Society of Chemistry). (**c**) Hand-held isotachophoresis instrument (dimensions: 7.6 × 5.7 × 3.8 cm) (Reprinted from [57] with permission of The Royal Society of Chemistry)

a sensitivity in the low nanomolar range [56]. In 2010, Backhouse et al. presented a hand-held isotachophoresis microchip device [240 g, (Fig. 4c)] with LIF detection, which was powered by a universal serial bus link connected to a laptop computer. A limit of detection of 100 pM for Alexa Fluor 647 was demonstrated [57].

Making the size of each optical component for LIF as small as possible and assembling them onto microfluidic systems is a straightforward way to develop a portable analyzer as described above. Another idea to minimize the instruments is to design and fabricate all optics directly on-chip, which means that the optics such as light source [58–60], lens [61, 62], and detector [63] are part of the microchip rather than minimizing them and interfacing them with microchip. This self-contained optical microfluidic, also called optofluidic, is reviewed in the Sect. 3 of this article.

2.3 Chemiluminescence

Chemiluminescence (CL) is a molecular luminescence spectrometry, originating from the release of energy in a chemical reaction: a metastable reaction intermediate or product in a singlet excited state relaxes to the ground state with the emission of a photon. By measuring the as-generated luminous intensity, one can conveniently determine the analyte concentration. Compared to other spectroscopic techniques, there is an absence of any excitation light source is the inherent advantage of CL detection, which not only eliminates the problematic background emission from interfering compounds or from the microfluidic substrate itself. CL is also more appropriate for miniaturization of the on-line detection system and to build a portable microfluidic instrument [64, 65]. Thus, CL is a promising alternative detection technique to be coupled to microfluidic analysis for a miniaturized device. However, CL is limited to some special reaction reagents, such as luminol and its derivatives [66], peroxyalate and oxalate [64], and ruthenium complexes and to direct oxidations [65]. Till now, CL detection coupled to capillary electrophoresis has been well reviewed [67–70]. Herein, we briefly introduce recent developments in the use of CL on microchips.

In 1998, CL detection coupled to microfluidic was first described by Mangru and Harrison [66]. Based on horseradish peroxidase (HRP)-catalyzed reaction of luminol with peroxide, they designed a postseparation detection scheme on a microchip by transporting peroxide into the separation channel from a bypass channel. A 7 nM detection limit for the fluorescein conjugate of HRP was achieved. To optimize the microchip layout and reaction conditions for different CL reagents, Liu et al. [71] compared two chip layouts (cross-combining with Y, and cross-combining with V) in PDMS and evaluated their performance with respect to two model CL systems that included the metal-ion-catalyzed luminol–peroxide reaction and the dansyl species conjugated peroxalate–peroxide reaction. A comparison between these chip layouts showed that the chip pattern of cross-combining with Y may be more promising for the luminol–peroxide CL system, while the chip pattern of cross-combining with V was preferred for the peroxalate–peroxide system.

Hashimoto et al. [64] simplified the chip layout without etching the manifolds. The detection point was set at the interface of separation channel and reaction reservoir. They successfully separated and detected dansyl amino acids (glycine and glycine) using peroxyoxalate CL reaction using CL detection-based microchip capillary electrophoresis. They further incorporated an immune reaction with CL detection to determine human serum albumin or immunosuppressive acidic protein as a cancer marker in human serum [72].

After CL detection was successfully introduced to the microchip and optimal channel design and reaction conditions were obtained, it was used for high-sensitivity detection, especially for single cells analysis. Zhao et al. [73] first reported that CL detection with microchip electrophoresis was used to determine the intracellular content of glutathione in single human red blood cells. CL detection showed about 100 times more sensitivity than LIF detection for single cell analysis.

The same group subsequently quantified Gly and Ala in single rat hepatocytes cells [74], taurine and amino acids in mice single fibrosarcoma cells [75], and sulfydryl compounds in single red blood cells [76]. They further developed a CL resonance energy transfer method to detect the amino acids in single human red blood cells [77]. Besides intracellular content analysis, the counting of intact cells by CL detection on microfluidic chip was designed by Wang et al. [78]. They counted CD4-positive ($CD4^+$) T-cells from whole blood by capturing $CD4^+$ T-cells with anti-CD4 antibody. The trapped cells were then bound to CD3 antibodies conjugated with HRP and incubated with chemiluminescent substrate. The emitted photons were proportional to the number of captured $CD4^+$ T-cells. The results can be provided automatically and digitally within seconds of running the test. Chemiluminescence detection on the microfluidic chip provides the highest potential to reduce the size of instruments compared with other optical detection systems due to the lack of an external light source, which makes it possible to produce low-cost point-of-care diagnosis microdevices. However, the crucial drawback of CL detection in microfluidics is its poor reproducibility because CL measurements are strongly affected by many factors, including temperature, pH, solvent, solution composition, etc. Furthermore, the number of CL reagents is limited.

2.4 Surface Plasmon Resonance and Localized Surface Plasmon Resonance

Surface plasmon resonance (SPR) is an optical method for probing the refractive index changes of medium (sample) in contact with the surface of a noble metal film in which coherent electron oscillations result in resonance. SPR is a particularly suitable technique for measuring rates and equilibrium constants of processes that involve molecular interactions and adsorption of biomacromolecules on surfaces. A light wave is coupled to the metal film to excite the surface plasmons. If analytes bind to the metal film surface, a change in the refractive index of the medium alters the characteristics of the light wave coupled to the surface plasmon. Measuring the alteration of the excitation light reflects the species and concentrations of the analytes. The basic principles, couplers and applications of SPR have been well documented by Homola [79]. An SPR spectroscope usually incorporates a flow cell to confine and control the sample flowing through the sensing surface. Microfluidic chips provide an easy manner for sensing very small amounts of samples without an optical tag when SPR or localized surface plasmon resonance (LSPR) is integrated onto microchips.

Coupling SPR to microfluidic systems is usually implemented by prism, waveguide, and metal grating. In the prism coupling mode, the walls of microfluidic channels are deposited with a layer of gold film for sensing the change of analytes flowing through the film [80–82]. The excited light is totally internally reflected by the prism, which is in contact with the metal film, and is introduced to the metal film through evanescent wave. Figure 5a shows a schematic of prism-coupled SPR for a microfluidic chip. This traditional coupling approach is simple, sensitive, and

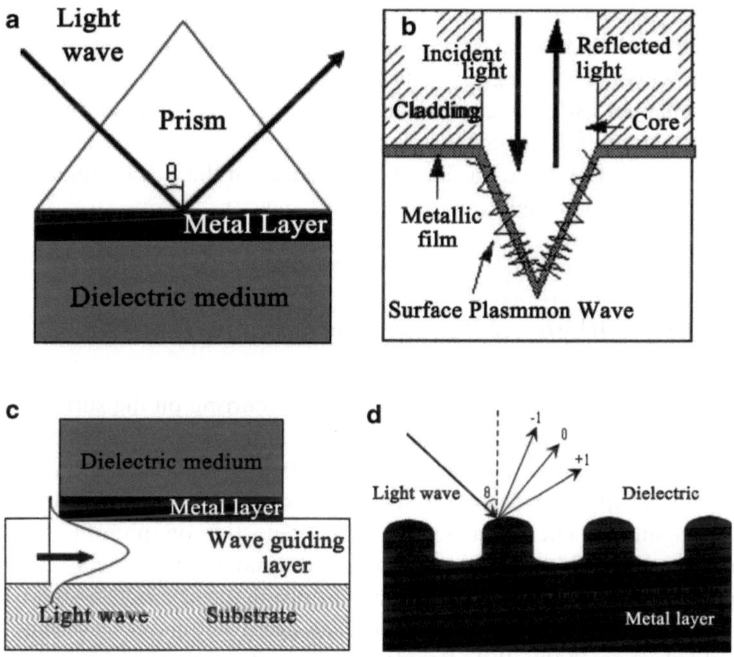

Fig. 5 Coupling schemes of incident light to surface plasmons: (**a**) prism coupling, (**b**) experimental setup for the fiber-optic SPR microsensor, (**c**) waveguide coupling (Reprinted from [83] with permission of Elsevier), and (**d**) grating coupling

robust, but it is less advantageous in miniaturization and integration. Optical fibers and waveguide have been developed as alternatives to the prism to generate evanescent waves and to excite the surface plasmon. Kurihara etched a fiber to form a gold-coated microtip at the end of the fiber and inserted it into a microchannel [83]. The coated microtip has features like a microprism (Fig. 5b). Fiber-based SPR on-chip has the problem of alignment and assembly. To overcome such issues, it was proposed that the planar optical waveguide coupling mode should be integrated on the microfluidic chip (Fig. 5c) [84, 85]. The waveguide-based SPR on-chip has the potential of high throughput analysis of multiple samples in parallel because multiple waveguide structures can be fabricated into one single platform. Another promising alternative method for the miniaturization is SPR-based grating, which is low-cost and can be mass-produced by disposable polymer materials instead of the glass chip used for prism or waveguide SPR [86, 87]. In this coupling mode, a metal grating profile coated on the channel diffracts the incident light wave, which is capable of matching the momentum of the plasmon and the wave and generating a surface plasmon (Fig. 5d). In this approach, noncontact optical excitation obviates the need for a prism and index-matching fluids and greatly simplifies instrument configuration.

The excitation of surface plasmons that occurs at the planar surfaces is denoted as SPR; that which occurs at the surface of nanometer-sized metallic structures is called LSPR. SPR-based systems require coating of metal film on the walls of the microchannel and finding an appropriate way to couple the light to the surface plasmon. However, LSPR merely requires the immobilization of metal nanoparticles in the channel [88], which substantially reduces the fabrication cost and the optical setup complexity. LSPR seems to be sensitive enough on the microfluidic chip because Cao et al. have measured the changes in the LSPR signal of a single Au nanoparticle by dark field microscopy [89].

2.5 Surface Enhanced Raman Spectroscopy/Imaging

Surface enhanced Raman spectroscopy (SERS) occurring on the surfaces of noble metal substrates or nanoparticles and is another highly sensitive detection method based on surface plasmon, without requiring labeling. In contrast to conventional Raman spectroscopy, the detection sensitivity of SERS is enhanced up to 6–10 orders of magnitude. The advantages of SERS detection on microfluidic chips are expected to yield reliable and reproducible results and to obtain precise quantitative measurement because the experimental conditions in the microchannel are easier to control than that of bulk SERS, which has been discussed in a review of chip-based SERS published in 2008 [90]. Besides probing the fingerprint of molecules, a SERS-coupled microfluidic device has been used for characterization of single Chinese hamster ovary cells, into which gold nanoparticles were introduced for Raman enhancement [91]. Herein, we only focus on progress published after 2007 and in which metal nanoparticles were exploited as the SERS enhancer.

The targets of present SERS on-chip are (1) to solve the problem of poor detection sensitivity caused by the small sample volume and (2) to develop robust methods to obtain a reliable and reproducible SERS signal. Aggregation of metal nanoparticles and mixing of enhancer with analytes are the straightforward ways to increase the SERS signal intensity and the uniformity of the SERS active site.

In general, SERS signal intensity is related to the number of molecules, Raman cross-section and the enhancement factors. Among these parameters, molecular quantity is the only controllable variable for the signal enhancement at given experimental conditions. Therefore, increasing the amount of absorbed molecules in the detection area can amplify the SERS signal. Various ways have been proposed for accumulating nanoparticles in microdevices and three are listed below:

1. Hampering nanoparticles by a dam or restricting them in reservoir: Choi et al. [92] concentrated rhodamine 6G-adsorbed gold nanoparticles by repeating a "filling–drying" cycle on a (compact disk) CD-based platform. They readily achieved a SERS signal of 1 nM rhodamine 6G. Wang et al. [93] developed a pinched and step microfluidic–nanochannel junction like an inverted dam, where nanoparticles and target molecules were stopped and assembled, and obtained electromagnetic enhancement factor of ~10^8. This technique was also used for monitoring the conformational transition of beta-amyloid peptide [94].

2. Use of electrokinetic force: Huh et al. [95] embedded microwells in a microchannel, the top and bottom of which were fabricated with a pair of electrodes. When electrically actuated, gold nanoparticles bound with DNA molecules were attracted into the microwells by the electric field, and the sensitivity of SERS was significantly increased. The authors reported the detection limit of 30 pM for Dengue sequences. Similarly, Cho et al. [96] reported accumulation of electrokinetically driven charge molecules on an Au surface and amplified the SERS signal by almost eight orders of magnitude.
3. Coupling a laser tweezer to a microdevice for aggregating metal nanoparticles: When nanoparticles flowed through focused red laser beam, they were trapped by the laser beam and assembled together. Another green laser was used for SERS excitation. Although the instrument was complex, it provided potential for controllable aggregation of nanoparticles and sequential analysis of samples in one run [97]. Based on above advances, we will not be surprised if magnetic force will be used for trapping metal enhancer in the near future.

SERS detection is restricted in quantitative analysis due to its poor reproducible results. A consistent and homogeneous mixing between the analytes and nanoparticles before they arrive the detection point is helpful in improving reproducibility. Therefore, various shapes of microchannels have been designed to increase the mixing effect, such as alligator-teeth, zigzag, 3D channel, etc [90]. Recently, Quang et al. [98] fabricated a pillar array in a PDMS microfluidic channel to efficiently mix sample and enhancer. Two confluent streams of analytes and silver nanocolloids were introduced into the pillar array channel and then the SERS spectra were recorded at the downstream end of the channel. Using this pillar array channel-based SERS, they achieved the quantitative analysis of dipicolinic acid and malachite green, and the detection limits were estimated to be 200 and 500 ppb, respectively.

3 Detection Inside the Chip

Optical detection inside the chip means that optical detection relevant units, such as light sources, filters, mirrors, waveguides, lenses, and detectors, are integrated on the chip. Those microfluidic chips integrated with optical components that are merged with or made of fluids are also called "optofluidic." It is obvious that the more functional units, including optical components, are integrated on the chip, the degree of miniaturization of the microdevice is higher. Compared with free-space optics systems, the integrated optical microfluidic system is alignment-free and technician-independent [99]. Since such integrated microsystem is compact and firm, it is insensitive to shock and vibration, and can be used under extreme environments. In addition, if the light is well guided inside the chip, it can minimize light loss and the emergence of stray light as much as possible, increase the utilization efficiency of probing light, and decrease the background. Most current progress on detection inside a chip is concentrated on developing various strategies for assembly of optical elements on the chip. For example, previous reviews have discussed the structure and fabrication of the optofluidic waveguide [100], optofluidic microlens [101], microscope on-chip [102] and so on. With the advance of manufacturing technologies, integrated optical devices in the near future would be less expensive, with response times much faster than ever before. The full integration of optical elements into microfluidic chip is becoming possible.

3.1 *Planar Optical Waveguide*

Optical waveguide is a physical structure that transmits light along its axis, which is generally composed of "core/cladding." A planar optical waveguide is a waveguide fabricated in a flat platform, and herein refers particularly to microfluidic chip. A waveguide is a key optical element for integrated optofluidics, and has the same importance as the fluidic channel in directing fluids. When the optical waveguide and the fluidic guide are merged together, the degree of integration is maximum. In microfluidic systems, planar waveguides have been used to enhance the detection sensitivity, such as focusing the light to the channel to increase the excitation power for fluorescence measurements [103], extending the optical path length in absorbance detection [104], and splitting the laser beam for interferometry sensing [105, 106]. The current status is that two types of waveguides have been assembled on microchips. One type is based on the refractive index, whereby the light is confined in the core material by total internal reflection because the core has a higher refractive index than that of the cladding. There is an electric field in the cladding layer, which typically penetrates 100–200 nm depths and decays exponentially outside the core. This field is called the evanescent wave. The other type of waveguide is the interference-based waveguide, in which the light is reflected back to the core by interference rather than internal reflection; therefore, the refractive index of the core need not be larger than that of the cladding.

3.1.1 Evanescent Wave-Based Waveguide

Depending on the materials used for core and cladding, waveguides based on total internal reflection are divided into three types: solid-state waveguides, liquid-core waveguides (LCWs), and liquid-core/liquid-cladding (L^2) waveguides.

Solid-state waveguides are usually buried in the chip and intersect the fluidic channel to excite the analytes and collect the response signal. Solid-state waveguides are fabricated with the same materials as the chip, such as silica, glass and polymer. Silica and glass have the advantages over polymer materials of being chemically inert, biocompatible, and optically clear, and are usually used for improving sensitivity in absorption and fluorescence detection. Petersen et al. buried a doped silica waveguide on microfluidic separation devices for UV/Vis absorbance measurements [20]. A ninefold higher sensitivity was achieved with this approach compared to a conventional CE instrument with a 75-μm inner diameter capillary. The authors further fabricated 128 waveguides on-chip for multipoint detection to measure particle velocities [107]. Cleary et al. fabricated a silica waveguide on a chip and realized time-correlated single-photon counting of nile blue with concentrations as low as 1.5 nM [108].

However, glass and silica materials suffer from fabrication and mass production difficulties due to their complex processing technology and high cost. A variety of recent developments in assembling waveguides on microdevices have concentrated on polymer materials such as PDMS [109, 110] and poly(methyl methacrylate) (PMMA) [111] for large-scale production and cost reduction. Very recently, Okagbare et al. [112] developed a cyclic olefin copolymer waveguide embedded in a PMMA fluidic chip. The evanescent wave outside the cyclic olefin copolymer waveguide was used to simultaneously excite fluorphores in 11 fluidic channels for high-throughput analyses. The limit of detection for an Alexa Fluor 647 dye solution was 7.1×10^{-20} mol at a signal-to-noise ratio of 2.

The solid state waveguide is assembled into the microchip but separately fabricated with the microchannels. The connection point of fluidic channel and solid waveguide is generally the interrogation area. However, LCWs share the same physical volume with the fluidic channel. The microchannel not only transports the sample, but also transmits the light, which greatly simplifies the design complexity and reduces light loss through the channel walls. If liquid (particularly water in a microfluidic chip, with refractive index of 1.33) is used as the core of waveguide, the refractive indices of the cladding should be smaller than that of water, based on the principle of total internal reflection. Teflon amorphous fluoropolymers (Teflon AF) with an indices of 1.29–1.31 is a suitable cladding material for a LCW in a microfluidic chip. Datta et al. [113, 114] coated Teflon AF on silicon and glass substrate as cladding layer, and fabricated the coated substrates into microchannels that were also LCWs to propagate light. Detection of nanomolar concentrations of fluorescent dye was achieved on the liquid-core waveguide integrated chips. In order to reduce the time-consuming and high-cost fabrication processes using glass and silica substrates, Wu et al. [115] coated Teflon on the PDMS substrate and assembled a liquid-core waveguide on a PDMS/glass hybrid microfluidic chip to sense nitrite. Cho et al. [116] developed a method to construct LCWs in PDMS microchannels by flowing Teflon AF solution through the channels. This coating approach avoids the difficulties of bonding between two Teflon-coated PDMS substrates [115]. A micro fluorescence-activated cell sorter in the Teflon-coated LCW was further demonstrated. The laser was introduced into the microchannel, i.e., the LCW, from its one end and the prestained cells in microchannel were all excited. Therefore, the cells can be sorted into a split channel at the downstream side [117]. Replacing a commercialized Teflon tube with a fabricated liquid core in channel is an alternative method for improving sensitivity in a microdevice. For example, Du et al. [118] coupled a Teflon AF capillary to the chip channel as the detection cell for microchip absorption detection and achieved an effective optical path length of 17.4 mm with a detection volume of only 90 nL. In 2010, Pan et al.

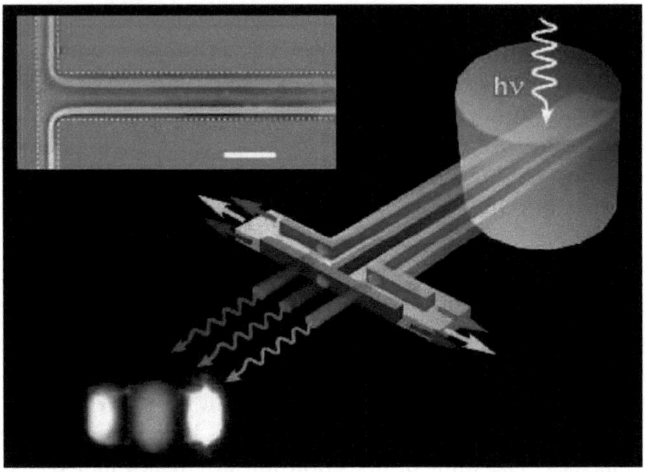

Fig. 6 Three parallel waveguides formed with liquid core and cladding in laminar flow systems. The direction of the light propagation can be altered by different flow rates of the adjacent fluids. *hν* refers to the energy of light; *h* is the Plank's constant; *ν* is the frequency of the light. (Reprinted from [121] with permission of The American Chemistry Society)

[27] integrated Teflon LCWs, photometer, LEDs and other components into a hand-held instrument. Teflon has long been the only option for low refractive index cladding for LCWs until nanoporous dielectrics were developed. Nanoporous materials are formed with a high-index solid material by adding air pores, therefore the refractive index can be adjusted by varying the air fraction. At the same time, the wall of nanoporous cladding can be chemically functionalized, overcoming some of the adhesion problems of Teflon [119]. Korampally et al. fabricated nanoporous film of low refractive index on the silicon microchannel and developed a miniaturized LCW [120].

L^2 waveguides are formed by two or more different laminar liquids of different refractive index flowing inside a fluidic channel. The index of the cladding liquid is smaller than that of the core liquid so that the light is guided in the channel by the total internal reflection that occurred at the interface of the liquids. By changing the flow rates of the liquids, the size and the direction of the waveguides can be reconstructed (Fig. 6) [121, 122]. The advantages of L^2 waveguides over solid-state waveguide are: (1) the optical performance is tunable by controlling the fluidic properties; and (2) the smooth interfaces between the core and cladding streams minimize optical loss. To couple an L^2 waveguide with a biosensor, Rosenauer et al. [123] fabricated a liquid-core/liquid-twin-cladding waveguide for on-chip fluorescence spectroscopy. The design included an inner and an outer cladding fluid outside of the core fluid; analyte was suspended in the inner cladding fluid and excited by the evanescent wave at the fluid interface of core and inner cladding. By tuning the flow rate ratios of the three different liquids, considerable reduction in sample volume and a low background noise were achieved. The detection limit of rhodamine B dye was 20 μM. The L^2 waveguide in a 2D plane suffers from optical loss at the top and bottom interfaces of core and channel walls. Lee et al. [124] therefore developed a 3D microfluidic L^2 waveguide by using microfluidic drifting technology [125], where the liquid core cannot come in contact with the channel walls because it is hydrodynamically focused in the center of the microchannel by the cladding fluids. In addition, the laminar flow-based L^2 configuration has been widely applied to construct other optical components, such as reconfigurable lenses [126], optical switches [127], and tunable micro-optofluidic prisms [128]. Although current L^2 waveguides may be restricted by the sensitivity of the stream interface to vibrations and diffusion effects, and limited on finding liquids with specific refractive indexes, the tunability of L^2 waveguides makes it possible to provide functionality and integration in future optofluidic devices.

3.1.2 Interference-Based Waveguide

The interference-based waveguide is a conceptually different approach for realizing LCWs in microfluidic chips. In contrast to the evanescent wave-based waveguides, interference-based waveguides do not require a cladding material with a lower index than that of the core material. Light is multiply reflected from a periodic dielectric cladding layer by wave interference. There are two types of interference-based waveguides: photonic crystal (PC) waveguides and antiresonant reflecting optical waveguides (ARROWs).

The integration of PC fibers on glass chip has been demonstrated by means of femtosecond laser exposure and chemical etching [129]. However, so far no microfabricated PC-based LCW has been reported. ARROWs have attracted more attention from researchers in the microfluidic chip field due to their relatively easy fabrication from accessible materials using current micro-electromechanical techniques. The main feature of ARROWs is that light is confined by Fabry–Perot antiresonant reflectors, rather than total internal reflection. Therefore, ARROWs can be constructed in a low refractive index core, such as an aqueous solution, surrounded by high refractive index cladding layer such as a silicon substrate. Planar integrated networks of intersecting solid and LCWs on a chip were used for single dye molecule detection [130] and single liposome particle detection [131]. Further information about integrated waveguiding can be found in reviews by Schmidt et al. [100, 132].

3.2 *Microlens*

The microlens is an important and key optical component for focusing and collimating light in the microfluidic optical detection system, and can be made up of a hard solid surface or interface (soft solid/liquid, liquid/liquid/...). Hard solid state lenses on-chip are usually fixed focal length lenses similar to the miniaturized traditional lenses used in the free-space detection system. They can be fabricated on-chip with the optical axial vertical to the chip plane, focusing light to the channel from top or bottom surface [133], as well as with optical axis parallel to the plane, introducing light from the side wall of the chip [25, 134]. Although solid lenses suffer from the complexity of fabrication and alignment, they are robust and vibration resistant. Recently, such microlenses have been fabricated on a SU-8/ PDMS hybrid microfluidic chip for optical manipulation of microparticles [135]. Other promising and popular microlenses on-chip utilize the curvature formed by the interface between two liquids, between liquid and soft solid, or between liquid and air. Such microlenses are also called optofluidic microlens, and they are dynamic optical elements and their focal lengths can be easily tuned by changing fluid characteristics such as species, concentration, and flow rate. Next we will focus the discussion on optofluidic microlenses.

The major advantages of these interface-based microlenses are that they are reconfigurable and optically smooth meniscus surfaces. Light can be modulated by simple adjustment of the flow conditions, thus eliminating the need for mechanical or electrical parts for manipulating the lens on the microchip.

Elastic PDMS is a commonly used polymer material for fabricating microfluidic chip and because it is readily curved by air pressure or liquid hydrodynamic, it has been used to shape a curvature and form a microlens by deforming with pneumatic force. For example, a PDMS wall with a thickness of 50 μm between chambers and microchannel can be distorted by applying liquid pressures flowing through chambers. The deformation of the PDMS wall and the index-matching fluid act as a double convex lens to focus both the excitation light source and the fluorescent emission signal. The focal length can be adjusted through different liquid pressures [136]. Changing the curvature of the PDMS surface can also be realized out-of-plane by exerting electrical force [137] and electromagnetic force [138]. However, the distortion of the surface is limited, and a large curvature of the PDMS surface is hard to form by external force. To address this issue, Shi et al. [139] fabricated a tunable in-plane optofluidic microlens set involving a static PDMS lens and an air–water interface that diverged the light and was reshaped by adjusting the flow rate of the injected water. The tunable divergence in the incident light beam, in turn tuning the overall focal length of the microlens, was converged by the static PDMS lens. The microlens achieved tunable focal distance of 11 mm and a ninefold light intensity enhancement at the focal point. In addition, the liquid–liquid interface promises another way of forming a large curvature microlens.

Liquid–liquid microlenses were generated for the first time on-chip by Mao et al. [140]. Subsequently, Rosenauer et al. designed a hydrodynamically adjustable convex lens with 3D light focusing ability instead of line focusing [141] (Fig. 7). The authors co-injected miscible fluids of different refractive indices and formed the interface in the channel. When the liquids flow through a 90° curve in a microchannel, a centrifugal effect also curves the fluidic interface and yields a reliable cylindrical microlens. The curvature of the fluidic interface can be altered by simply changing the flow rate. Higher flow rates generate a microlens with larger curvature and, hence shorter focal length. Tang et al. [142] reported another L^2 lens formed by laminar flow of three streams of fluids (Fig. 8). The refractive index of

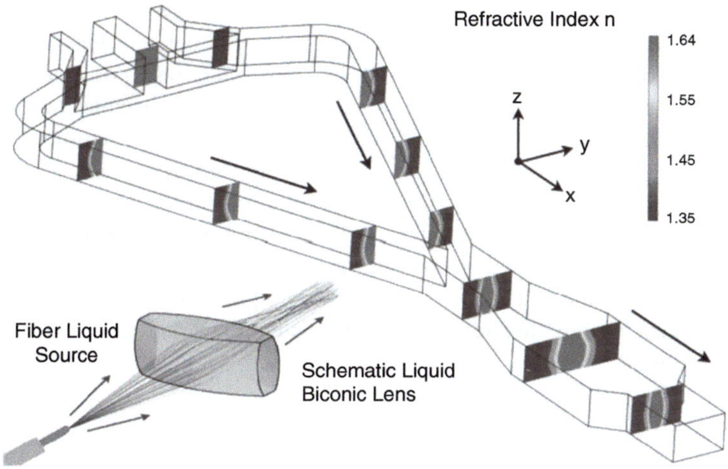

Fig. 7 A 3D light-focusing optofluidic microlens. The displayed cross-sections show the shaping of the lens and the refractive index step defined by the choice of core and cladding fluids (Reprinted from [141] with permission of The Royal Society of Chemistry)

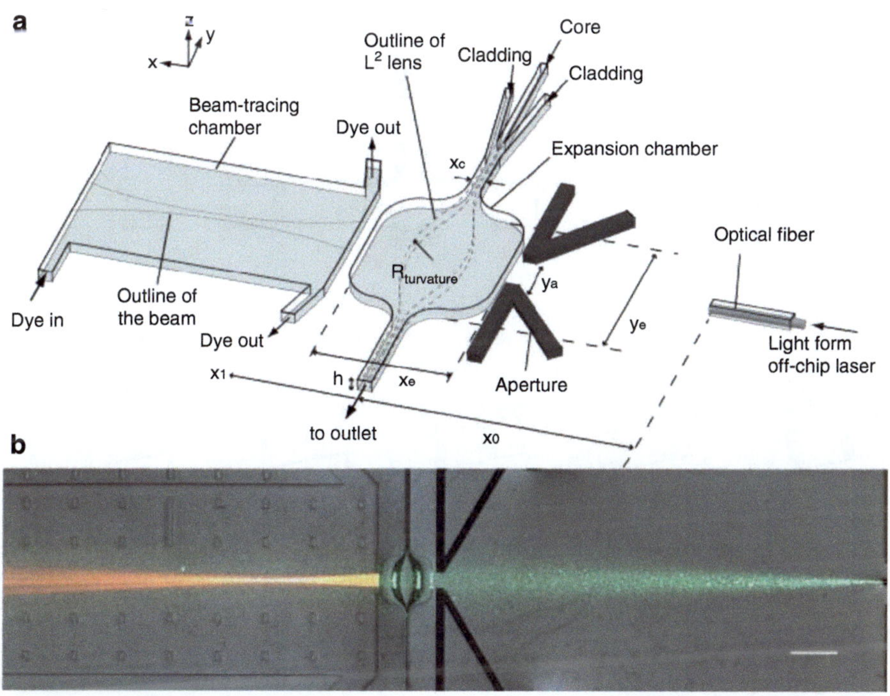

Fig. 8 (**a**) Experimental setup for focusing light exiting an optical fiber through the liquid-core/liquid-cladding (L^2) lens. (**b**) Bright-field image of the L^2 lens. The *green* laser beam from the fiber is visible in front of the aperture because PDMS contains nanoparticles of silica that scatter light. The focused beam in the beam-tracing chamber is *orange* due to the fluorescence of the dye. (Reprinted from [142] with permission of The Royal Society of Chemistry)

the central (core) stream was higher than that of the sandwiching (cladding) streams. When the streams entered an expansion chamber, the lens was formed. The curvature of the interface between the liquid core and the liquid cladding can be modulated by adjusting the relative flow rates, resulting in variable focal distance. As a result of the change of the expansion chamber to a circular chamber, interfaces with a perfect arc shape and a short focal length can be obtained [143, 144]. However, diffusion of solute across the fluid–fluid interface deteriorates the smoothness of the interface. Therefore, Huang's group developed in-plane [126] and out-of-plane [145] lenses on the microchip that were dependent on the liquid gradient refractive index (L-GRIN), and focus light due to the gradient of the refractive index within the liquid medium, rather than due to curved refractive interfaces (Fig. 9).

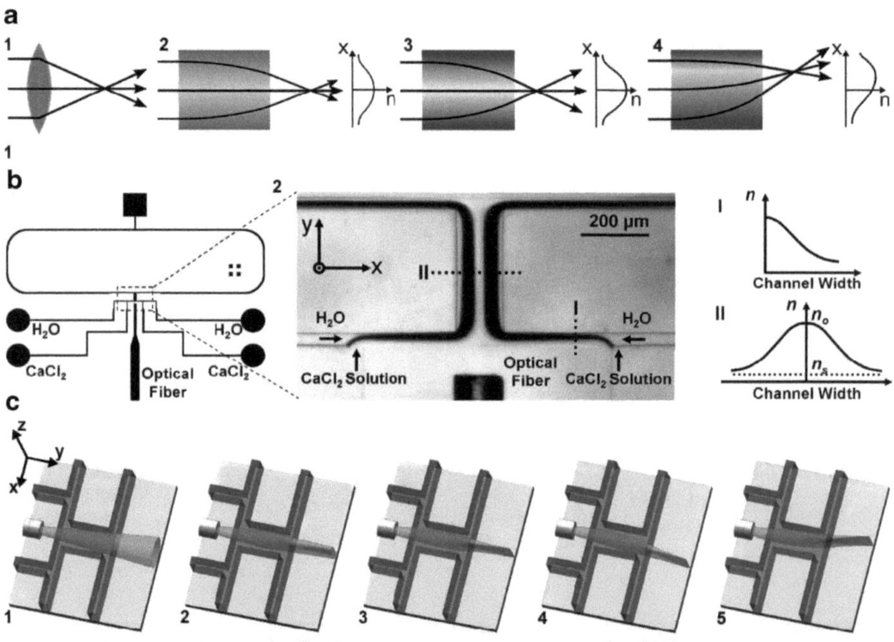

Fig. 9 Principle and design of the L-GRIN lens. (**a**) Comparison between the classic refractive lens (*a1*) and GRIN lens (*a2*). Change of the refractive index contrast in GRIN lens can result in change of focal distance (*a2, a3*), and shift of optical axis can result in change of output light direction (*a4*). (**b**) Schematic of the L-GRIN lens design (*b1*), microscopic image of the L-GRIN lens in operation (*b2, left*), refractive index distribution at two locations (*b2, right*) (**c**) Drawings showing two operation modes of the L-GRIN lens: the translation mode with variable focal length including no-focusing (*c1*), a large focal distance (*c2*), and a small focal distance (*c3*); and the swing mode with variable output light direction (*c3–c5*) (Reprinted from [126] with permission of The Royal Society of Chemistry)

3.3 Optofluidic Laser

To achieve the ultimate goal of integration of many optical components onto a compact chip, laser integration on-chip is a key challenge, although conventional bulky lasers have been widely applied in microfluidic systems through free-space coupling [7, 8]. The miniaturized laser had been independently developed before the microfluidic chip was put forth. However, only when the microfluidic device and microphotonic system are synergetically combined, can the optofluidic chip be successful. Therefore, we will not attempt to discuss light sources that are miniaturized and assembled on the microchip but have nothing to do with fluids, although they have made contributions to the development of laser miniaturization. For example, the laser diode, photodetector, and waveguide have all been integrated on a silicon chip [146], as well as an organic semiconductor film laser assembly on a PMMA substrate [147]. The dye laser is currently the optofluidic laser most usually integrated on microfluidic chips [148]. The advantages of optofluidic dye lasers are that they can be fabricated on the same materials as the microfluidic system and readily integrated with other fluidic functionalities, and that the characteristics of dye lasers are amenable to tuning by changing the dye solution. Since the first optofluidic dye laser was reported in 2003 [149], rapid and significant progress has been obtained in terms of various configurations and materials. Lasers consists of three principal elements: gain medium, pump source, and resonator cavity. Since the pump sources for the present optofluidic dye laser are all external light sources such as lasers, and the gain mediums are various dyes in different solvents, both of which are less related to chip structure, we will summarize the development of optofluidic lasers with respect to the resonator, which is an optical feedback volume that repeatedly transmits a resonant light through the gain medium to create a very intense, coherent laser. This optical feedback system is conventionally called the optical resonator or optical cavity.

The simplest optical cavity comprises only two facing plane mirrors, denoted as a Fabry–Perot cavity. Helbo et al. [149] first deposited a pair of opposing metal mirrors on the top and bottom of a microchannel acting as a resonator in which rhodamine 6G solution flew through as the gain medium (Fig. 10). Song and Psaltis have recently developed a pneumatically tunable optofluidic dye laser [150], in which a LCW filled with dye solution is the resonator at the ends of which two air chambers act as the mirrors. The wavelength is chosen by inflating the air-gap etalon with compressed air.

A ring resonator is the second simple configuration for a cavity consisting of three or more mirrors to loop the light in the resonator. Galas et al. fabricated a ring resonator using four mirrors (air chamber) to generate a tunable dye laser [151]. Suter et al. [152] presented side-coupled ring geometry to generate an optofluidic ring resonator laser, which can be produced in PDMS substrates with contact molding (Fig. 11). A kind of variation of the ring cavity structures is the droplet-based whispering gallery mode. A droplet can be treated as an optical microcavity when the index of refraction of enclosed dye solution is higher than that of

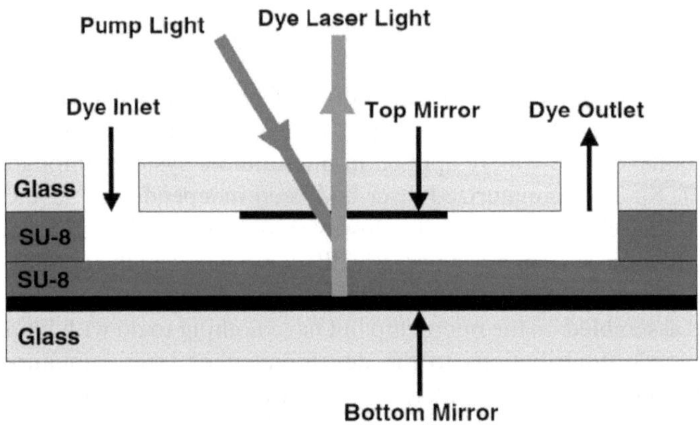

Fig. 10 Cross-sectional outline of the microcavity fluidic dye laser. The laser dye is pumped through a microfluidic channel. The laser cavity is formed inside the microfluidic channel by metal mirrors, deposited on the top and bottom glass wafers. The microcavity fluidic dye laser is pumped optically by an external laser, and output is emitted through the semitransparent top mirror (Reprinted from [149] with permission of IOP Publishing)

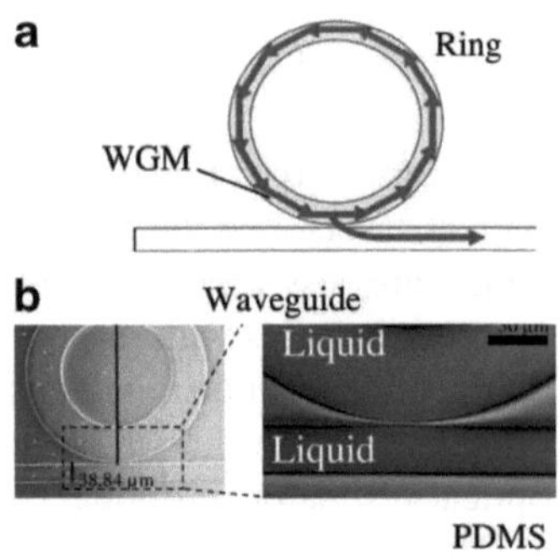

Fig. 11 (**a**) Cross-sectional outline of the microcavity fluidic dye laser; *WGM* whispering gallery mode. (**b**) Photograph of a glass/SU-8 chip with a microcavity fluidic dye laser (Reprinted from [152] with permission of The Optical Society of America)

the carrier liquid. Light emitting from the dyes is confined and amplified in the drops by total internal reflection. This method could provide a multicolor laser in a train of alternating dye drops and is capable of switching the wavelength at high frequency [153].

Besides the mirror-based optical cavity, there is another type of optical feedback structure involving diffraction grating, called distributed feedback (DFB). The group of Kristensen demonstrated a 130th order Bragg grating-based optofluidic dye laser [154] to achieve a single mode laser. However, lower-order DFB gratings are expected to minimize the intrinsic scattering losses and decrease the threshold of pump energies. Therefore, the15th order [155] and third order [156] Bragg gratings were successively fabricated on a microfluidic chip. Song et al. [157] have systematically studied third, second, and first order DFB lasers, where the second order lasers exhibited the lowest pump threshold. Vannahme et al. [158] embedded a first order DFB optofluidic dye laser in a foil for mass production and cost reduction.

At least two issues should be improved to fulfill the promise of optofluidic dye lasers as on-chip coherent light sources. Firstly, new strategies should be developed to widen the laser wavelength range and to reduce the tuning time. Secondly, the optical pump source should be integrated onto the same laser chip, which should demonstrate the lowest laser threshold.

3.4 Optical Detector

Optical detectors have been integrated on microfluidic chips, such as silicon photodiode, organic photodiodes (OPDs), and complementary metal–oxide–semiconductor (CMOS) chips. The former two are point detectors without spatial resolution, whereas CMOS detectors can capture an image. In the early stage, integrated detectors on microfluidic chips were mainly silicon detectors due to the mature fabrication technique of semiconductors, e.g., a microavalanche photodiode on PDMS [159], a photodiode built on a silicon substrate [160], and a hydrogenated amorphous silicon photodiode assembly on glass [161]. Generally speaking, the fabrication procedure for a detector on-chip becomes less and less complex, and the cost of detector trends to become lower and lower. For example, if the substrates are changed from silicon to glass and then to polymer, the fabrication temperature is decreased from a high temperature to ambient temperature. Recently, Pereira et al. presented an example by integrating a PIN photodiode on a microfluidic chip to perform chemiluminescent detection, and the detection results were comparable with laboratory analysis for detection of HRP [162].

Silicon photodiodes could render relatively high sensitivity for detection of low analyte concentrations but are too expensive and complicated to fabricate as an integral part of a disposable microfluidic chip. In contrast to the silicon counterpart, OPDs are a promising potential for future point-of-care technology because they are inexpensive and easily fabricated, and are attracting more and more attention from researchers in the field of lab-on-chip.

Hofmann et al. [163] showed the feasibility of integration of thin-film OPDs on microchips for microscale chemiluminescence. The detector was fabricated by a layer-by-layer vacuum deposition method. The quantitation of hydrogen peroxide yielded a detection limit of 1 mM. They further implemented the integration of OPD on PDMS instead of silicon as the substrate material [164] and applied it to analyze antioxidant capacity [165]. To reduce the background and increase the detection limit of the OPD detector on chips, Pais [166] placed a second polarizer layer before fluorescence entered into the OPD detector by preventing the leaking of excitation light into the detector. The detectable concentrations of rhodamine 6 G and fluorescein dyes were as low as 100 nM and 10 μM, respectively. Wojciechowski et al. [167] developed a chemiluminescence immunoassay-based miniaturized biosensor, which was composed of a disposable microscope slide and a hand-held reader. The OPD substrate was printed on the glass slide. This OPD system is approaching the final requirements of point-of-care, i.e., inexpensive, easy to use, highly sensitive, and robust.

Besides the various photodiodes mentioned above integrated on microfluidic chips, the CMOS is another type of detector that can be assembled on microchips and offers the great advantage of the ability to observe the image of sample object [63]. One of the most important applications of an integrated CMOS detector on-chip is the lensless optofluidic microscopy (OFM). In 2005, Lange et al. [168] attached a CMOS chip at the bottom of a microfluidic chip to record the projection image of *Caenorhabditis elegans*. But, the quality of this imaging method needs to be further improved by choosing appropriate apertures. Yang et al. [169, 170] were inspired to develop a high-resolution optofluidic microscope by covering a nanohole array layer on the imaging chip. They measured a resolution limit of 490 ± 40 nm, comparable resolution to a conventional microscope. The principle of lensless microscopy is shown in Fig. 12. The authors extended this method to color OFM by integrating a RGB CMOS color sensor chip substrate [171]. Laurberg et al. then exploited OFM to 3D spatial resolution and on-chip particle image velocimetry [172].

Yang's group further presented a non-nanohole method for reconstruction of a high-resolution image from a low-resolution image processed with a pixel super-resolution algorithm [173]. This method further compacts the OFM by eliminating the nanoaperture array, and follows the consistent requirement of

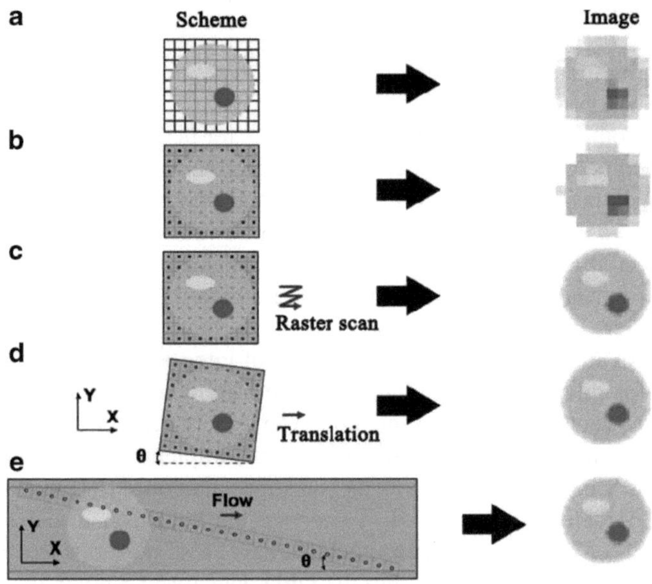

Fig. 12 Principles of nanohole microscopy. (**a**) Direct projection imaging scheme. (**b**) By placing the specimen on a grid of apertures, we can obtain a sparse image. (**c**) Raster-scanning the specimen over the aperture grid. In this case, the image resolution is limited by the aperture size. (**d**) The scanning scheme can be simplified into a single-pass flow of the specimen across the grid. (**e**) The aperture grid can be simplified by substitution with a long linear aperture array (Reprinted from [170] with permission of The National Academy of Science, USA)

microfluidic chip in miniaturization as small as possible. The rapid development of OFM is following the advances in optofluidics but also contributes to them.

4 Conclusions

Optical detection still maintains its predominant application in microfluidics due to its characteristics of being noninvasive, easily coupled, quick to respond, and highly sensitive. As discussed, although conventional free-space detection strategy is less amenable to being miniaturized and compacted in a portable microdevice, it is highly familiar to analytical chemists, and is still helpful when a proof-of-principle needs to be quickly demonstrated on microfluidic chips and extreme sensitivity is a particular concern. As the most promising alternative to free-space detection, optofluidics is becoming an active field in lab-on-a-chip, merging fluidics with optics in the same microchip, in which optics are generated by fluids and fluids are controlled by optics. Current optofluidics is developing and optimizing single optics, but has potential to minimize and integrate all components onto the same microscale device, which includes analytical units and optical pieces. In optofluidics, various optical components such as light source, filters, microlens, detectors, and waveguides can be made of fluids; therefore, the space in the chip is greatly saved and the complexity of fabrication is simplified. The microchannel both handles the analytes and transmits light to probe the sample and collect analytical results. Future optofluidic chips undoubtedly can integrate more components and be mass-produced to meet the requirements of point-of-care diagnostics.

Acknowledgments The authors would like to thank Xingbo Shi, Xinliang Liu, Lichun Sun, and Haixia Ding for helpful discussions. HWG gratefully acknowledges the support of National Natural Science Foundation of China (20705007, 21075033, 20703016).

References

1. Manz A, Fettinger JC, Verpoorte E, Ludi H, Widmer HM, Harrison DJ (1991) TrAC Trends Anal Chem 10:144–149
2. Einav S, Gerber D, Bryson PD, Sklan EH, Elazar M, Maerkl SJ, Glenn JS, Quake SR (2008) Nat Biotechnol 26:1019–1027
3. Walsh CL, Babin BM, Kasinskas RW, Foster JA, McGarry MJ, Forbes NS (2009) Lab Chip 9:545–554
4. Gerber D, Maerkl SJ, Quake SR (2009) Nat Methods 6:71–74
5. Dossi N, Susmel S, Toniolo R, Pizzariello A, Bontempelli G (2009) Electrophoresis 30:3465–3471
6. Llopis X, Pumera M, Alegret S, Merkoci A (2009) Lab Chip 9:213–218
7. Yi CQ, Zhang Q, Li CW, Yang J, Zhao JL, Yang MS (2006) Anal Bioanal Chem 384: 1259–1268
8. Viskari PJ, Landers JP (2006) Electrophoresis 27:1797–1810
9. Gotz S, Karst U (2007) Anal Bioanal Chem 387:183–192
10. Myers FB, Lee LP (2008) Lab Chip 8:2015–2031

11. Pennathur S, Fygenson DK (2008) Lab Chip 8:649–652
12. Schwarz MA, Hauser PCS (2001) Lab Chip 1:1–6
13. Lu Q, Copper CL, Collins GE (2006) Anal Chim Acta 572:205–211
14. Collins GE, Lu Q, Pereira N, Wu P (2007) Talanta 72:301–304
15. Salimi-Moosavi H, Jiang Y, Lester L, McKinnon G, Harrison DJ (2000) Electrophoresis 21:1291–1299
16. Billot L, Plecis A, Chen Y (2008) Microelectron Eng 85:1269–1271
17. Noda T, Takao H, Yoshioka K, Oku N, Ashiki M, Sawada K, Matsumoto K, Ishida M (2006) Sens Actuators B 119:245–250
18. Ma B, Zhou X, Wang G, Dai Z, Qin J, Lin B (2007) Electrophoresis 28:2474–2477
19. Ou J, Glawdel T, Ren CL, Pawliszyn J (2009) Lab Chip 9:1926–1932
20. Petersen NJ, Mogensen KB, Kutter JP (2002) Electrophoresis 23:3528–3536
21. Gustafsson O, Mogensen KB, Ohlsson PD, Liu Y, Jacobson SC, Kutter JP (2008) J Micromech Microeng 18:055021
22. Llobera A, Wilke R, Büttgenbach S (2004) Lab Chip 4:24–27
23. Snakenborg D, Mogensen KB, Kutter JP (2003) Proc Micro Total Anal Syst:841–844
24. Mogensen KB, Kutter JP (2009) Electrophoresis 30:S92–S100
25. Ro KW, Lim K, Shim BC, Hahn JH (2005) Anal Chem 77:5160–5166
26. Llobera A, Demming S, Wilke R, Büttgenbach S (2007) Lab Chip 7:1560–1566
27. Pan JZ, Yao B, Fang Q (2010) Anal Chem 82:3394–3398
28. Prabhakar A, Mukherji S (2010) Lab Chip 10:748–754
29. van der Sneppen L, Ariese F, Gooijer C, Ubachs W (2009) Annu Rev Anal Chem 2:13–35
30. Waechter H, Litman J, Cheung AH, Barnes JA, Loock H-P (2010) Sensors 10:1716–1742
31. Waechter H, Bescherer K, Durr CJ, Oleschuk RD, Loock H-P (2009) Anal Chem 81:9048–9054
32. Yao B, Luo G, Wang LD, Gao YD, Lei GT, Ren KN, Chen LX, Wang YM, Hu Y, Qiu Y (2005) Lab Chip 5:1041–1047
33. Yao B, Yang HH, Liang QL, Luo G, Wang LD, Ren KN, Gao YD, Wang YM, Qiu Y (2006) Anal Chem 78:5845–5850
34. Ren KN, Liang QL, Mu X, Luo GA, Wang YM (2009) Lab Chip 9:733–736
35. Fu JL, Fang Q, Zhang T, Jin XH, Fang ZL (2006) Anal Chem 78:3827–3834
36. Whiting CE, Dua RA, Duffy CF, Arriaga EA (2008) Electrophoresis 29:1431–1440
37. Harrison DJ, Manz A, Fan ZH, Ludi H, Widmer HM (1992) Anal Chem 64:1926–1932
38. Harrison DJ, Fluri K, Seiler K, Fan ZH, Effenhauser CS, Manz A (1993) Science 261:895–897
39. Ocvirk G, Tang T, Harrison DJ (1998) Analyst 123:1429–1434
40. Haab BB, Mathies RA (1999) Anal Chem 71:5137–5145
41. Liu SR, Shi YN, Ja WW, Mathies RA (1999) Anal Chem 71:566–573
42. Dittrich PS, Manz A (2005) Anal Bioanal Chem 382:1771–1782
43. Schiro PG, Kuyper CL, Chiu DT (2007) Electrophoresis 28:2430–2438
44. Liang H, Cheng XL, Ma YF (2009) Anal Chem 81:2059–2066
45. Sun W, Marchuk K, Wang G, Fang N (2010) Anal Chem 82:2441–2447
46. Gai H, Li Y, Silber-Li Z-H, Ma Y, Lin B (2005) Lab Chip 5:443–449
47. Le NCH, Yokokawa R, Dao DV, Nguyen TD, Wells JC, Sugiyama S (2009) Lab Chip 9:244–250
48. Shi X, Meng X, Sun L, Liu J, Zheng J, Gai H, Yang R, Yeung ES (2010) Lab Chip 10:2844–2847
49. Bu X, Chen H, Gai H, Yang R, Yeung ES (2009) Anal Chem 81:7507–7509
50. Chen H, Gai H, Yeung ES (2009) Chem Commun:1676–1678
51. Yu LF, Huang HQ, Dong XL (2008) Electrophoresis 29:5055–5060
52. Klepárník K, Horký M (2003) Electrophoresis 24:3778–3783
53. Hellmich W, Pelargus C, Leffhalm K, Ros A, Anselmetti D (2005) Electrophoresis 26:3689–3696

54. Hellmich W, Greif D, Pelargus C, Anselmetti D, Ros A (2006) J Chromatogr A 1130:195–200
55. Renzi RF, Stamps J, Horn BA, Ferko S, VanderNoot VA, West JAA, Crocker R, Wiedenman B, Yee D, Fruetel JA (2005) Anal Chem 77:435–441
56. Novak L, Neuzil P, Pipper J, Zhang Y, Lee SH (2007) Lab Chip 7:27–29
57. Kaigala GV, Bercovici M, Behnam M, Elliott D, Santiago JG, Backhouse CJ (2010) Lab Chip 10:2242–2250
58. Kim JH, Kim YH, Shin KS, Kim BK, Kim YM, Lee YH, Moon SI, Kim TS, Kang JY, Yang EG, Kim SS, Ju BK, Park JO (2005) A miniaturized fluorescence detection system with an integrated organic light emitting diode. In: Proceedings 18th IEEE international conference on micro electro mechanical systems, 30 Jan–3 Feb, Miami, FL, pp 806–809
59. Miyaki K, Guo YL, Shimosaka T, Nakagama T, Nakajima H, Uchiyama K (2005) Anal Bioanal Chem 382:810–816
60. Xu J, Xiong Y, Chen SH, Guan YF (2009) Prog Chem 21:1325–1334
61. Jang JM, Shin HJ, Hwang SW, Yang EG, Yoon DS, Kim TS, Kang JY (2005) Sens Actuators B 108:993–1000
62. Roulet JC, Volkel R, Herzig HP, Verpoorte E, de Rooij NF, Dandliker R (2001) Opt Eng 40:814–821
63. Adams M, Enzelberger M, Quake S, Scherer A (2003) Sens Actuators A 104:25–31
64. Hashimoto M, Tsukagoshi K, Nakajima R, Kondo K, Arai A (2000) J Chromatogr A 867:271–279
65. Su RG, Lin JM, Qu F, Chen ZF, Gao YH, Yamada M (2004) Anal Chim Acta 508:11–15
66. Mangru SD, Harrison DJ (1998) Electrophoresis 19:2301–2307
67. Lara FJ, Garcia-Campana AM, Velasco AI (2010) Electrophoresis 31:1998–2027
68. Hashimoto M, Tsukagoshi K, Nakajima R, Kondo K, Arai A (1999) Chem Lett:781–782
69. Garcia-Campana AM, Lara FJ, Gamiz-Gracia L, Huertas-Perez JF (2009) TrAC Trends Anal Chem 28:973–986
70. Huang XY, Ren JC (2006) TrAC Trends Anal Chem 25:155–166
71. Liu BF, Ozaki M, Utsumi Y, Hattori T, Terabe S (2003) Anal Chem 75:36–41
72. Tsukagoshi K, Jinno N, Nakajima R (2005) Anal Chem 77:1684–1688
73. Zhao SL, Li XT, Liu YM (2009) Anal Chem 81:3873–3878
74. Zhao SL, Huang Y, Liu YM (2009) J Chromatogr A 1216:6746–6751
75. Ye FG, Huang Y, Xu Q, Shi M, Zhao SL (2010) Electrophoresis 31:1630–1636
76. Zhao SL, Huang Y, Ye FG, Shi M, Liu Y (2010) J Chromatogr A 1217:5732–5736
77. Zhao SL, Huang Y, Shi M, Liu RJ, Liu YM (2010) Anal Chem 82:2036–2041
78. Wang ZAK, Chin SY, Chin CD, Sarik J, Harper M, Justman J, Sia SK (2010) Anal Chem 82:36–40
79. Homola J (2008) Chem Rev 108:462–493
80. Furuki M, Kameoka J, Craighead HG, Isaacson MS (2001) Sens Actuators B 79:63–69
81. Wheeler AR, Chah S, Whelan RJ, Zare RN (2004) Sens Actuators B 98:208–214
82. Liu C, Cui D, Li H (2010) Biosens Bioelectron 26:255–261
83. Kurihara K, Ohkawa H, Iwasaki Y, Niwa O, Tobita T, Suzuki K (2004) Anal Chim Acta 523:165–170
84. Stöcker P, Menges B, Langbein U, Mittler S (2004) Sens Actuators A 116:224–231
85. Levy R, Ruschin S (2007) Sens Actuators B 124:459–465
86. Zhang N, Liu H, Knoll W (2009) Biosens Bioelectron 24:1783–1787
87. Unfricht DW, Colpitts SL, Fernandez SM, Lynes MA (2005) Proteomics 5:4432–4442
88. Huang C, Bonroy K, Reekmans G, Laureyn W, Verhaegen K, De Vlaminck I, Lagae L, Borghs G (2009) Biomed Microdevices 11:893–901
89. Cao C, Sim SJ (2009) Lab Chip 9:1836–1839
90. Chen LX, Choo J (2008) Electrophoresis 29:1815–1828
91. Zhang X, Yin H, Cooper JM, Haswell SJ (2008) Anal Bioanal Chem 390:833–840
92. Choi D, Kang T, Cho H, Choi Y, Lee LP (2009) Lab Chip 9:239–243
93. Wang M, Jing N, Chou I-H, Coté GL, Kameoka J (2007) Lab Chip 7:630–632

94. Chou I-H, Benford M, Beier HT, Coté GL, Wang M, Jing N, Kameoka J, Good TA (2008) Nano Lett 8:1729–1735
95. Huh YS, Chung AJ, Cordovez B, Erickson D (2009) Lab Chip 9:433–439
96. Cho H, Lee B, Liu GL, Agarwal A, Lee LP (2009) Lab Chip 9:3360–3363
97. Tong L, Righini M, González MU, Quidant R, Käll M (2009) Lab Chip 9:193–195
98. Quang LX, Lim C, Seong GH, Choo J, Do KJ, Yoo S-K (2008) Lab Chip 8:2214–2219
99. Hunt HC, Wilkinson JS (2008) Microfluid Nanofluid 4:53–79
100. Hawkins AR, Schmidt H (2008) Microfluid Nanofluid 4:17–32
101. Nguyen N-T (2010) Biomicrofluidics 4:031501
102. Psaltis D, Quake SR, Yang C (2006) Nature 442:381–386
103. Wang SL, Fan XF, Xu ZR, Fang ZL (2005) Electrophoresis 26:3602–3608
104. Du WB, Qun F, Fang ZL (2004) Chem J Univ Chin 25:610–613
105. Ymeti A, Greve J, Lambeck PV, Wink T, van Hovell S, Beumer TAM, Wijn RR, Heideman RG, Subramaniam V, Kanger JS (2007) Nano Lett 7:394–397
106. Crespi A, Gu Y, Ngamson B, Hoekstra HJWM, Dongre C, Pollnau M, Ramponi R, van den Vlekkert HH, Watts P, Cerullo G, Osellame R (2010) Lab Chip 10:1167–1173
107. Mogensen KB, Kwok YC, Eijkel JCT, Petersen NJ, Manz A, Kutter JP (2003) Anal Chem 75:4931–4936
108. Cleary A, Glidle A, Laybourn PJR, García-Blanco S, Pellegrini S, Helfter C, Buller GS, Aitchison JS, Cooper JM (2007) Appl Phys Lett 91:071123
109. Sheridan AK, Stewart G, Ur-reyman H, Suyal N, Uttamchandani D (2009) IEEE Sens J 9(12):1627–1632
110. Kee JS, Poenar DP, Neuzil P, Yobas L (2008) Sens Actuators B 134:532–538
111. Xu F, Datta P, Wang H, Gurung S, Hashimoto M, Wei S, Goettert J, McCarley RL, Soper SA (2007) Anal Chem 79:9007–9013
112. Okagbare PI, Emory JM, Datta P, Goettert J, Soper SA (2010) Lab Chip 10:66–73
113. Datta A, Eom I-Y, Dhar A, Kuban P, Manor R, Ahmad I, Gangopadhyay S, Dallas T, Holtz M, Temkin H, Dasgupta PK (2003) IEEE Sens J 3:788–795
114. Manor R, Datta A, Ahmad I, Holtz M, Gangopadhyay S, Dallas T (2003) IEEE Sens J 3:687–692
115. Wu CW, Gong GC (2008) IEEE Sens J 8:465–469
116. Cho SH, Godin J, Lo Y-H (2009) IEEE Photonics Technol Lett 21:1057–1059
117. Cho SH, Chen CH, Tsai FS, Godin JM, Lo Y-H (2010) Lab Chip 10:1567–1573
118. Du WB, Fang Q, He QH, Fang ZL (2005) Anal Chem 77:1330–1337
119. Risk W, Kim H, Miller R, Temkin H, Gangopadhyay S (2004) Opt Express 12:6446–6455
120. Korampally V, Mukherjee S, Hossain M, Manor R, Yun MS, Gangopadhyay K, Polo-Parada L, Gangopadhyay S (2009) IEEE Sens J 9:1711–1718
121. Ligler FS (2009) Anal Chem 81:519–526
122. Wolfe DB, Conroy RS, Garstecki P, Mayers BT, Fischbach MA, Paul KE, Prentiss M, Whitesides GM (2004) Proc Natl Acad Sci USA 101:12434–12438
123. Rosenauer M, Vellekoop MJ (2009) Appl Phys Lett 95:163702
124. Lee KS, Kim SB, Lee KH, Sung HJ, Kim SS (2010) Appl Phys Lett 97:021109
125. Mao X, Waldeisen JR, Huang TJ (2007) Lab Chip 7:1260–1262
126. Mao X, Lin S-CS, Lapsley MI, Shi J, Juluri BK, Huang T (2009) Lab Chip 9:2050–2058
127. Nguyen NT, Kong TF, Goh JH, Low CLN (2007) J Micromech Microeng 17:2169–2174
128. Song C, Nguyen N-T, Asundi AK, Tan S-H (2010) Opt Express 35:327–329
129. Maselli V, Grenier JR, Ho S, Herman PR (2009) Opt Express 17:11719–11729
130. Yin DL, Deamer DW, Schmidt H, Barber JP, Hawkins AR (2006) Opt Lett 31:2136–2138
131. Yin DL, Lunt EJ, Rudenko MI, Deamer DW, Hawkins AR, Schmidt H (2007) Lab Chip 7:1171–1175
132. Schmidt H, Hawkins AR (2008) Microfluid Nanofluid 4:3–16
133. Kuo J-N, Hsieh C-C, Yang S-Y, Lee G-B (2007) J Micromech Microeng 17:693–699
134. Camou S, Fujita H, Fujii T (2003) Lab Chip 3:40–45

135. Guo HL, Zhao P, Xiao GZ, Zhang ZY, Yao JP (2010) IEEE J Sel Top Quantum Electron 16:919–926
136. Hsiung SK, Lee CH, Lee GB (2008) Electrophoresis 29:1866–1873
137. Lee JY, Choi ST, Lee SW, Kim WB (2009) Proc SPIE 7426:742603
138. Lee SW, Lee SS (2007) Appl Phys Lett 90:121129
139. Shi JJ, Stratton Z, Lin SCS, Huang H, Huang TJ (2010) Microfluid Nanofluid 9:313–318
140. Mao X, Waldeisen JR, Juluri BK, Huang TJ (2007) Lab Chip 7:1303–1308
141. Rosenauer M, Vellekoop MJ (2009) Lab Chip 9:1040–1042
142. Tang SKY, Stan CA, Whitesides GM (2008) Lab Chip 8:395–401
143. Song C, Nguyen N-T, Tan S-H, Asundi AK (2009) Lab Chip 9:1178–1184
144. Song CL, Nguyen NT, Tan SH, Asundi AK (2009) J Micromech Microeng 19:085012
145. Huang H, Mao X, Lin SCS, Kiraly B, Huang YP, Huang TJ (2010) Lab Chip 10:2387–2393
146. Ohira K, Kobayashi K, Iizuka N, Yoshida H, Ezaki M, Uemura H, Kojima A, Nakamura K, Furuyama H, Shibata H (2010) Opt Express 18:15440–15447
147. Vannahme C, Klinkhammer S, Kolew A, Jakobs P-J, Guttmann M, Dehm S, Lemmer U, Mappes T (2010) Microelectron Eng 87:693–695
148. Li Z, Psaltis D (2008) Microfluid Nanofluid 4:145–158
149. Helbo B, Kristensen A, Menon A (2003) J Micromech Microeng 13:307–311
150. Song W, Psaltis D (2010) Appl Phys Lett 96:081101
151. Galas JC, Torres J, Belotti M, Kou Q, Chen Y (2005) Appl Phys Lett 86:264101
152. Suter JD, Lee W, Howard DJ, Hoppmann E, White IM, Fan XD (2010) Opt Lett 35:2997–2999
153. Tang SKY, Li Z, Abate AR, Agresti JJ, Weitz DA, Psaltis D, Whitesides GM (2009) Lab Chip 9:2767–2771
154. Balslev S, Kristensen A (2005) Opt Express 13:344–351
155. Li ZY, Zhang ZY, Emery T, Scherer A, Psaltis D (2006) Opt Express 14:696–701
156. Gersborg-Hansen M, Kristensen A (2006) Appl Phys Lett 89:103518
157. Song W, Vasdekis AE, Li Z, Psaltis D (2009) Appl Phys Lett 94:051117
158. Vannahme C, Christiansen MB, Mappes T, Kristensen A (2010) Opt Express 18: 9280–9285
159. Chabinyc ML, Chiu DT, McDonald JC, Stroock AD, Christian JF, Karger AM, Whitesides GM (2001) Anal Chem 73:4491–4498
160. Webster JR, Burns MA, Burke DT, Mastrangelo CH (2001) Anal Chem 73:1622–1626
161. Kamei T, Paegel BM, Scherer JR, Skelley AM, Street RA, Mathies RA (2003) Anal Chem 75:5300–5305
162. Pereira AT, Pimentel AC, Chu V, Prazeres DMF, Conde JP (2009) IEEE Sens J 9:1282–1290
163. Hofmann O, Miller P, Sullivan P, Jones TS, de Mello JC, Bradley DDC, de Mello AJ (2005) Sens Actuators B 106:878–884
164. Wang X, Hofmann O, Das R, Barrett EM, Demello AJ, Demello JC, Bradley DDC (2007) Lab Chip 7:58–63
165. Wang X, Amatatongchai M, Nacapricha D, Hofmann O, de Mello JC, Bradley DDC, de Mello AJ (2009) Sens Actuators B 140:643–648
166. Pais A, Banerjee A, Klotzkin D, Papautsky I (2008) Lab Chip 8:794–800
167. Wojciechowski JR, Shriver-Lake LC, Yamaguchi MY, Fureder E, Pieler R, Schamesberger M, Winder C, Prall HJ, Sonnleitner M, Ligler FS (2009) Anal Chem 81:3455–3461
168. Lange D, Storment CW, Conley CA, Kovacs GTA (2005) Sens Actuators B 107:904–914
169. Heng X, Erickson D, Baugh LR, Yaqoob Z, Sternberg PW, Psaltis D, Yang C (2006) Lab Chip 6:1274–1276
170. Cui XQ, Lee LM, Heng X, Zhong WW, Sternberg PW, Psaltis D, Yang CH (2008) Proc Natl Acad Sci USA 105:10670–10675
171. Pang S, Cui X, DeModena J, Wang YM, Sternberg PW, Yang CH (2010) Lab Chip 10:411–414
172. Vig AL, Marie R, Jensen E, Kristensen A (2010) Opt Express 18:4158–4169
173. Zheng G, Lee SA, Yang S, Yang C (2010) Lab Chip 10:3125–3129

Top Curr Chem (2011) 304: 203–260
DOI: 10.1007/128_2011_153

Published online: 24 May 2011

Integrated Microfluidic Systems for DNA Analysis

Samuel K. Njoroge, Hui-Wen Chen, Małgorzata A. Witek, and Steven A. Soper

Abstract The potential utility of genome-related research in terms of evolving basic discoveries in biology has generated widespread use of DNA diagnostics and DNA forensics and driven the accelerated development of fully integrated microfluidic systems for genome processing. To produce a microsystem with favorable performance characteristics for genetic-based analyses, several key operational elements must be strategically chosen, including device substrate material, temperature control, fluidic control, and reaction product readout. As a matter of definition, a microdevice is a chip that performs a single processing step, for example microchip electrophoresis. Several microdevices can be integrated to a single wafer, or combined on a control board as separate devices to form a microsystem. A microsystem is defined as a chip composed of at least two microdevices. Among the many documented analytical microdevices, those focused on the ability to perform the polymerase chain reaction (PCR) have been reported extensively due to the importance of this processing step in most genetic-based assays. Other microdevices that have been detailed in the literature include those for solid-phase extractions, microchip electrophoresis, and devices composed of DNA microarrays used for interrogating DNA primary structure. Great progress has also been made in the areas of chip fabrication, bonding and sealing to enclose fluidic networks, evaluation of different chip substrate materials, surface chemistries, and the architecture of reaction conduits for basic processing steps such as mixing. Other important elements that have been developed to realize functional systems include miniaturized readout formats comprising optical or electrochemical transduction and interconnect technologies. These discoveries have led to the development of fully autonomous and functional integrated systems for genome processing

S.K. Njoroge, H.-W. Chen, and M.A. Witek
Department of Chemistry, Louisiana State University, Baton Rouge, LA 70803, USA

S.A. Soper (✉)
Department of Chemistry, Louisiana State University, Baton Rouge, LA 70803, USA
and
Department of Mechanical Engineering, Louisiana State University, Baton Rouge, LA 70803, USA
e-mail: chsope@lsu.edu

that can supply "sample in/answer out" capabilities. In this chapter, we focus on microfluidic systems that are composed of two or more microdevices directed toward DNA analyses. Our discussions will primarily be focused on the integration of various processing steps with microcapillary electrophoresis (μCE) or microarrays. The advantages afforded by fully integrated microfluidic systems to enable challenging applications, such as single-copy DNA sequencing, single-cell gene expression analysis, pathogen detection, and forensic DNA analysis in formats that provide high throughput and point-of-analysis capabilities will be discussed as well.

Keywords Continuous flow PCR · DNA analysis · DNA microarrays · Genetic analysis · Integrated microsystems · Microcapillary electrophoresis · Microfluidics · Micro-PCR devices · Solid-phase extraction

Contents

1 Introduction 205
 1.1 The Human Genome Project 205
 1.2 Molecular Processing Pipeline for DNA Analyses 206
2 Microfluidics and DNA Analysis 207
 2.1 Cell Lysis 207
 2.2 Nucleic Acid Extraction, Purification, and Preconcentration 209
 2.3 Microfluidic Polymerase Chain Reactors 212
 2.4 Microfluidic Thermal Heating Methods 218
3 Analysis Methods of Reaction Products 222
 3.1 Microcapillary Electrophoresis 222
 3.2 DNA Microarrays 229
4 Integrated Microfluidic Systems 234
 4.1 Integrated Systems with μCE Readout 235
 4.2 Integrated Systems with Microarray Readout 247
5 Concluding Remarks 252
References 253

Abbreviations

AOM	Aluminum oxide membranes
BPU	Biochemical processing unit
CAE	Capillary array electrophoresis
CFPCR	Continuous flow polymerase chain reaction
COC	Cyclic olefin copolymer
ddNTPs	Dideoxy nucleoside triphosphates
DNA	Deoxyribonucleic acid
dNTPS	Deoxyribonucleotide triphosphate
DRIE	Deep reactive ion etching
EDTA	Ethylenediaminetetraacetic acid
ELFSE	End-labeled free-solution electrophoresis

EOF	Electroosmotic flow
FPC	Flexible printed circuits
FSCE	Free solution conjugate electrophoresis
HEC	Hydroxyethyl cellulose
HGP	Human genome project
ITO	Indium–tin oxide
LDR	Ligase detection reaction
LPA	Linear polyacrylamides
MDR-TB	Multidrug resistant tuberculosis
MHEC	Methyl hydroxyethyl cellulose
PC	Polycarbonate
PCR	Polymerase chain reaction
PDMS	Poly(dimethylsiloxane)
PEO	Poly(ethylene oxides)
PMMA	Poly(methyl methacrylate)
PPC	Photoactivated polycarbonate
RNA	Ribonucleic acid
RTD	Resistance temperature detector
RT-PCR	Reverse transcription polymerase chain reaction
SGE	Slab gel electrophoresis
SPE	Solid-phase extraction
STRs	Short tandem repeats
TE	Thermoelectric
TTE	*tris*-Taps-EDTA
μCE	Microcapillary electrophoresis
μTAS	Micro-total analysis systems

1 Introduction

1.1 The Human Genome Project

The completion of the human genome sequence in 2003 was one of the most important scientific accomplishments in human history [1] and marked a significant milestone for the human genome project (HGP). This achievement has led to compelling genomic and proteomic research discoveries with unprecedented impacts in areas such as forensic DNA analysis [2–5], medical diagnostics [6, 7], infectious disease management [8–11], and chemical and biological sciences [12, 13]. Some of the important tools for DNA processing include solid-phase extraction (SPE) and purification of DNA, polymerase chain reaction (PCR) or other thermally induced amplification strategies, electrophoresis, and DNA microarrays. Although many conventional benchtop tools currently exist to process DNA samples, efforts are being heavily invested into further automating the processing strategy, reducing the cost of performing the assay, and increasing the sample throughput. In this chapter, we will discuss the use of microfluidics, in particular integrated systems, for processing a variety of DNA samples.

1.2 Molecular Processing Pipeline for DNA Analyses

Complete nucleic acid analyses (RNA or DNA) for a variety of applications can be accomplished using commercial benchtop instruments, and typically consists of several molecular processing steps including:

1. Lysis of cells to release the nucleic acids of interest
2. Purification and isolation of the nucleic acids from other cellular components (e.g., cell debris and proteins)
3. Amplification of trace amounts of nucleic acids to generate sufficient copy numbers for detection
4. Analysis of unique regions within the genetic material using a combination of techniques

To complete an entire assay, a well-equipped laboratory and significant technical expertise are commonly required, with intervention at several stages of the processing pipeline to manipulate samples and/or reagents. In addition, the total time required for sample processing can be several hours to several days.

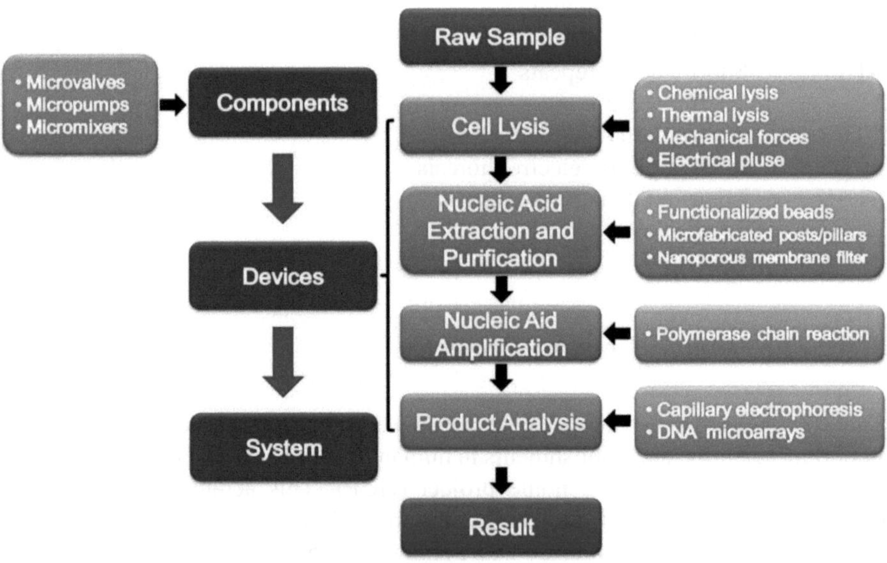

Fig. 1 Flow diagram showing the molecular processing steps required for analyzing nucleic acids. The main steps include cell lysis, nucleic acid extraction and purification, PCR amplification, and analysis methods for identifying the resultant products. For each of the processing steps shown, a device will have poised on it a component for one of these functional steps, and each device may be comprised of various components (such as pumps, valves, and micromixers) in order to carry out the desired operation. A system comprises two or more devices, meaning that it will have integrated onto it multiple processing steps. The ultimate goal is to incorporate all of the molecular processing steps onto a single platform to provide sample in/answer out capabilities with no operator intervention

Derived from the concept of micro-total analysis systems (μTAS) or lab-on-a-chip platforms first proposed by Manz et al. [14] in the early 1990s, integrated microfluidic systems have emerged that incorporate several molecular processing steps into a single platform with sample-to-answer capabilities. These systems are particularly compelling for DNA/RNA analyses. To create such a system, a series of discrete devices performing specific molecular functions such as cell lysis, nucleic acid extraction and purification, nucleic acid amplification, and other supporting analysis techniques (e.g., capillary electrophoresis, microarrays) must be interconnected with minimal dead volumes due to the ultrasmall samples processed (picoliters to nanoliters). Fluids are manipulated using on-chip or off-chip components, such as micropumps, microvalves, and micromixers (see Fig. 1). As noted previously, microsystems are composed of two or more microdevices and, in many cases, microdevices consist of components such as on-chip valves, mixers, and/or pumps. In most cases, DNA/RNA processing requires multiple processing steps and therefore devices need to be combined, either in a single wafer format or a modular format, to form the system targeted for a given genetic analysis (see Fig. 1).

2 Microfluidics and DNA Analysis

The capability of handling a volume of liquid as small as a few nanoliters and even a few picoliters, the common sample size in most microfluidic systems, can be utilized to permit DNA extraction following cell lysis and thermal reactions without creating sample dilution, minimize reagent usage, provide process automation, allow in-the-field analyses, and minimize possible contamination. Microfluidic systems also offer rapid, accurate, and cost-effective analyses. Performing sequencing or genotyping using microfluidic platforms can also lead to significant increases in throughput. For example, DNA sequencing read lengths of 600–800 bases can be achieved in 25 min using microcapillary electrophoresis (μCE) with a separation channel length of 20 cm (Sanger sequencing) [15], whereas the same separation would require 1–2 h in a capillary array electrophoresis (CAE) system [16, 17]. In Sects. 2.1–2.3, we will give a brief introduction to the various microdevices that have been fabricated to handle steps in the processing of nucleic acids, such as cell lysis, extraction and/or purification of the nucleic acids, their amplification, determination of sequence variations, and readout.

2.1 Cell Lysis

Cell lysis is the first step in most DNA analyses and involves disassembly of the cell membranes and release of the genomic material and other cellular contents.

A variety of lysis methods, including chemical lysis [18, 19], thermal lysis [20], and lysis by mechanical forces [21, 22], or electrical pulse [23–26], have been successfully demonstrated in microfluidic devices.

Transitioning the chemical lysis methods commonly used in macroscale workups to microfluidic devices is straightforward. Chemical lysis methods involve mixing the target cells with lytic agents, such as sodium dodecyl sulfate or guanidium hydrochloride and hydroxide, that can solubilize the lipid membranes. One issue associated with the use of chemical lysis is that lytic agents can interfere with downstream processing, such as PCR, and therefore must be removed from the sample before subsequent reactions, increasing the microfluidic design complexity. Carol and coworkers [18] reported a polydimethylsiloxane (PDMS) microfluidic device for on-chip cell lysis based on local hydroxide electrogeneration. In this device, hydroxide ions porated the cell membrane, leading to cell lysis. During lysis, hydrogen ions, which were simultaneously generated on-chip, reacted with excess hydroxide ions creating a neutral pH lysate and eliminating the need for a final washing step.

Thermal lysis, which involves disrupting the cell membranes by heating cells to near boiling temperatures, is another method that can be incorporated into a microfluidic device as long as the microfluidic material can withstand the temperature required to lyse the cells. The advantage of thermal lysis is that no interfering reagents are required that may interfere with downstream reactions [27]. The device design can be further simplified by lysing cells in the initial denaturation step of downstream PCRs [28]. But, thermal methods are not applicable for certain cell types, such as Gram-positive bacteria.

Mechanical forces, such as sonication, can be integrated to the microdevice to disrupt cells via gaseous cavitation. In this process, air pockets form from dissolved gases in the aqueous media and collapse rapidly, creating high pressure and high temperature environments sufficient to break cell membranes. This method is suitable for hard-to-lyse cells or spores, but can generate considerable amounts of heat and free radicals [29]. Belgrader et al. [21] reported a minisonicator combined with a spore lysis cartridge. *Bacillus* spores were sonicated in the presence of glass beads and were successfully lysed to release DNA for PCR amplification in ~30 s.

Electrical pulse methods represent another method for cell lysis and are based upon electroporation of the membrane. In the electroporation process, the application of high electric field pulses causes the formation of small pores in cell membranes [30]. However, the use of high electric fields can lead to heating and gas generation. To overcome this limitation, a microfluidic electroporation device for the lysis of human carcinoma cells was demonstrated by Lu et al. [23] In their design, a straight microchannel was constructed in glass, where the side-walls consisted of gold saw-tooth-shaped electrodes supported by the polymer, SU-8. Using pressure-driven flow, cells were directed through the channel and electroporated by the saw-tooth electrodes (see Fig. 2). The magnitude of the electric field was in the range of a few kilovolts per centimeter, while the AC voltage was >10 V peak-to-peak, minimizing heat generation and bubble formation.

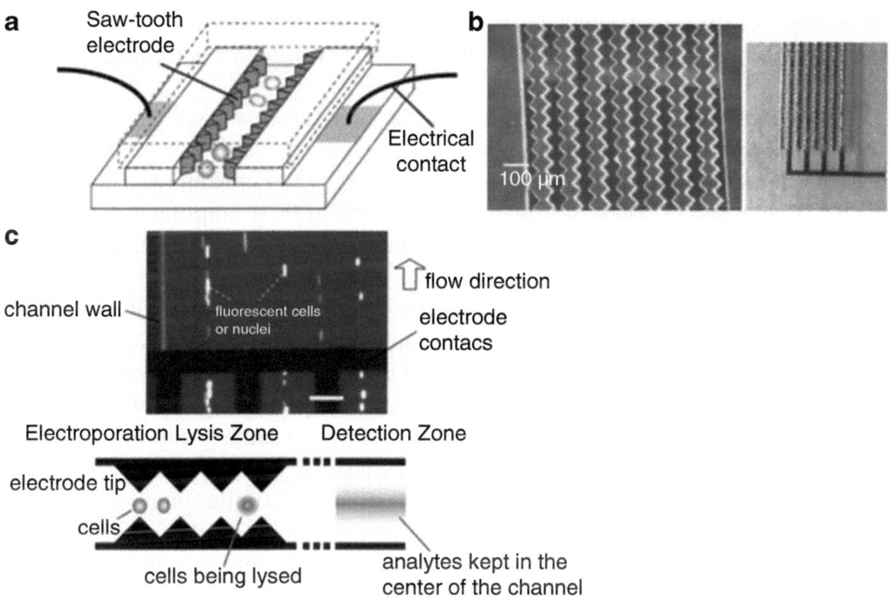

Fig. 2 (**a**) Microelectroporation device for cell lysis. (**b**) Device at various steps of the fabrication process: after metallization and electrode-mold formation (*left*) and after electroplating (*right*). (**c**) Dielectrophoresis (DEP) effect observed in the flow channels (*top*). Saw-tooth microelectrodes acting as a DEP device for focusing intracellular materials after electroporation (*bottom*). Reproduced from [23] with permission

2.2 Nucleic Acid Extraction, Purification, and Preconcentration

Following cell lysis, DNA extraction, purification, and preconcentration are usually achieved by microsolid phase extraction (micro-SPE) devices. This step is essential in order to purify and isolate the genomic materials from other cellular components, contaminants, and chemicals introduced in the cell lysis step that might potentially interfere with downstream enzymatic reactions. In addition, the nucleic acids may be enriched in this phase of the processing strategy to preconcentrate the targets to a level that is amenable for further downstream processing.

A variety of well-established macroscale SPE methods for nucleic acid extraction have been successfully transferred to microscale devices [10, 31–57]. Although the physical principles of these methods may be different (e.g., chaotropic interactions, electrostatic interactions, affinity interactions, etc.), micro-SPE protocols typically consist of three steps: (1) selective adsorption of nucleic acids onto a solid phase; (2) removal of contaminants by a washing step; and (3) elution of the preconcentrated nucleic acids from the solid support using water or a low salt buffer [31]. Like their macroscale counterparts, micro-SPE devices possess a loading level of target material that is dependent upon the available surface area within the extraction bed and, thus, are manufactured either by packing the solid phase (typically consisting of silica beads) into the device or by directly fabricating microstructures inside the device to increase the available load capacity of the device.

As noted above, micro-SPE devices can be fabricated by packing silica beads, sol–gel immobilized silica beads, photopolymerized monoliths, or modified magnetic particles into microfabricated channels [10, 31–43]. For example, Landers and his research group [32] demonstrated the extraction of PCR-amplifiable DNA from lysed white blood cells using silica particles packed into a capillary tube; the DNA recovery was found to be ~70% in a 10 min processing time. The load of DNA into the device was found to be on the order of 10–30 ng/mg of DNA. To circumvent the high backpressure introduced by flowing silica beads into a microchannel and to improve reproducibility, a silica bead/sol–gel hybrid matrix was packed into a glass microchip (see Fig. 3a) [31, 33]. Other matrices, such as a sol–gel monolith [35] or photopolymerized monoliths [36] were explored by the same group in an effort to overcome the aging and shrinkage problems associated with the silica bead/sol–gel hybrid matrix. Klapperich and coworkers [37–39] used a similar photopolymerized monolith as a solid-support matrix to confine silica beads within a cyclic olefin copolymer (COC) microchip to extract a variety of samples, such as lambda

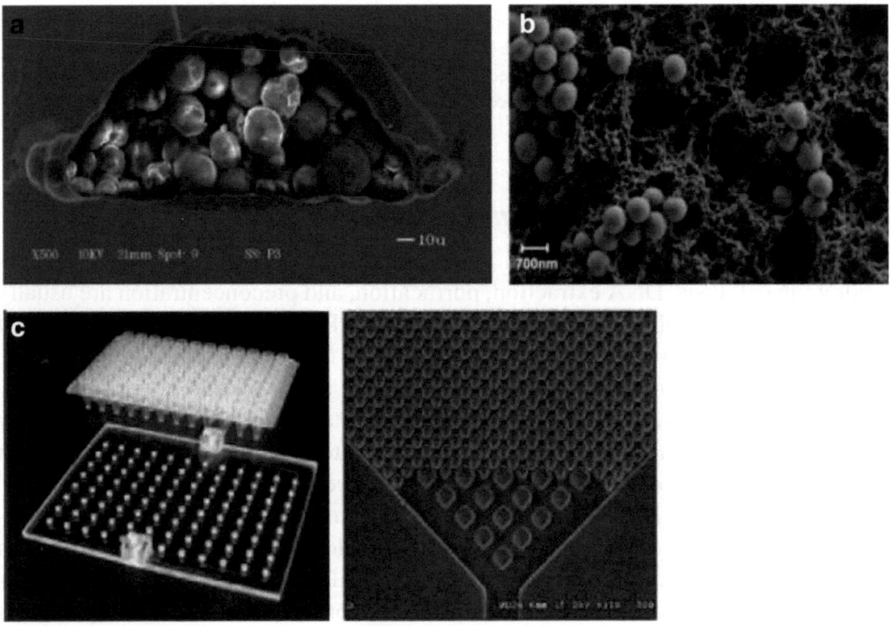

Fig. 3 (**a**) SEM image of the cross-section of a glass microchip channel packed with silica bead/sol–gel hybrid at 500× magnification. Reproduced from [33] with permission. (**b**) SEM image of a porous polymer monolith filled with silica beads at10,000× magnification. Reproduced from [39] with permission. (**c**) Photograph of a 96-well polycarbonate solid phase extraction microfluidic plate and a commercial 96-well titer plate (*left*). SEM image of the micropillars that were fabricated in the purification bed (*right*). Reproduced from [47] with permission

(λ)-DNA, Gram-positive and Gram-negative bacterial genomic DNA, inoculated human blood and urine samples (see Fig. 3b).

Issues associated with SPE beds comprised of silica beads or polymerized polymer monoliths include: (1) residual chaotropic reagents, such as guanidinium or sodium iodide salts, present in the initial elution phase and interfering with downstream amplification steps; (2) packing the matrix, which requires a postfabrication process that can be tedious and demands experienced engineers to accomplish; and (3) aging and shrinkage of the packing material, which can affect the efficiency and reproducibility of the extraction process.

Alternatively, one can produce a SPE surface directly in the device via microfabrication techniques. Micropost or micropillar structures can be fabricated inside the extraction chamber to increase the available surface area, thus enhancing the load of target material as well as the interaction probability between the solution-borne nucleic acids and the SPE bed. Christel et al. [44] fabricated, using deep reactive ion etching (DRIE) and anodic bonding, a micro-SPE device in silicon that contained high aspect ratio (aspect ratio = structure height divided by structure diameter) micropillars with a total surface area of 3.5 mm^2. The binding capacity of DNA was found to be 40 ng/cm^2 with a 50% extraction efficiency for short (500 bp) and medium sized (48 kbp) DNAs. Cady et al. [45] extended this work to *Escherichia coli* cell lysates. To circumvent the tedious DRIE and bonding processes required for fabricating these high aspect ratio microstructures in silicon, Soper and coworkers[46] developed a micro-SPE device made from photoactivated polycarbonate (PPC). This micro-SPE device contained high aspect ratio (5/1) micropillars, hot embossed from a LIGA-fabricated nickel molding tool. The SPE bed possessed a total active surface area of 2.3×10^7 μm^2. Nucleic acids were selectively immobilized onto the PPC surface, which contained carboxylic acid groups generated using UV radiation, using an immobilization buffer containing polyethylene glycol, NaCl, and ethanol. After cleanup using ethanol, the purified and concentrated nucleic acids were eluted from the PPC surface using water or PCR buffer. The load capacity and recovery of *E. coli* genomic DNA were estimated to be 790 ng/cm^2 and 85 $\pm$ 5%, respectively. This work was followed by a report on the fabrication of a high-throughput device consisting of 96 micro-SPE beds, each containing an array of 3,800 20-μm diameter micropillars (see Fig. 3c) [47, 48]. Both genomic DNA and total RNA could be extracted and purified from bacterial cells seeded into mammalian blood samples.

Another approach for the SPE of nucleic acids is the use of commercially available nanoporous aluminum oxide membranes (AOM). In a high salt concentration buffer, genomic DNA or RNA will aggregate and bind to the nanoporous membrane; both nanofiltration and electrostatic interactions contribute to the retention and purification of the target DNA or RNA. The retained DNA/RNA can be recovered using a PCR buffer. Kim et al. [49] investigated the extraction of genomic DNA from blood samples with a recovery of ~90% using an AOM sandwiched between PDMS microchannels. The AOM SPE device was later integrated to a microchamber PCR device, demonstrating successful amplification of both DNA from a bacterial sample and RNA from virus samples [50]. The advantages of this method included the ability to use high flow rates to shorten processing time and low protein absorption onto the AOM. However, the handling of the thin and brittle AOM remains a challenge.

In addition to genomic DNA or RNA purification and preconcentration before an amplification step, some applications, for example the purification of dye terminator Sanger sequencing products, require high quality DNA free of background species such as salts, unincorporated primers, dNTPs, and dye-labeled ddNTPs prior to the electrophoresis step. Soper and coworkers [51] demonstrated the use of a PPC micro-SPE device for the purification of Sanger sequencing products to provide high quality DNA free from background species. PPC micro-SPEs were successfully coupled to capillary gel electrophoresis [54] that also contained a continuous flow Sanger extension thermal cycler [51, 52].

Mathies and his group [53] purified Sanger extension products using a microchamber containing a sparsely crosslinked polyacrylamide gel copolymerized with complementary oligonucleotide probes appended onto the target DNA products. DNA elution was achieved by thermal denaturation of the hybrids. This microchamber was coupled to a Sanger extension chamber and microchip electrophoresis to form an integrated Sanger sequencing bioprocessor. With a 400-fold reduction in sequencing reagents and 10- to 100-fold reduction in DNA template required compared to benchtop approaches, 556 continuous bases were sequenced using this bioprocessor with 99% accuracy [54].

2.3 *Microfluidic Polymerase Chain Reactors*

Since the discovery of the PCR in 1986 by Mullis et al., [58] PCR has become a crucial tool in basic molecular biology discovery, genome sequencing, clinical research, in vitro diagnostics, and evolutionary studies [59]. PCR is an enzymatic reaction that allows any nucleic acid fragment to be generated in vitro and in high abundance. Theoretically, the amount of product doubles during each PCR cycle, as shown by the following equation:

$$N = N_0 2^n, \tag{1}$$

where N is the number of amplified DNA molecules, N_0 is the initial copy number of DNA molecules and n is the number of amplification cycles [60, 61]. Experimentally, the amplification efficiency (E) can range from 0 to 1, and therefore the true copy number produced is given by:

$$N = N_0(1 + E)^n. \tag{2}$$

In PCR, denaturation and annealing are nearly instantaneous events that occur as soon as the correct temperature is reached (e.g., ~94 °C for denaturation; 50–60 °C for annealing) and the extension step is limited only by the kinetics of the polymerase enzyme. Implementation of the thermostable *Taq* polymerase as a substitute for the Klenow fragment of *E. coli* DNA polymerase I [62] has made it possible to automate the PCR amplification step by using various thermal cycles carried out by a block thermal cycler. Investigators have shown that *Taq* DNA polymerase has an extension rate of 60–100 nucleotides/s at 72 °C [63]. For efficient amplification, a device with low heat capacity that can transfer heat quickly to the sample and quickly draw away the heat when cooling is preferable. Most conventional thermal cyclers have large thermal masses resulting in high power requirements and slow heating and cooling rates with long reaction times, typically exceeding 1.5 h in spite of the fact that kinetically, a 500 bp PCR product should be produced in as little as 5 s.

Due to the intrinsically small PCR chamber volumes and mass, exquisite heat transfer capabilities can be realized in microfluidic polymerase chain reactors (micro-PCR) devices that can significantly reduce the processing time compared to conventional benchtop thermal cyclers. Short thermal cycling times can be realized and still provide amplification efficiencies comparable to their macroscale counterparts with designs that are sometimes not conducive to macroscale formats. Micro-PCR devices have adopted design formats such as chamber type PCR devices, in which the PCR cocktail and target are mixed inside a microchamber and the chamber is then cycled between the various temperatures required for the amplification process. Another design approach uses the continuous-flow format, in which the PCR device consists of isothermal zones brought to equilibrium prior to the amplification process. The PCR cocktail is shuttled between these isothermal zones either electrokinetically or hydrodynamically to affect the thermal processing. Another format that has been employed for microscale PCR devices is a thermal convection-driven PCR device in which a temperature gradient is applied to a closed reaction chamber and the fluid is shuttled through the temperature gradient using a Rayleigh–Bénard convection cell. In Sects. 2.3.1–2.3.4, we will briefly introduce these PCR designs.

2.3.1 Chamber-Type Micro-PCR Devices

In chamber-type micro-PCR devices, a static PCR cocktail containing the target is repeatedly cycled between three different temperatures: one for denaturation, a second for renaturation, and a third for polymerase extension, which is similar to that used in a conventional PCR thermal cycler. Chamber-type micro-PCR devices consist of either a single chamber or multiple chambers configured on a single wafer with the appropriate heating modalities to allow thermal cycling. The primary advantage of these types of microthermal cyclers is the low thermal masses that must be heated/cooled, providing faster reaction times compared to block thermal cyclers.

2.3.2 Single-Chamber Micro-PCR Devices

In 1993, Northrup et al. [64] reported the first PCR microfluidic device, which consisted of a 50-μL well structure serving as the reaction chamber and was fabricated in silicon using wet chemical etching. Twenty amplification cycles were carried out, with the cycling time four times faster than a conventional benchtop PCR device. In 1994, Wilding et al. [65, 66] developed a silicon/glass hybrid device that held 5–10 μL of reaction mixture in a chamber, whose performance was improved by surface passivation through salinization of the micro-chamber surface[67] and which was heated using an external copper block. Single-chamber PCR devices have been widely investigated since these initial reports [68–77]. However, single-chamber micro-PCR devices possess low throughput.

2.3.3 Multichamber Micro-PCR Devices

Multiple PCR chambers have been fabricated on a single microfluidic chip and explored for high throughput PCRs [78–83]. An example of a multichamber micro-PCR device, the micro-DNA amplification and analysis device, (μ-DAAD) consisted of 16μ-DAADs in parallel with each μ-DAAD consisting of four micro-reactors fabricated on a 4″ silicon wafer (see Fig. 4). Multichamber micro-PCR devices [84] have been demonstrated for DNA amplifications of five gene sequences related to *E. coli* from three different DNA templates and detected by TaqMan chemistry with a limit of detection (LOD) of 0.4 copies of target DNA.

2.3.4 Continuous Flow PCR Devices

Another configuration for micro-PCR devices employs a flow-through format with a "time–space conversion" concept, in which the sample is continuously transitioned through isothermal zones for denaturation, annealing, and extension. This is in contrast to chamber-type PCR devices, in which heating and cooling occurs on a static sample with the entire device heated and cooled to the desired temperatures [79, 83, 85, 86]. The continuous flow PCR (CFPCR) approach allows for short reaction times because the small-volume fluid elements can be heated or cooled to the required temperature within 100 ms [59]. In 1998, Kopp et al. [87] reported the first CFPCR microdevice (see Fig. 5). This device consisted of 20 thermal cycles comprised of a serpentine channel design whose dimensions were 40 μm in width and 90 μm in depth, with a total length of 2.2 m, producing a pressure drop of ~14.5 PSI. For DNA amplification, 10 μL of a PCR mixture was hydrostatically pumped at volumetric flow rates ranging from 5.8 to 72.9 nL/s with a flow-through time of 18.7–1.5 min [87]. The channel walls were silanized with dichlorodimethylsilane to reduce possible adsorption of the polymerase enzyme (*Taq* polymerase) and DNA

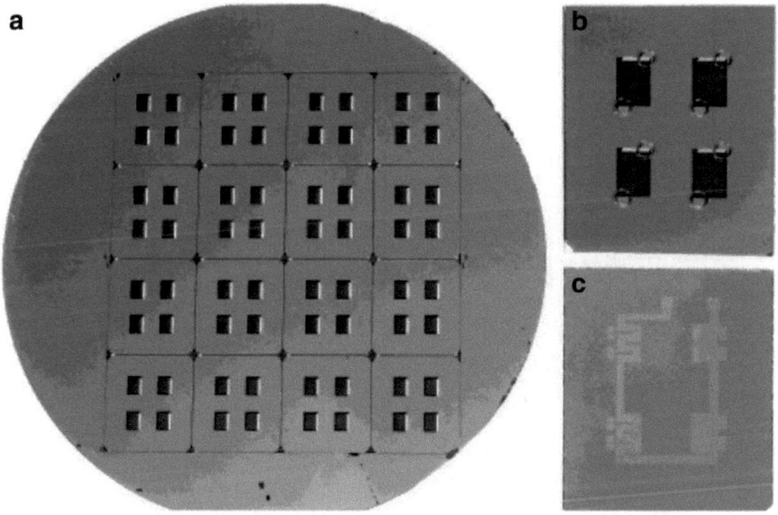

Fig. 4 Photographs of μ-DAAD production steps. (**a**) Front side of a 4″ silicon wafer populated with etched microreactors; 16 μ-DAAD are processed in parallel, each consisting of four microreactors. (**b**) Front-side view of a single μ-DAAD (16 × 1 mm^2) after bonding a cover plate and dicing. DNA arrays are printed onto the bottom of the microreactor cavities, but cannot be seen in this image because of their small size. Holes of 1 mm in diameter are drilled in the cover glass for the filling of the μ-DAAD reactors with reagent. (**c**) Back-side view of the device with platinum heater coil and thermoresistors placed at the corresponding area of the microreactor. Reproduced from [79] with permission

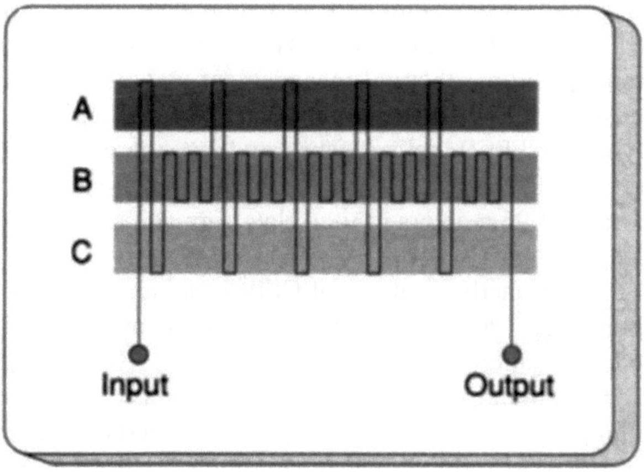

Fig. 5 Chip for CFPCR. Three well-defined isothermal zones are poised at 95 °C (*A*), 77 °C (*B*), and 60 °C (*C*) by means of thermostated copper blocks. The sample is hydrostatically pumped through a single channel etched into a glass wafer. The channel passing through the three temperature zones defines the thermal cycling process. Reproduced from [87] with permission

onto the glass surface. A zwitterionic buffer and nonionic surfactant were used as the PCR buffer additives to impart a dynamic coating [88].

Using a 20-cycle spiral microchannel hot embossed into a polycarbonate (PC) substrate configured for performing CFPCR [89], the PCR cycle time was reduced

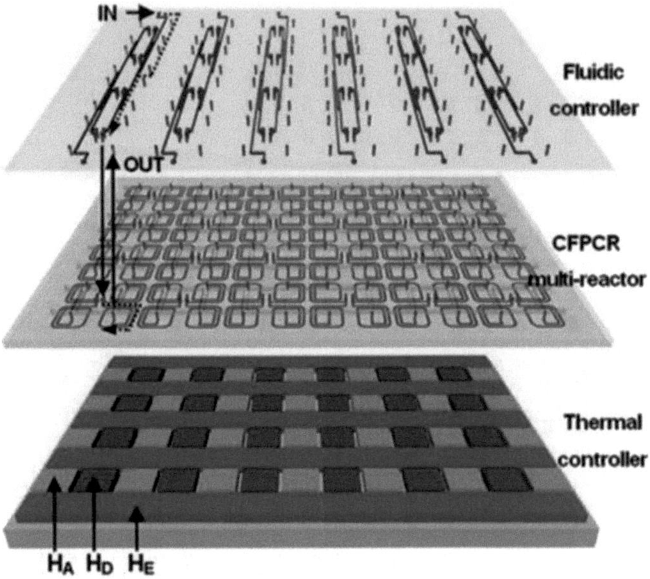

Fig. 6 High-throughput CFPCR multireactor platform consisting of three functional units: a fluidic controller for distributing reagents and analyte to the reactors, a CFPCR multireactor, and a distributed temperature controller. H_D denotes the denaturation heaters (90–95 °C), H_A annealing heaters (50–70 °C), and H_E extension heaters (70–77 °C). Reproduced from [92] with permission

to the kinetic limit set by the polymerase incorporation rate; 500- and 997-bp fragments were amplified in a total time of 1.7 min (5.2 s/cycle) and 3.2 min (9.7 s/cycle), respectively. The amplification efficiency was further optimized through proper thermal management using numerical models and experiments to evaluate the effects of different combinations of temperature distribution in a typical CFPCR device fabricated by hot embossing PC substrates [90]. Chen et al. [91–93] reduced the footprint of each spiral reactor to 8 mm by 8 mm and arranged 96 reactors in titer-plate format (12 × 8) for high throughput processing (Fig. 6).

The attractive features of CFPCR devices consist of: (1) very rapid heat transfer during the PCR, with run times on the order of minutes; (2) low possibility of contamination (closed architecture); [4, 10, 94] and (3) facile integration with various liquid transport processes, such as magneto-hydrodynamic (MHD) actuation [95]. Additional advantages include reduced sample consumption and reagents (lower cost) and simple integration to other DNA processing devices [11, 96]. A limitation of this approach is the fixed cycle number that can be employed by the chip, which is dictated by the device layout. To overcome this drawback, Chen et al. [97] demonstrated the use of a microfabricated PC chip for DNA amplification in a continuous flow (CF) mode using electrokinetically driven synchronized pumping (Fig. 7). A 500-bp fragment from λ-DNA was obtained with a total time of amplification of ~18.1 min for 27 cycles.

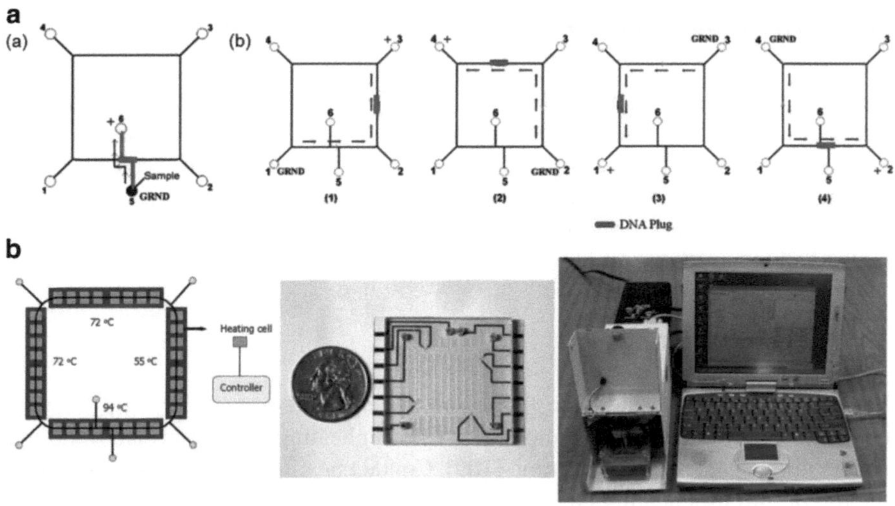

Fig. 7 (**a**) Principle of electrokinetic synchronized cyclic CFPCR process. Sample injection (*a*): DNA was filled into reservoir *5* and a voltage was applied to the electrodes in reservoirs *5* (*GRND* indicates ground) and *6* (+ indicates high voltage input). Sample moved across the reactor channel to fill the crossed-T injector. Sample cycling (*b*): Following injection, the sample is shuttled through the various isothermal zones by moving the position of the applied electric field in a cyclic fashion as denoted in diagrams *1–4*. (**b**) Schematic view and photographs of the electrokinetically synchronized CFPCR microchip. The actual microchip, fabricated via replication technology into PC, is shown in the *middle photograph* next to the quarter. Poised on the PC chip are electrode contacts for applying the voltaged in an automated fashion to the various reservoirs. Reproduced from [97] with permission

The arrangement of the three temperature zones on most rectangular serpentine channel CFPCR devices consists of denaturation, extension, and annealing in that order. Although this arrangement can establish a smooth temperature gradient, amplification efficiency may degrade because the melted single-stranded DNA (ssDNA) is likely to form double strands with the template DNA or with their complementary fragments when passing through the extension zone. To circumvent this problem, a novel three temperature zone arrangement in a "circular" format consisting of denaturation, annealing, and extension has been exploited [95, 98–100].

The PCR channel for these CFPCR devices can also consist of either capillary tubes [99, 100] or an on-chip annular channel [95, 101]. The serpentine channel formats on a monolithic chip can utilize thermal insulation with the aid of air gaps [102, 103] or by utilizing glass chips with a low thermal conductivity [104, 105]. Recently, a novel spiral channel configuration was also used to perform CFPCR on a single PC wafer with a circular arrangement of three temperature zones, allowing for a compact footprint and a minimal number of heaters for temperature control [106, 107]. CFPCR microfluidics can also use a unidirectional PCR or oscillatory flow [108].

2.4 *Microfluidic Thermal Heating Methods*

The choice of a heating method for micro-PCR devices is important in achieving efficient temperature ramping rates. The diversity of materials exhibiting differences in thermal mass means that different heating methods may be required. At present, temperature cycling on microfluidic devices can be performed either with contact or noncontact heating methods. Summarized below are various heating methods that have been employed in microfluidic devices.

2.4.1 Contact Heating

For contact heating, heaters are fabricated directly within the microchip or are in contact with the outside of the microchip. Contact heating utilizes an electrothermal conversion to heat the PCR solution [109]. Contact heating can be achieved through the use of thin film heating elements, which are mainly fabricated using deposition techniques; through the use of metal heating blocks, which primarily consist of inserting a heating cartridge into the metal blocks; or by utilizing Peltier elements.

2.4.2 Thin Film Heating

Heating elements can be fabricated on-chip using thin film deposition. Platinum [78, 79, 110–114] is the most commonly used material for heating elements due to its ability to withstand high temperatures, good chemical stability, and ease of micromanufacturing. Some other metals, alloys, or inorganic compounds have also been used as thin film heaters in micro-PCR devices, such as Al [70, 81, 82, 115], Ni [104], W [101], Ag/graphite inks [85], Ag/Pd [102, 116], Ni/Cr [117], Cr/Al [82], Al_2N_3 [118], and indium–tin oxide (ITO) [119–121]. Microheaters fabricated by Pt thin film deposition often require a thin layer of Ti serving as an adhesion layer. The Ti layer exhibits a high diffusion rate at high temperatures, which can deteriorate the Pt heater [122]. Commercial thin film resistive heaters [85, 106, 107] have proven to be efficient and robust for achieving fast PCR cycling, in contrast to conventional PCR devices.

2.4.3 Metal Heating Blocks

Conventional PCR instruments typically utilize contact heating, which involves a metal heating block in contact with the sample container, to cycle the temperature of the PCR solution that is held within a thin-walled polypropylene tube. In spite of their large thermal mass and slow temperature ramping rates, metallic heating blocks and Peltier-based thermo-electric (TE) ceramic heating blocks are widely applied in micro-PCR devices [69, 84, 87, 123–128]. To achieve fast thermal transition, two [8, 98, 123, 129] or more TE devices can be coupled to thermally cycle the PCR solution, and a total of six TE devices have been used in a portable miniaturized thermal cycling system [84]. The temperature of the peltiers could be independently controlled and programmed to be at different temperature levels necessary for effective annealing and denaturation. To ensure good thermal contact between the TE element and the cycled region of the device, supporting substances with higher thermal conductivity, such as mineral oil [8, 129] or a metallic thin wafer [130], can be added to the interface of the TE element. Normally, the TE cell consists of an array of parallel P–N junctions and each parallel P–N junction establishes its own temperature differential for a given voltage. The P–N junction is formed by joining P-type and N-type semiconductors (e.g. silicon) and Peltier-based TE use the differing behavior of charge carriers in P and N type semiconductor to move heat. Consequently, a radial temperature gradient on the hot surface of the TE cell is created, which causes nonhomogeneity of the surface temperature of the TE cell and compromises the efficiency of the PCR [109]. To achieve a homogenous temperature distribution across the surface of the TE cell, an oxygen-free thin copper wafer is necessary to redistribute the surface temperature [126]. Other reliable contact heaters are resistive heating coils [99, 127, 131] and single-sided flexible printed circuits (FPCs) [132].

It is important to note that the thermal-cycling rate is limited by the thermal mass of the heating element itself and of the entire micro-PCR device as well. Moreover, in the case of an external contact element, localized heating is ultimately limited in terms of lateral resolution by the thermal conductivity of the substrate material. In the case of on-chip integrated heaters, these devices still require tedious and complicated micromanufacturing processes, which restrict the flexibility to reconfigure the PCR design [133].

2.4.4 Noncontact Heating

The inherent problem with contact thermal heaters is their relatively large thermal mass. More thermal mass is added to the PCR device when contacting the chamber containing the PCR solution, which hinders fast thermal cycling rates. For the integration of PCR with μCE, thermal management becomes difficult because contact resource is regarded as part of the PCR chip and not part of the electrophoresis chip itself. These restrictions have triggered interest in the development of noncontact thermal cycling in which the heating is remote from the microfluidic device and not in physical contact with the PCR chamber [109].

2.4.5 Noncontact Heating Based on IR Radiation

Noncontact heating using IR radiation was first reported by Oda et al. [134] in 1998. In their work, an IR light, which used a single and inexpensive tungsten lamp as the noncontact heat source, was used for heating glass microchambers. The authors achieved temperature ramping rates of 10 °C/s for heating and 20 °C/s for cooling. In 2000, Hühmer and Landers [71] reported IR-mediated fused-silica capillary cycling with nanoliter volumes (160 nL), with improved heating and cooling rates of 65 and 20 °C/s, respectively. Giordano et al. [74] developed a novel polyimide (PI) PCR microchip, which utilized IR-mediated thermal cycling for the amplification of a 500-bp λ-phage DNA fragment in a 1.7-μL chamber with a total reaction time of only 240 s for 15 cycles. In 2003, Ferrance et al. [72] presented IR-mediated PCR amplification of genomic DNA using primers defining a 380-bp fragment of the β-globin gene followed by electrophoretic analysis on a single glass chip for the analysis of Duchenne muscular dystrophy (DMD) in less than 15 min for 35 cycles.

2.4.6 Noncontact Heating Based on Hot Air Cycling

Wittwer et al. [135, 136] developed noncontact heating for PCR based on hot air cycling. In their work, temperature cycling was performed without physical contact between the heating source and the reaction chamber by rapidly switching streams of air set to the desired temperature. Due to the low thermal mass of air, a high temperature ramping rate could be obtained, which has been further improved by several research groups [68, 112, 113, 137–139].

2.4.7 Noncontact Heating Based on Laser-Mediated Heating

The tungsten lamp is a noncoherent source with large focus projection, which limits the heating efficiency when applied to microchips with a small cross-section. Laser-mediated noncontact heating utilizes a photothermal effect produced by a diode laser coherent light source to heat an absorbing target. Tanaka et al. [140] used a diode laser to control the temperature of a chemical reaction by heating an absorbing target of a black ink point placed on top of a glass microchip cover plate above the reaction channel. The integrated glass microchip with noncontact IR laser-mediated heating has been demonstrated for fast and localized temperature control under flowing conditions with ultrafast heating and cooling rates of 67 and 53 °C/s, which is 30 times faster than a conventional device and 3–6 times faster than electrothermal miniaturized thermal cyclers [133]. This heating method may be very attractive and desirable due to its high resolution for spatially localized heating, ease of manipulation along the chip, and its property of being a point light source. Unfortunately, this heating method has not been applied for temperature control in microfluidics.

2.4.8 Noncontact Heating Based on Microwave Irradiation

Noncontact heating utilizing a focused microwave source was demonstrated by Fermér et al. [141]. In their work, a single-mode microwave cavity was used to heat 100 μL of PCR mixture in a 0.5-mL polypropylene tube for 25 cycles. Most recently, microwave-induced milliliter-scale PCR (see Fig. 8) was reported [142] for real-time PCR analysis. Although the amount of amplified nucleic acid product

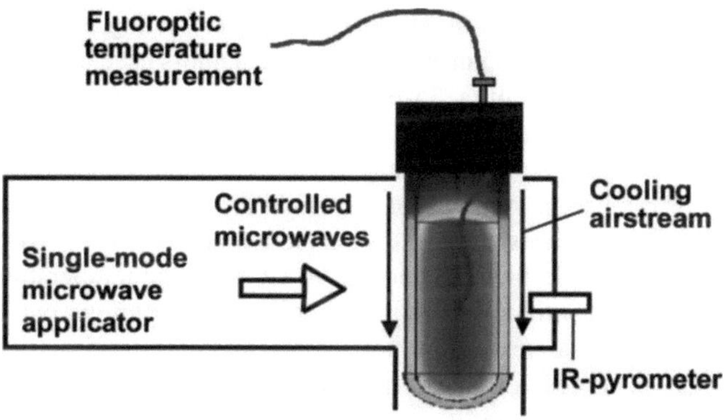

Fig. 8 Experimental set up for microwave-heated PCR that is used to perform milliliter-scale PCR utilizing highly controlled microwave thermal cycling. Reproduced from [142] with permission

after 33 cycles indicated incomplete amplification, which was attributed to temperature "over-shooting" at the denaturation phase and subsequent deactivation of the *Taq* polymerase [143], microwave heating was quite promising due to the following properties [141]:

1. Efficiency of the optimized microwave conditions nearly reached 70% that of conventional PCR
2. Irradiation energy was used to heat only the PCR solution and not a heating block or the sample containment tube
3. Temperature ramping time was substantially shortened
4. Required temperature was reached almost instantaneously and simultaneously, allowing for shortening of the incubation time
5. Modern microwave cavities can deliver a uniform field density without "hot-spots"

2.4.9 Other Noncontact Heating Methods

Other noncontact heating methods have also been described, such as noncontact heating using a halogen lamp as a low power radiation source for rapid temperature ramping in a silicon microreaction chamber [144]. This method achieved a rate of 4 °C/s for heating and 4 °C/s for cooling. Another noncontact heating method is based on induction heating and was first reported by Pal et al. [145]. Induction heaters are much simpler to fabricate, and heating and cooling rates of 6.5 and 4.2 °C/s can be achieved by optimizing the heater dimensions and frequency. The advantage of this method is that accurate positioning of the reaction mixture with respect to the heater is not necessary, deposition steps to pattern thin-film heaters on the chip are not required, and elaborate percentage/integrator/differentiator (PID) control is not needed [144].

Convectively driven PCR is an alternative thermal cycling process, which was first reported in 2002 by Krishnan et al. [146]. The authors used a Rayleigh–Bénard convection cell consisting of a 35-μL cylindrical cavity to perform the PCR amplification of the β-actin gene (295-bp fragment). Rayleigh–Bénard convection is generated by buoyancy-driven instability in a confined fluid layer heated from below [147]. The inherent structure of Rayleigh–Bénard convection-steady circulatory flow between surfaces employs two fixed temperature zones to facilitate the convection-driven sample flow. In contrast to CFPCR, the temperature cycling is achieved as the fluid continuously shuttles vertically between the two temperature zones poised for annealing/extension (top, 61 °C) and denaturation (below, 97 °C). Therefore, there is no need of an external force to drive the fluid through different temperature zones, simplifying its operation and allowing the implementation of the desired number of thermal cycles.

3 Analysis Methods of Reaction Products

Following amplification, the identification of amplification products must be performed to read the results and/or to confirm that the correct product was generated. Analysis techniques that can be used should provide short analysis times, high sensitivity and specificity, and favorable LODs. A variety of analysis methods have been successfully demonstrated for microdevice examples, including μCE and DNA microarrays. While there are a number of alternative techniques for reading successful PCRs, we will restrict our discussion to these two particular techniques.

3.1 Microcapillary Electrophoresis

The operational advantages of μCE are the high separation efficiency, providing high specificity, and relatively short analysis times [148–151]. In narrow channels that possess the capacity to efficiently dissipate Joule heating, electric field strengths as high as 1 kV/cm can be used. Consequently, the analysis time is typically one order of magnitude shorter than that found in conventional slab gel electrophoresis (SGE). Another feature of μCE is that the operational conditions can be kept constant, defined, and reproducible by filling the separation channel with fresh electrolyte prior to each analysis. DNA separations employing μCE have undergone extensive development since its description and demonstration by Manz et al. [14] and Harrison et al. [152] in the early 1990s. These reports were followed by demonstrations of μCE for the high resolution separation of ssDNAs in chips consisting of many separation channels. For example, Woolley and Mathies [16, 153] reported DNA sequencing separations using a glass μCE device. DNA was introduced electrokinetically through an injection cross and separated on a 5-cm-long gel-filled microchannel in only 120 s (see Fig. 9). A brief description

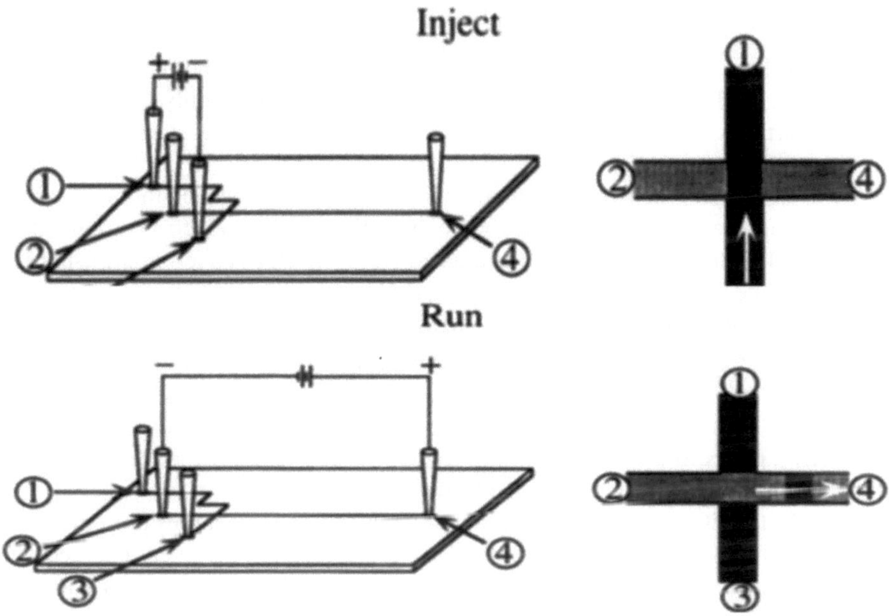

Fig. 9 Diagram of an electrophoresis chip indicating the injection procedure. The injection channel connects reservoirs *1* and *3* and the separation channel connects reservoirs *2* and *4*. In the injection mode (*top*), a field is applied between reservoirs *1* and *3* causing the DNA to migrate through the gel-filled intersection toward reservoir *1*. In the run mode (*bottom*), a field is applied between reservoirs *2* and *4* causing the DNA fragments in the intersection region to migrate toward reservoir *4* through the gel and into the separation channel. The actual device had 15 electrophoresis units integrated on each chip. Reproduced from [16] with permission

of μCE separation devices used for sorting DNAs will be described in Sects. 3.1.1–3.1.5 as well as a brief description of the fundamentals associated with CE in general.

3.1.1 Electrophoretic Mobilities

When a voltage is applied across a separation channel, the analyte migrates with an electrophoretic mobility and direction that generally depends on its chemical properties in the background electrolyte (BGE) and the applied electric field strength, E. The electric field strength is a ratio of the applied voltage (V) and the total capillary length (L_c) as shown in the following equation:

$$E = \frac{V}{L_c}. \tag{3}$$

By relating the velocity, v_{ep}, to E, the electrophoretic mobility (μ_{ep}) can be determined using:

$$\mu_{\text{ep}} = v_{\text{ep}}E^{-1} = q/(6\pi r\eta), \tag{4}$$

where q is the net charge on the analyte, r is the Stokes radius, and η is the viscosity of the BGE. It should be noted that the above equation is valid only for spherically shaped particles and must be modified for those molecules that adopt alternative configurations, such as DNAs.

Two electrically dependent phenomena contribute to the net mobility of an analyte, i.e., the intrinsic electrophoretic mobility and the electro-osmotic flow (EOF) [154]. The mobility can be evaluated from the electropherogram using the migration time, t_{m}, of the analyte migrating a distance L_{d} from the injection point to a detection point using:

$$\mu_{\text{app}} = \frac{L_{\text{d}}L_{\text{c}}}{t_{\text{m}}V} = \frac{L_{\text{d}}/t_{\text{m}}}{E}, \tag{5}$$

where μ_{app} refers to the apparent mobility, which accounts for the electrophoretic mobility of the analyte as well as the EOF.

In μCE, the EOF magnitude and direction depend on the charges present on the surface of the microchannel. For example, glass walls are negatively charged due to deprotonated silanol groups, which induce an EOF that travels from anode to cathode. The μ_{app} of an analyte inside a fused silica capillary or a microchannel that is superficially charged by contact with a solution typically consists of two contributions. The first is the electrophoretic movement of the analyte with respect to the electrolyte, which is characterized quantitatively by the effective mobility, μ_{eff}. The second contribution is the EOF of the liquid with respect to immobile, charged surfaces, which is characterized by μ_{eof} The net effective electrophoretic mobility (μ_{eff}) can be evaluated from the apparent mobility μ_{app} as follows:

$$\mu_{\text{eff}} = \mu_{\text{app}} - \mu_{\text{eof}}. \tag{6}$$

The μ_{eof} can be determined by using a neutral marker to determine t_{eof} from:

$$\mu_{\text{eof}} = \frac{L_{\text{d}}/t_{\text{eof}}}{E}. \tag{7}$$

We should note that in most DNA electrophoretic separations, the EOF is typically suppressed using either linear polyacrylamides covalently attached to the separation channel wall or dynamic coatings. Therefore, due to the fact that DNAs are polyanionic, they will exclusively migrate from cathode to anode, requiring that the electrophoresis analysis should be operated with the injection end cathodic.

3.1.2 Separation Selectivity

Separation selectivity, S, in electrophoresis is defined as the effective mobility difference between two migrating components [155], and is expressed as:

$$S = \frac{\Delta\mu_{\mathrm{eff}}}{\mu_{\mathrm{av}} + \mu_{\mathrm{eof}}}, \tag{8}$$

where μ_{av} is the average of the effective mobilities. For closely migrating analytes, the average effective mobilities may be replaced by the effective mobility of one component [156].

3.1.3 Resolution

The resolution of two components can be determined from [157]:

$$R = \frac{L_{\mathrm{c}}(\mu_{\mathrm{eff1}} + \mu_{\mathrm{eff2}})}{4\mu_{\mathrm{eff1}}\left[((\mu_{\mathrm{eff1}} E_{\mathrm{inj}} t_{\mathrm{inj}})^2/12) + (2DL/\mu_{\mathrm{eff1}} E)\right]^{1/2}}, \tag{9}$$

where L_{c} is the channel length, μ_{eff1} and μ_{eff2} are the effective mobilities of the two components of interest, E_{inj} is the electric field applied to perform injection, t_{inj} is the injection time, and D is the average diffusion coefficient of the components. Depending on which of two regimes is operating, the resolution depends on either the length of the channel or the square root of the length. In the first regime, band broadening due to electrokinetic injection dominates and the resolution scales with length. In the second regime, diffusion contributes primarily to band broadening, resulting in a square root dependence of R on length. The contribution of the electrokinetic injection to band broadening can be reduced by microfabrication and in conjunction with control of voltages on channel arms for determining the shape and size of the injected plug for short separation lengths [158].

3.1.4 DNA Separation Matrices

Size-based separations of homogeneous polyelectrolytes, such as DNA, are not possible in free solution electrophoresis [159]. This is due to the proportionality of the friction hydrodynamic force and total charge of the molecule to its length. The friction hydrodynamic forces exerted on the free-drained polymer coil while it moves as well as the accelerating electrostatic force both increase proportionally with the addition of a nucleotide to the chain. This is why one must typically use a sieving media, such as a gel or an entangled polymer solution, to obtain size-based separations of DNA using electrophoresis.

DNA electrophoretic mobilities are highly dependent on the nature of the matrix in which the separation takes place. For example, in free solution, the DNA molecules migrate with a mobility that is independent of size [160, 161], because the charge per unit mass is the same for all DNA molecules. However, the mobility becomes dependent on molecular mass in sieving gels. The mobility in sieving gels is related to the fractional volume $f(C)$ of the matrix available to the migrating DNAs according to:

$$\frac{\mu}{\mu_0} = f(C), \tag{10}$$

where, μ is the mobility observed in the matrix, μ_0 is the free solution mobility and $f(C)$ is the available fractional volume in the gel or entangled polymer solution [162]. If the sieving mechanism in the matrix is similar to that described by Ogston [163], $f(C)$ is related to gel concentration, C, as shown in the following:

$$\frac{\mu}{\mu_0} = f(C) = \mathrm{e}^{-KC}, \tag{11}$$

where K, the retardation coefficient, is dependent on the macromolecule size [164, 165]. Plots of the logarithm of μ/μ_0 as a function of gel concentration are known as Ferguson plots [166]. Linear Ferguson plots indicate that the separation takes place in the Ogston regime, where the average pore radius of the gel is larger than the radius of gyration of the migrating macromolecule. DNA mobilities in various sieving media are determined by the interplay of three factors: (1) the relative size of the DNA molecule with respect to the effective pore size of the matrix; (2) the effect of the electric field on the matrix; and (3) specific interactions of DNA with the matrix during electrophoresis [167].

Many different polymers have been investigated as separation matrices, including liquefied agarose, cellulose derivatives, dextrans, linear polyacrylamides (LPA) and their derivatives, poly(ethylene oxides) (PEO) and polyvinyl alcohols. The solution properties of these polymers are described elsewhere [168–176]. Most of these polymers provide good separations in certain DNA mass ranges [174, 175, 177]. For example, Njoroge et al. [178] used a 4% w/v LPA gel of ~6.6 MDa as a sieving matrix in a poly(methyl methacrylate) (PMMA) μCE device and achieved favorable resolution of PCR fragments of forensic *Alu* elements varying over a large size range (199–887 bp). The LPA was suspended in 1× TTE that contained 0.05% (w/v) methyl hydroxyethyl cellulose (MHEC) for EOF suppression in the PMMA device, and 7 M urea to denature the DNA. Using this matrix, PMMA microchip separations provided near-baseline resolution for all 20 components of a 900-bp sizing ladder. The separation of large DNA has been improved in sparsely crosslinked "nanogels" by ~10.4% compared to separation in LPA [179]. Nanogels are synthesized by incorporating a low percentage (~10^{-4} mol%) of *N,N*-methylene bisacrylamide crosslinker in LPA of high molar mass [180].

For polymers of comparable molar mass, hydrophilic polymers form more highly entangled networks because each molecule occupies a relatively large volume in solution [176]. The major differences between DNA separations in slab gels and entangled polymer solutions is that the entangled polymers are free to move about in the solution [181], creating dynamic pores in the matrix as they change interacting partners. The movements of the polymer chains give rise to a process called constraint release, which increases the mobility of the DNA [182–184].

3.1.5 End-Labeled Free-Solution Electrophoresis

Mayer, Slater, and Drouin [185] were the first to quantitatively examine end-labeled free-solution electrophoresis (ELFSE), also referred to as free-solution conjugate electrophoresis (FSCE), for the potential of sorting DNA in free solution without requiring a sieving matrix. Their theoretical model was based on two assumptions: The first was that the velocity of a hybrid ELFSE molecule (made of a charged component, the DNA fragment, and an uncharged component, the drag-tag) is given by the total electrical force applied to the molecule divided by the total friction coefficient of a free-draining molecule. The second assumption was that, in free solution, the Einstein relation [186] should relate the diffusion coefficient, $D(M)$, to the electrophoretic mobility, $\mu(M)$, of a DNA molecule of size M [see (14)].

In free solution, a DNA molecule of M bases behaves as a free-draining coil, and its $\mu(M)$ is independent of M. However, labeling the DNA with a molecular species having a different charge-to-friction ratio can lead to a size-dependent mobility. By neglecting the eventual electrostatic and hydrodynamic interactions between the DNA and the friction-generating label, the free-solution mobility of the end-label–DNA complex simply equals its charge-to-friction ratio, which is calculated from the following equation:

$$\mu(M) \approx \frac{\rho(M-\beta)}{\xi(M+\alpha)} = \mu_0 \frac{(M-\beta)}{(M+\alpha)}, \tag{12}$$

where α is the friction due to the end-label, ξ is the friction of one base, $-\beta$ is the effective charge carried by the end-label (the negative sign in front of β arises from the fact that DNA is negatively charged), ρ is the electric charge carried by one base, and $\mu_0 = \rho/\xi$, which is the free-solution mobility of a normal free draining DNA molecule. This equation shows that the mobility of an end-labeled DNA complex is now a function of the size of the DNA fragment when $\alpha \neq -\beta$ and the separation is possible with end-labels having a charge-to-friction ratio that is different from the charge-to-friction ratio of DNA (in practice, $M > \beta$ is also required to ensure that all the molecules migrate in the same direction). In ELFSE, the $\mu(M)$ of a DNA fragment can also be determined from the following equation:

$$\mu(M) = \frac{\mu_0}{1+\alpha/M}. \tag{13}$$

In contrast to classical gel electrophoresis, complexes containing long DNA fragments will have higher velocities than complexes containing short DNA fragments [185]. By considering the Brownian diffusion as a source of band broadening only in free solution, the Einstein relation [186] should relate $D(M)$ to $\mu(M)$ as shown in the following expression:

$$D(M) = \frac{\mu(M)k_B T}{\rho(M-\beta)} = \frac{\mu_0 k_B T}{\rho(M+\alpha)}, \tag{14}$$

where k_B is the Boltzmann constant and T is the temperature. By assuming that consecutive end-labeled DNAs have the same intensity and a Gaussian shape, the minimum migration distance, $L_{min}(M)$, required to separate molecules differing in length by a single base is approximately derived from:

$$L_{\min}(M) \approx \pi r^2 \frac{S\mu(M)D(M)}{E\left(\frac{\partial\mu(M)}{\partial(M)}\right)^2}\left(1+\left[(1+\gamma^2)^{1/2}\right]\right), \tag{15}$$

$$\gamma = \frac{\omega_0 E}{(S)^{1/2}D(M)} \times \frac{\partial\mu(M)}{\partial(M)},$$

where ω_0 is the peak width, and S is a numerical factor of order unity and depends on the efficiency of the detection method [185].

As a proof of concept, Heller et al. [187] used ELFSE for the separation of double-stranded DNA (dsDNA) fragments by free-solution CE. The drag-tag was added to the DNA fragments using a method similar to that previously used to study DNA migration in polyacrylamide gels and polymer solutions [188–190]. The results shown in Fig. 10 confirmed the predictions [185] that in the absence of EOF, the end-labeled molecules with longer DNA segments migrate faster than the shorter ones, and that higher resolution could be observed with a larger label or, in this case, with two labels instead of one. Using the original Mayer et al. [185] ELFSE theory to fit the data, the relative friction coefficient, α, for a streptavidin drag-tag can be estimated by the slope of the straight line obtained from plotting μ_0/μ^{-1} versus $1/M$ (see inset in Fig. 10). Using the data of Fig. 10, it was estimated that adding a single streptavidin drag-tag generated a friction equivalent to ~23 uncharged nucleotides, whereas adding two streptavidins generated a friction equivalent to 54 uncharged nucleotides (inset in Fig. 10a). The potential of this technique has been demonstrated by the separation of DNA sequencing fragments [191] and detection of single-nucleotide polymorphisms using electrophoresis [192].

Sinville et al. [193] combined an allele-specific ligase detection reaction (LDR) with FSCE (LDR–FSCE) for multiplexed electrophoretic screening of low-copy number mutations in a high abundance of wild-type DNA using PMMA microchip devices. To conduct FSCE separations of LDR-drag tags products, the LDR discriminating primers were reconfigured to allow for the addition of drag-tags onto their 5′ terminus. Using a dynamically coated separation channel, the PMMA

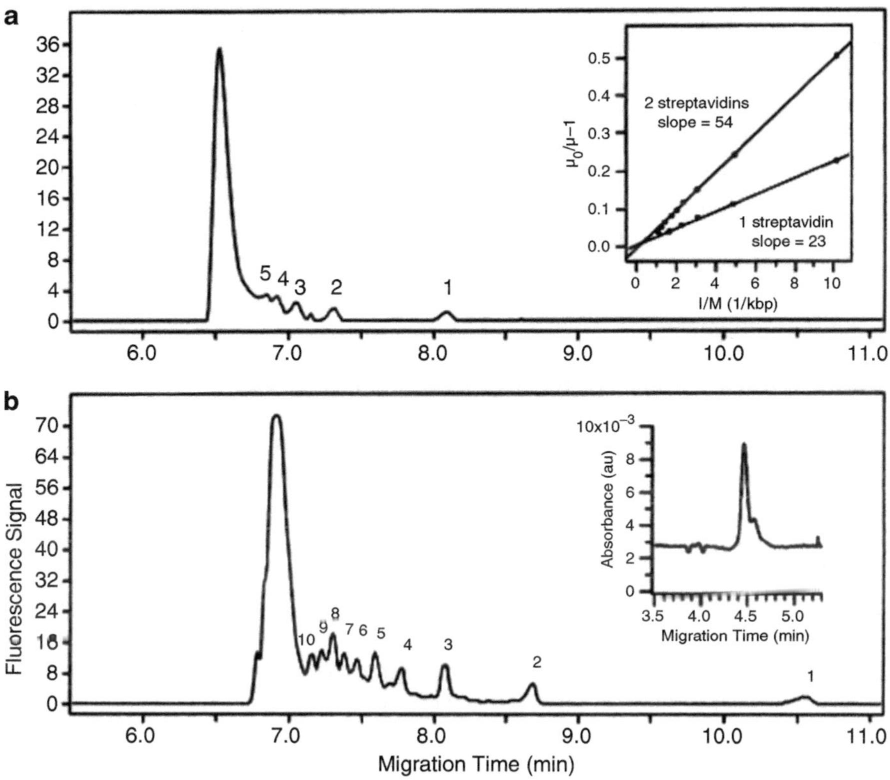

Fig. 10 Separation of a 100-bp dsDNA ladder with (**a**) one or (**b**) two streptavidin molecules used as ELFSE drag-tags. The peaks marked *1–10* represent the 100–1,000 bp dsDNA fragments, respectively. *Inset* in (**a**): plot of μ_0/μ^{-1} vs. $1/M$. *Inset* in (**b**): polydispersity of streptavidin as measured by CE. Reproduced from [187] with permission

μCE device achieved simultaneous detection of four mutant alleles in ~85 s, which was >7.5 times faster than a commercial capillary system.

3.2 DNA Microarrays

DNA microarrays were first developed in the early 1990s [194–196] and have become an important tool for high-throughput DNA analysis. A DNA microarray consists of a collection of oligonucleotide probes attached to a solid support in an orderly manner, typically a two-dimensional array. The probes readily hybridize to amplified gene fragments (targets) that are complementary to a specific probe. Readout of successful hybridization events is accomplished using a fluorescent dye or other such label attached covalently to the target. The mRNA expression levels or DNA sequence variations from hundreds to thousands of genes can be interrogated simultaneously. A few recent reviews of microarray technology are recommended to interested readers [197–201].

The basic elements required for the DNA microarray are the solid substrate, the attachment chemistry of the probe to the solid support, the approach adopted to "spot" the probes at particular locations of the two-dimensional array, the method employed to bring the solution target to the appropriate location of the array (passive or active), and the readout modality. A brief discussion of some of these important elements is given in Sects. 3.2.1–3.2.3.

3.2.1 Substrate Materials for Microarray Construction

A variety of solid substrates have been explored for microarrays, such as glass [202, 203], polymers [69, 98, 204–212], gold [213], optical fibers [214], and microbeads [215, 216]. Several issues must be considered in choosing the appropriate substrate, including the level of scattering and fluorescence background generated from the substrate, its chemical stability and complexity, the amenability to modification or derivatization of the substrate, loading capacity, and the degree of nonspecific interactions [202]. Glass has been widely adopted as a substrate material due to its favorable optical properties, which are highly desired for signal readout of the microarray using fluorescence. However, the microarray fabrication process involved for glass uses siloxane chemistry to tether oligonucleotide probes to the glass. These siloxane linkages are susceptible to hydrolytic cleavage, especially at extreme pH values. Recently, polymers have been used as alternative microarray substrate materials because of their diverse properties that can be selected to suite different immobilization strategies of probes to the substrate, and the ability to microfabricate structures in a low cost, mass-production mode for single-use applications. Polymers that have been used for microarray applications include PDMS [204, 205], PC [69, 98, 206], PMMA [207–211], and polystyrene [212].

The use of microbeads as substrates for the immobilization of oligonucleotide probes has also been reported. Fan and coworkers [215] described a dynamic DNA hybridization approach using paramagnetic beads as a transportable solid support. DNA targets were immobilized onto beads via streptavidin–biotin linkages for interrogation with probes that were transported via pneumatic pumping. Their experiment showed that beads containing DNA targets could be sequentially interrogated up to 12 times with no measurable change in the hybridization signal. Ali et al. [216] demonstrated a chip-based array composed of avidin-coated agarose microbeads for the discrimination of single-nucleotide mismatches. In their work, the biotinylated oligonucleotide probes containing microbeads were selectively arranged in micromachined cavities localized on silicon wafers, and the fluorophore-conjugated DNA target was a complement to the probe. The microcavities possessed *trans*-wafer openings allowing for both fluid flow through the bead chambers and optical analyses at numerous bead sites. Hybridization times on the order of minutes, with point mutation selectivity factors greater than 10,000 and an LOD of 10^{-13} M were achieved using this microbead array.

3.2.2 Surface Modification for the Immobilization of Probes

Various surface modification strategies have been used to attach probes to different solid supports. Oligonucleotide probes can be electrostatically attached to polylysine-derivatized or amino-silanized glass slides, representing a noncovalent approach [217]. Probes can also be covalently linked to the surface of the array by brief exposure to UV light [218]. Biotinylated DNA probes can be attached to streptavidin-coated magnetic beads [215], or thiol-terminated oligonucleotide probes can be immobilized to gold.

Methods for the end attachment of chemically modified oligonucleotide probes to a solid substrate have been reported as well. For example, Joos et al. [203] developed the covalent attachment of amine-terminated oligonucleotide probes to a glass substrate. Glass slides were derivatized with aminophenyl or aminopropyl silanes, and amine-terminated oligonucleotides were attached to the silanized glass with a crosslinking reagent such as glutardialdehyde. Using this approach, up to 90% of the attached oligonucleotides were available for hybridization.

Lenigk and coworkers [206] demonstrated the use of bifunctional linkers for the immobilization of amine-terminated oligonucleotide probes. In their work, a PC surface was coated with a photosensitive polymer (SurModic's photoreactive reagents) followed by 60 s UV irradiation to generate functional groups that allowed amine-terminated oligonucleotide probes to be covalently attached onto the surface. Detection of four pathogenic bacteria surrogate strains from multiple samples was accomplished using this device.

Wang et al. [207] reported the covalent attachment of amine-terminated oligonucleotide probes to a chemically modified PMMA substrate. In their protocol, the PMMA surface was aminated using a *N*-lithioethylenediamine solution, where methyl-ester functional groups were replaced by *N*-lithioethylenediamine. After aminolysis, the surface was activated with a homo-bifunctional crosslinker, glutardialdehyde, via a Schiff's base reaction and was converted to an aldehyde-terminated surface, which allowed for the covalent attachment of oligonucleotide probes. They found that the oligonucleotide coupling chemistry allowed reuse of the array >12 times without significant hybridization signal loss.

McCarely et al. [219, 220] described a simplified photomodification protocol of PMMA and PC substrates through direct and controlled UV exposure of the substrates in an oxygen-rich environment to yield surface carboxylic acid moieties. Patterns of carboxylic acid sites could be formed by exposure of the polymers in air to UV irradiation at 254 nm with a power density of 15 mW/cm^2 for ~60 min without significant physical damage to the polymer surface. The so-formed chemical patterns allowed for further functionalization to yield arrays or other structured architectures through covalent attachment chemistry.

Soper et al. [211] presented the fabrication of DNA microarrays onto PMMA surfaces using a UV modification protocol as shown in Fig. 11. Briefly, the PMMA surface was first activated via exposure to UV irradiation, which produced carboxylic acid functional groups onto its surface. EDC/NHS coupling chemistry was then used to facilitate the formation of succinimidyl ester intermediates on the surface, which allowed for the covalent attachment of amine-terminated

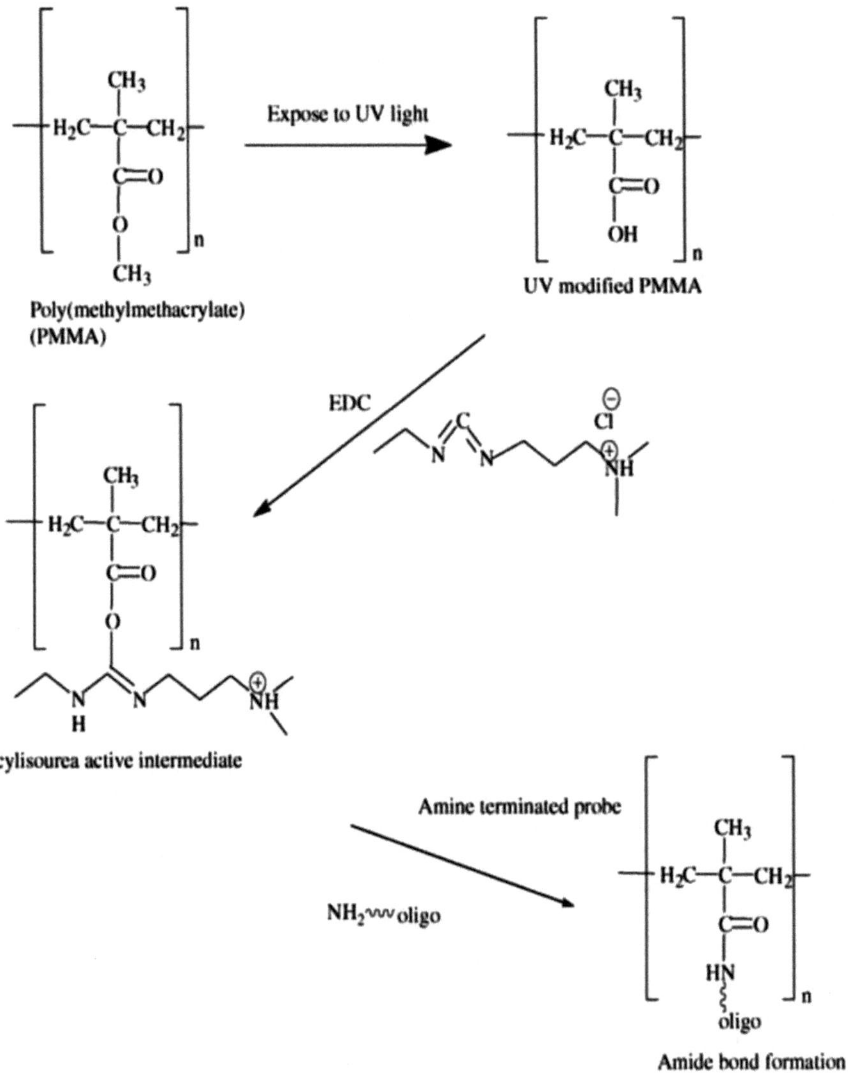

Fig. 11 UV activation of PMMA forming surface-confined carboxylic acid groups with the subsequent attachment of 5′ amine-containing oligonucleotides. Reproduced from [211] with permission

oligonucleotide probes. Recently, this same group enhanced the density of surface carboxylate groups by utilizing oxygen plasma treatment to create low density arrays on the surface of a free-standing, air-embedded PMMA waveguide [209].

3.2.3 Hybridization Efficiency Improvements

Merging microarrays to microfluidics is a step toward building integrated microfluidic systems for genetic analysis. In addition, it can provide significant reductions in target and probe hybridization reaction times resulting from diffusion-limited hybridization kinetics. Compared to a conventional two-dimensional hybridization array in a 1×1 cm format, hybridization occurring within a microfluidic channel significantly reduces the diffusional distances between the target molecules and the probes immobilized onto the surface. Shuttling back-and-forth, the hybridization mixture inside a microchannel can further facilitate mass transport and, thus, reduce hybridization time. Examples demonstrating improvements in hybridization efficiency using microfluidics have been reported [204, 205, 207, 221–224].

Liu and Rauch [221] investigated DNA hybridization in a microfluidic channel fabricated from a variety of plastic materials. By oscillating the hybridization mixture in the microfluidic channel, maximum signal was observed within a hybridization time of 15 min. Wang et al. [207] reported a low density array constructed inside a PMMA microfluidic device and used flow-through feed of the hybridization mixture, which successfully reduced the hybridization time from ~5 h to 1 min and reported an LOD of 10 pM for the identification of low abundance point mutations (one mutant in 10,000 wild-type DNA molecules) found in a K-*ras* oncogene. Erickson et al. [204] developed a theoretical model for electrokinetically controlled DNA hybridization in microfluidic devices, which predicted that reducing the height of the microchannel would effectively accelerate the diffusion-limited reaction kinetics and reduce the time required for the hybridization reaction to reach steady state. Following numerical simulations, the experimental results indicated that all processes from sample dispensing to hybridization detection could be completed within 5 min inside a glass–PDMS microchannel with a height of 8 μm. Yuen and coworkers [205] fabricated a microfluidic device consisting of two interconnected reaction chambers molded in PDMS on a standard microscope slide for closed-loop fluidic circulation and mixing. Fluid samples were circulated and mixed by the rotation of a magnetic stirring bar driven by a standard magnetic stirrer. A two- to fivefold increase in hybridization efficiency was observed with fluid circulation. Wei et al. [222] described the use of discrete sample plugs in a hybrid glass/PMMA microfluidic device for droplet hybridization (see Fig. 12). In this case, plugs were shuttled back and forth inside a channel sweeping over the probes, which were thoroughly mixed by the natural recirculating flows, significantly reducing the hybridization reaction volume to 1 μL. The total reaction time was 500 s, and the LOD was 19 amol.

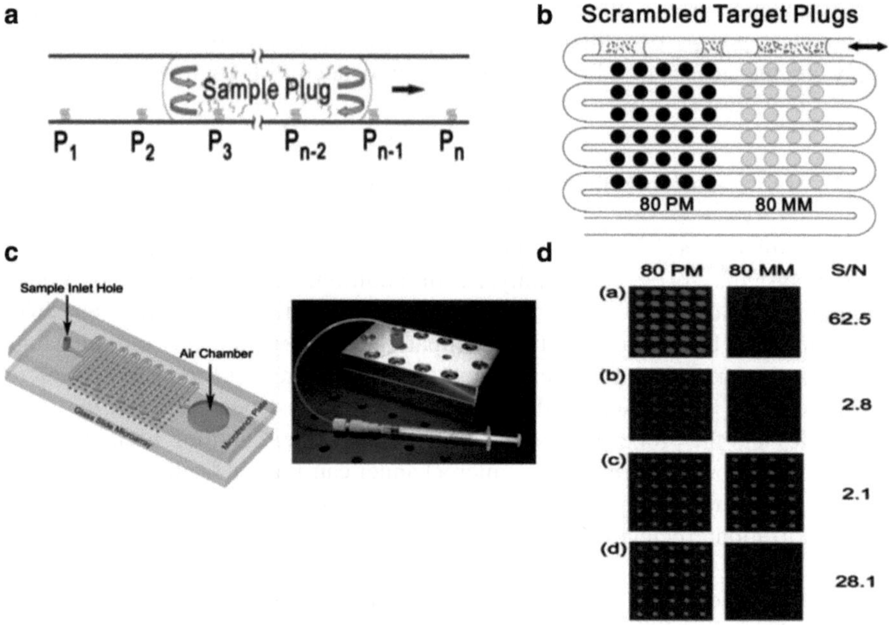

Fig. 12 (**a**) Droplet shuttle hybridization in a microchannel. $P_1, P_2 \ldots P_n$ refer to the probe spots. (**b**) Illustration showing that scrambled discrete plugs sweep over different probes in the channel. *80 PM* and *80 MM* denote the perfect match and mismatch probes, respectively. (**c**) Microtrench plate is stacked on a glass microarray slide (*right*) with a photograph of the assembled device (*left*). (**d**) Signal-to-noise (*S/N*) ratios for various hybridization formats: *1* shuttle hybridization at 500 s, sample volume 1 μL; *2* static microfluidic hybridization at 500 s, sample volume 10 μL; *3* flat glass hybridization at 500 s, sample volume 30 μL; and *4* flat glass hybridization at 18 h, sample volume 30 μL. The target concentration is 90 nM. The *left column* presents the fluorescence images with 80 PM probe, and the *right column* presents the fluorescence images with 80 MM probe. Reproduced from [222] with permission

4 Integrated Microfluidic Systems

The integration of sample pretreatment with analytical processing steps for the analysis of biological samples has remained the primary goal of μTAS as described by Manz and coworkers [14] over two decades ago. Many of these visions are becoming a reality and some of these systems will be described here. As noted previously, genetic analysis systems are defined as a single wafer or a collection of wafers seamlessly interconnected that possess two or more processing steps for the analysis of a genetic sample. Genetic analysis encompasses a large number of different types of applications, for example, DNA forensics where unique genetic markers are used for human identification either at crime scenes or in battlefields. For in vitro diagnostics, mutations in certain gene fragments can be detected and used to discover the presence of a disease in a particular patient, and also provide information to the clinician on how to treat that patient. In Sects. 4.1 and 4.2 we will discuss systems that include the sample preprocessing functions followed by either μCE or microarray readout of the preprocessing steps.

4.1 Integrated Systems with μCE Readout

The analytical challenge in coupling μCE with front-end sample preprocessing is the integration of pressure-driven flows required for the front-end with electrokinetically driven flows required for the back-end. In addition, the presence of any residual hydrodynamic flow during the separation can degrade separation efficiency due to the parabolic flow profile that is generated, compared to plug flow for electrokinetic flow. As such, the integration challenge in microfluidics is interfacing PCR with μCE separation because of sample transfer from hydrodynamic to electrokinetic flow. The quality of a separation depends, in part, on successful injection processes. The contribution of the injection process to the height equivalent of a theoretical plate H can be estimated from the following equation [225]:

$$H = \frac{L_{\text{sample}}^2}{Kl}, \tag{16}$$

where L_{sample} is the length of the sample plug along the separation channel, K is the injection profile factor, and l is the separation length (distance between injection and detection points). The injection profile factor reflects the peak shape of sample plug during injection and depends on the injected volume, the injector geometry, and the method of injection, and can vary between 1 for an exponential-decay injection and 12 for an ideal sample plug [226, 227]. This implies that the contribution of a specific injector to the total band broadening may vary by more than a factor of 10.

The injection step usually involves several intersecting channels and many injection designs have been developed and reviewed [228]. Most designs use cross- or double-T intersections, proposed initially by Harrison et al. [229, 230] and Effenhauser et al. [231, 232], respectively. For each injection design, different injection modes can be employed depending on the electric field sequences and distributions [233]. The three major limitations for electrokinetic injection are:

1. It is biased toward small DNA fragments with high electrophoretic mobility thereby complicating injection optimization. For example, unpurified PCR products contain high concentrations of salt and an excess of short unreacted primers, which can lead to long loading times.
2. The injected sample is small ($<$1% of the total sample volume), which imposes unavoidable restrictions on the sensitivity and quantitation of the analysis [234, 235].
3. The injection time becomes highly variable when one is injecting limiting amounts of analyte from nanoliter volumes [5, 54].

The injection methods used in μCE can either be stacking- or extraction-based methods. Stacking methods such as mediated stacking [236], gated injection [237], field amplified sample injection [238], the staggered-T configuration [239], field-amplified sample injection [238], and pressure-driven injection [10, 86] are widely adopted in μCE because they are simple to implement. Extraction-based techniques, such as membrane filtration [240, 241], SPE [32, 242], liquid–liquid extraction [243], and bioaffinity purification [53, 244] offer a more scalable platform for complete sample injection.

One of the first reports of an integrated microchip for DNA analysis was reported in 1998 by Burns and colleagues [245], which included a resistive heating region for DNA amplification, a sample loading region, and a gel-based separation region. This work utilized a single-strand displacement assay [246] with other sample analysis steps on one device. The system (see Fig. 13a) was fabricated entirely in Si using standard photolithography and consisted of devices for nanoliter sample injection, mixing of reagents, amplification or enzymatic digestion of the DNA sample, gel electrophoresis and, finally, fluorescence detection. The only components not situated on the chip were the fluorescence excitation source, pressure source, and control circuitry. Figure 13b shows the fluorescence image of a 50-bp ladder separated using μCE with a polyacrylamide gel. The system was demonstrated to perform an isothermal amplification of DNA using strand displacement amplification with the DNA target stained with SYBR green.

Waters et al. [20] reported a system that was micromanufactured in glass and consisted of a rather simple architecture: the microchip used a reservoir machined into the glass as a reaction chamber to perform, sequentially, cell lysis and a multiplex PCR amplification. The chip was fabricated using photolithography and wet chemical etching of the glass substrate. The monolithic chip was placed directly onto a commercial thermal cycler to allow thermal processing of the sample. Following PCR, the products were directly injected using a cross-T injector in a μCE channel and separated in a 1% hydroxyethyl cellulose (HEC) sieving matrix for size separation. An intercalating dye was used for the fluorescence detection of the generated DNA fragments.

Obeid and coworkers [128] reported a microsystem fabricated on two glass plates (each 40 × 45 × 0.55 mm), where a continuous channel network was etched into the bottom plate by standard photolithography and wet chemical etching, followed by thermal fusion-bonding of the two plates to form a closed structure. This system (see Fig. 14a) used a continuous flow concept to demonstrate functional integration of reverse transcription (RT) and PCR (RT-PCR) with operator selection of the number of amplification cycles to secure the results shown in Fig. 14b. The RT phase of the measurement involved the synthesis of DNA using mRNA templates and was performed before DNA amplification to allow quantification of mRNAs. The integration of RT and PCR processes within a monolithic chip is often problematic as RT components can interfere with the subsequent PCR [59]. Obeid and colleagues[128] tackled this problem by reducing the flow rate at which reverse transcription was performed so that at the intersection of the RT and PCR channels, the RT mixture constituted only ~10% of the total PCR volume.

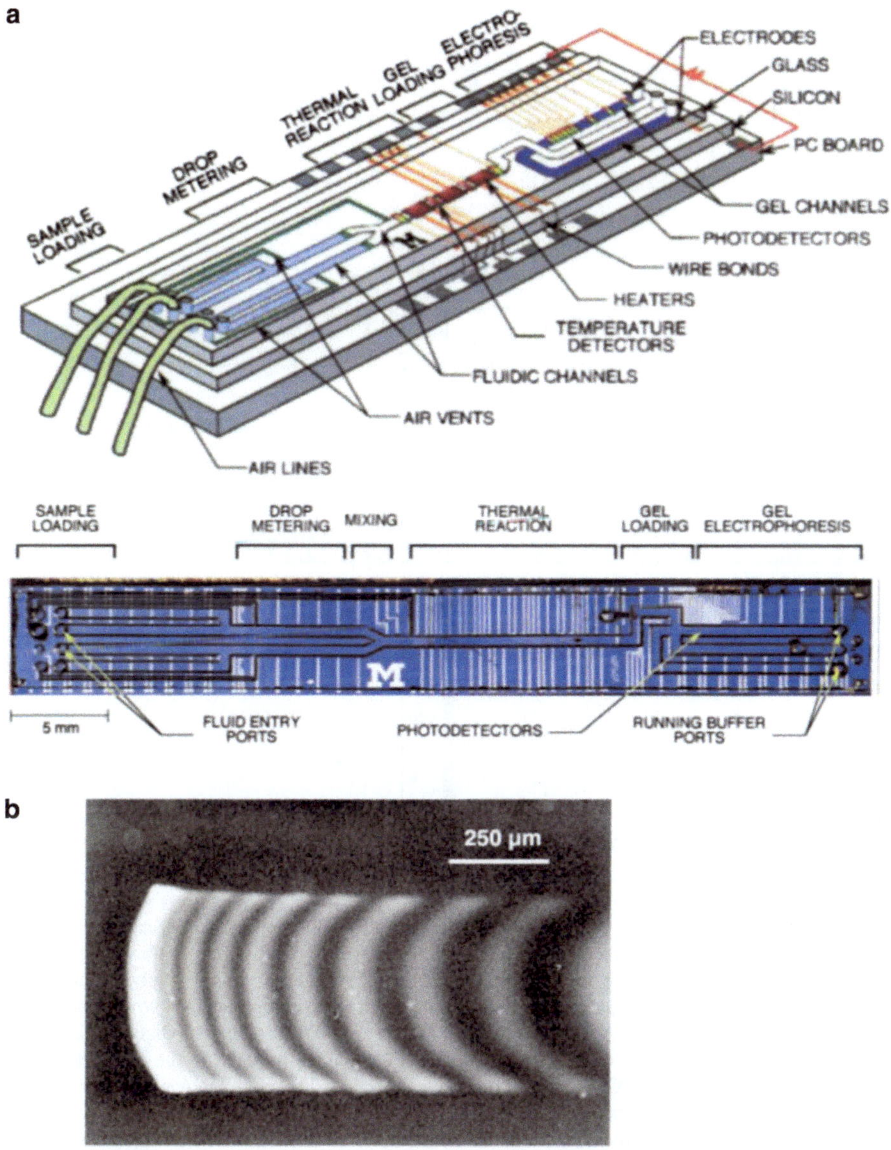

Fig. 13 (**a**) Integrated DNA analysis system developed by Burns and coworkers [245]. The device was made on a Si substrate and included the following elements: fluidic channels, heaters, temperature sensors, and a fluorescence detector to monitor the generation of DNA products. The following functions were incorporated into the system: mixing of solutions, amplifying or digesting DNA, separating the products using μCE, and detecting these products directly on-chip. *Blue* liquid sample, *green* hydrophobic surfaces, *purple* polyacrylamide gel. The *bottom panel* shows an optical micrograph of the device from above. (**b**) Optical micrograph of a 50-bp DNA ladder sorted in a 500-μm-wide polyacrylamide gel. Reproduced from [245] with permission

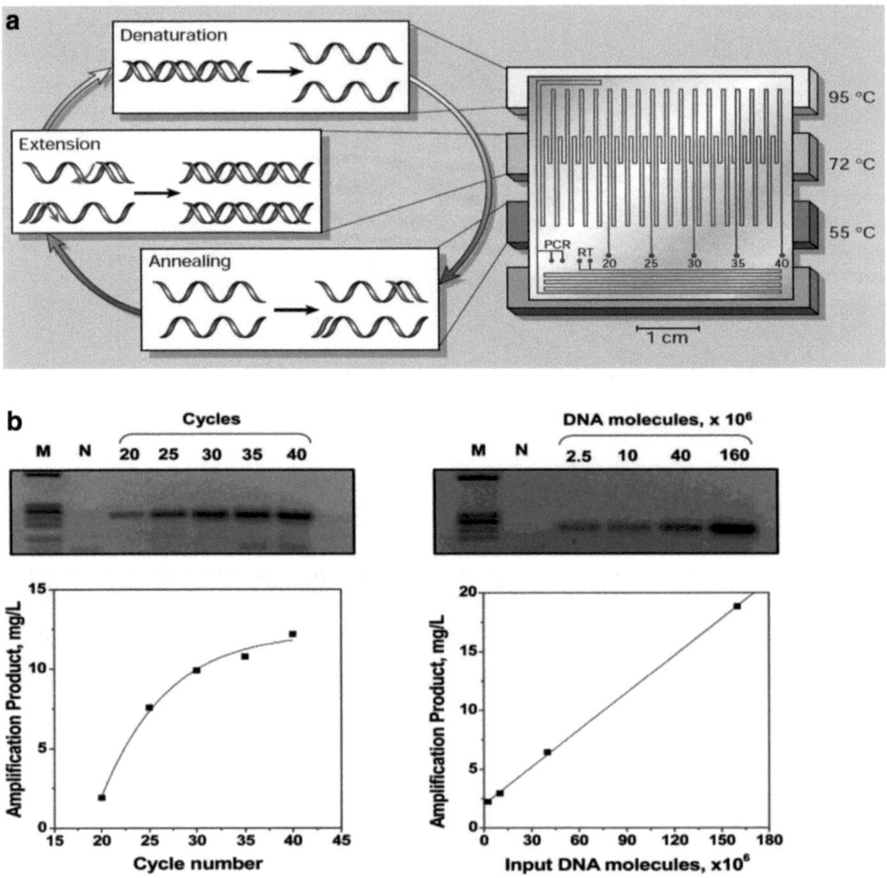

Fig. 14 (**a**) Microfabricated device for CFPCR and RT-PCR in which DNA can be amplified quickly using the PCR. When a sample is added at either of the PCR inputs, it flows over heating blocks whose temperatures are set to induce the three steps of PCR consecutively: denaturation, annealing of the PCR primers, and extension of the primers into complete DNA strands producing two copies of the original DNA strand. Samples can be extracted from the chip between 20 and 40 PCR cycles at the points indicated. (**b**) *Left*: Effect of the cycle number on the concentration of amplification product in CFPCR with on-chip cycle selection. The input DNA was 10^7 molecules. *Right*: Study of the concentration of the amplification product as a function of the input DNA molecules. All products were collected at 30 cycles. A 0.23-kbp PSA DNA template was used. *M* DNA markers. *N* negative (no DNA template). Reproduced from [128] with permission

The system was demonstrated by performing an amplification of the β-globin gene from human genomic DNA isolated from whole blood. The RT-PCR products were analyzed off-chip using agarose gel (2%) electrophoresis and ethidium bromide staining.

Soper and colleagues [208] investigated the effect of carryover from a primary PCR on a subsequent LDR in terms of LDR yield and fidelity using a continuous flow system microstructured on a PC substrate. Using this PC chip, three steps were carried out: (1) amplification of the gene fragments from genomic DNA using PCR; (2) mixing of the resultant PCR product(s) with an LDR cocktail via a Y-shaped passive micromixer (see Fig. 15); and (3) ligation of two primers for detection of mutant DNA in an excess of wild-type sequences (1:1,000 mutant:wild type). It was found that a post-PCR treatment to deactivate *Taq* polymerase prior to the LDR phase of the assay was not essential [247]. The PCR/LDR continuous flow microsystem was demonstrated by detecting single nucleotide polymorphisms in *KRAS* genes, which carry high diagnostic value for colorectal cancers. The PCR/LDR products were analyzed off-chip using polyacrylamide gel electrophoresis and ethidium bromide poststaining.

Wang et al. [52] reported a CFPCR chip that consisted of a 20-loop spiral microfluidic channel hot-embossed into a PC substrate. The CFPCR chip was interfaced to a solid phase reversible immobilization (SPRI) chip made from UV-activated PC for purification of Sanger DNA sequencing ladders produced in the CFPCR chip. The CFPCR chip had three well-defined temperature zones poised at 95, 55, and 60°C for denaturation, renaturation, and DNA extension, respectively. The system was demonstrated for Sanger cycle sequencing using dye-terminator chemistry. The CFPCR-SPRI system could prepare a sample for electrophoretic sorting in less than 30 min. Following CFPCR-SPRI processing, the purified products were collected into a microtiter plate and directly introduced into a conventional capillary gel electrophoresis machine for separation and automated base calling. Average read lengths of 682 bases with a 99% calling accuracy were reported.

Lagally et al. [79] developed a fully integrated genomic analysis microsystem that included microfabricated heaters, temperature sensors, and 200 nL PCR chambers directly connected to μCE. The system was microstructured using a hydrofluoric acid etching procedure for glass as described by Simpson et al. [248]. The system was also equipped with PDMS valves and hydrophobic vents to provide controlled and sensorless sample positioning into PCR chambers connected to electrophoretic separation channels. The use of microfabricated heaters and temperature sensors created uniform heating over the entire PCR chambers and fast thermal response times, while minimizing power requirements. The heating and cooling rates for the PCR were 20 °C/s and the 20 PCR cycles were completed in 10 min. The ramp rates were within the range reported for devices in which the heating elements are in contact with the PCR chambers [20, 64, 66, 75, 117, 249, 250], but slower than rates reported for noncontact devices [134, 251].

Koh et al. [9] demonstrated a microsystem fabricated using a poly(cyclic olefin) substrate, which was produced by first creating the desired pattern on a glass plate using photolithography and wet etching. A layer of metal (100 μm–1 mm) was then electroplated onto the surface of the glass plate to create a molding tool (an electroform) from which microstructures were embossed into a poly(cyclic olefin) resin using compression molding. The resulting microsystem contained a cell thermal lysis unit along with integrated valves, which could withstand pressures up to 100 psi, and a PCR device that was directly connected to a CE microchannel to sort fluorescently labeled PCR products generated from different strains of

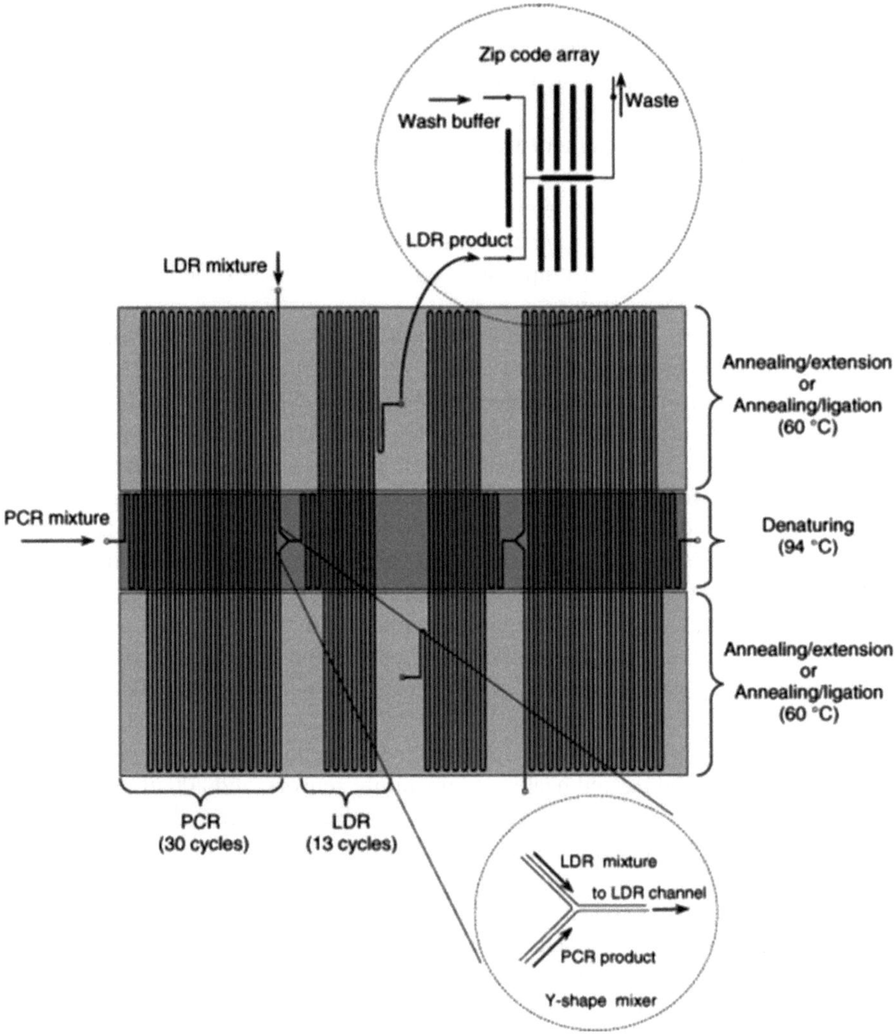

Fig. 15 Topographical layout of a CFPCR/CFLDR device. The microchip possessed channels that were 50 μm in width and 100 μm in depth with a 400 μm interchannel spacing. The total length of the thermal cycling channel was 2.28 m and consisted of a 30-cycle PCR (1.57 m long) and a 13-cycle LDR (0.71 m long). The *top inset* represents a microscope image of the turns of the CF thermal cycling channel. The *bottom inset* is an enlarged schematic of the Y-shaped passive micromixer for mixing the PCR product with the LDR cocktail. Three different Kapton film heaters were attached to the appropriate positions on the CFPCR/CFLDR chip for providing the required isothermal zones. Thermocouples were inserted between the microchip cover plate and the film heaters for monitoring the temperatures. Reproduced from [208] with permission

E. coli (a thiazole orange intercalating dye was admixed to label the PCR products for fluorescence detection). The thermal cycling device also used printed-ink contact resistors, which allowed heating at 12 °C/s. The system was used to identify

E. coli O157:H7 organisms. An LOD of ~6 copies of target DNA was reported using this system.

Easley et al. [10] developed a microfluidic genetic analysis (MGA) system fabricated on a glass substrate. The MGA was capable of accepting whole blood as a crude biological sample for the detection of *Bacillus anthracis* (anthrax) in 750 nL of whole blood from living asymptomatic infected mice, with the endpoint generation of a genetic profile. To demonstrated the sample-in-answer-out capability of the integrated monolithic chip (see Fig. 16), the blood was mixed with a lysis buffer and a volume equivalent to 750 nL (containing 15–45 ng of murine DNA) of whole blood was loaded onto the device with subsequent DNA extraction, a process which was completed in <10 min. DNA extractions used silica beads (5–30 μm) that were packed into the SPE domain (see Fig. 16d) against an etched weir filter by using an applied vacuum and replaced before each analysis. Flow rates employed for all extractions were 4.16 μL/min. Upon capture of the purified DNA in a 550-nL PCR microchamber, IR-mediated thermal cycling was performed to generate a 211-bp amplicon found on plasmid pX01 of *B. anthracis* (11 min processing time for the PCR). This was followed by pressure-injection of postamplified products into the separation domain of the chip along with a DNA sizing ladder and, then, electrokinetic injection into a separation channel for electrophoretic sorting, which was carried out in <180 s. The entire sample processing could be completed in <30 min (see Fig. 17).

Thaitrong et al. [252] fabricated a PCR-CE microsystem manufactured on a glass substrate, whose layout is shown in Fig. 18a. In this work, two types of injectors were compared: a sidearm injector (see Fig. 18b) that utilized a cross-T injection mode, and an in-line injector (see Fig. 18c) that employed a ~1-mm μCE section of crosslinked DNA affinity-capture matrix synthesized in situ by co-photopolymerizing acrylamide with 5′ acrydite-modified oligonucleotide capture probes [253]. For the in-line injector strategy, the PCR products were captured and preconcentrated before thermal release and injection. This strategy was demonstrated with PCR-μCE multiplexed analysis of *E. coli* K12 and *E. coli* O157:H7, where the in-line injector eliminated band broadening and increased the injection efficiency to ~100%. In addition, the on-chip generated PCR amplicons processed on this system with an oligonucleotide affinity capture gel exhibited ~3.6-fold increase in the signal-to-noise ratio, a sixfold increase in resolution, and yielded separation efficiencies in the range of 2×10^5 plates/m compared to the sidearm injector [254, 255].

There is need for portable field testing systems to perform DNA typing, microbial forensics, and ensure public health. For example, point-of-use systems are needed to determine the presence of microorganisms that are potentially harmful to humans [256]. These applications would benefit from a portable microsystem that could provide robust, quantitative analyses for the detection of infectious diseases or analyze DNA forensic samples in a timely manner.

Along these lines, Lagally and coworkers [111] reported an integrated portable genetic analysis microsystem to perform pathogen detection and genotyping directly from whole *E. coli* and *Staphylococcus aureus* cells. The system contained

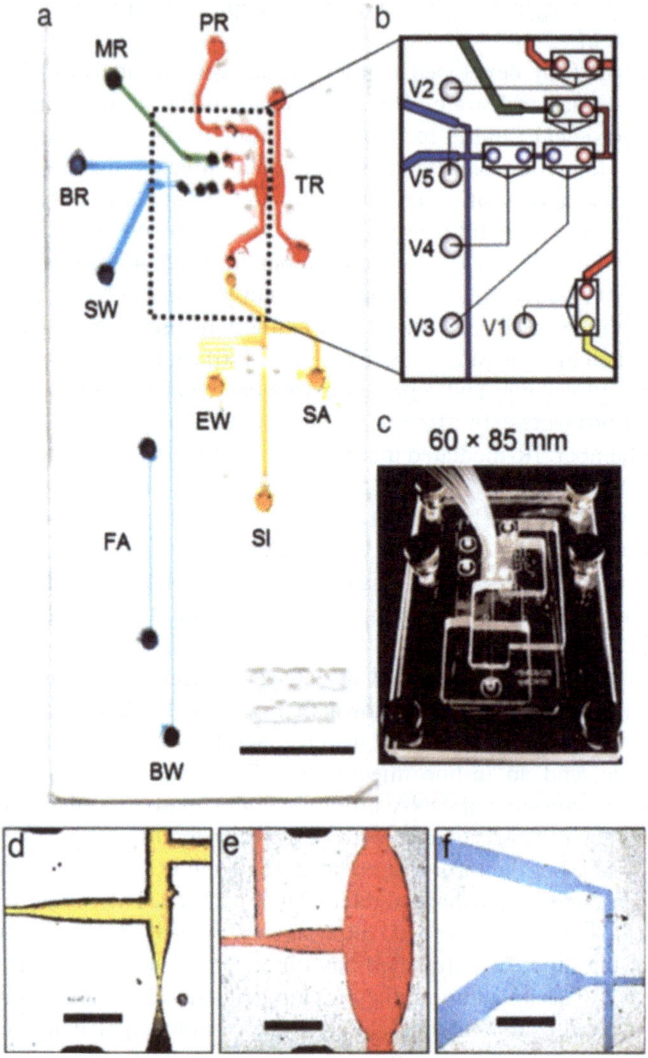

Fig. 16 Microfluidic genetic analysis (MGA) system. (**a**) Dyes are placed in the channels for visualization (*Scale bar*: 10 mm). Domains for DNA extraction (*yellow*), PCR amplification (*red*), injection (*green*), and separation (*blue*) are connected through a network of channels and vias. SPE reservoirs are labeled for sample inlet (*SI*), sidearm (*SA*), and extraction of waste (*EW*). Injection reservoirs are labeled for the PCR reservoir (*PR*), marker reservoir (*MR*), and sample waste (*SW*). Electrophoresis reservoirs are labeled for the buffer reservoir (*BR*) and buffer waste (*BW*). Additional domains patterned onto the device included the temperature reference (*TR*) chamber and fluorescence alignment (*FA*) channel. The flow control region is outlined by a *dashed box*. Device dimensions are 30.0 × 63.5 mm with a total solution volume <10 μL (*Scale bar*: 10 mm). (**b**) Flow control region. Valves are shown as *open rectangles*. *V1* separates the SPE and PCR domains. *V2* and *V5* are inlet valves for the pumping injection, *V3* is the diaphragm valve, and *V4* is an outlet valve. (**c**) Device loaded into the manifold. (**d**) Intersection between SI and SA inlet channels, with the EW channel tapering to increase flow resistance (*Scale bar*: 1 mm).

a single 200-nL PCR amplification chamber directly connected to a μCE separation microchannel, with turns to increase its length. The chip was coupled with a compact electrical control and laser-induced fluorescence (LIF) detector. The system contained microfabricated Ti/Pt heaters with gold leads on the reverse side of the chip, Ti/Pt resistance temperature detectors (RTDs) fabricated within the PCR chamber, and membrane valves to provide controlled sample positioning within the 200-nL PCR chamber (see Fig. 19). In this work, two novel PDMS microvalves were assembled on the top surface of the system. These microvalves exhibited dead volumes as low as 8 nL and were actuated with small pressures and vacuums [257]. Pt electrodes were also placed within the system, allowing application of high voltages without the need for external electrodes. The microsystem was the size of a microscope slide and was placed into a portable analysis instrument containing all of the necessary electronics, optics, and control hardware for conducting genetic analyses (form factor was 20.3 cm × 25.4 cm × 30.5 cm). This microsystem was used to perform a triplexed PCR targeting genes that encode for 16S ribosomal RNA, the flagellar antigen, and the shiga toxin in *E. coli*. Fluorescein-labeled PCR products were electrophoretically analyzed in <10 min.

A portable field testing system for performing human identification via DNA typing directly at a crime scene can be used to rapidly identify potential suspects. In the work of Liu et al. [5], a portable forensic genetic analysis system consisting of two identical PCR-CE systems symmetrically arranged on a 4″. wafer for the amplification and separation of short tandem repeats (STRs) was reported. The structure of the system was similar to that developed by Lagally et al. [111] (see Fig. 19), but the design was modified to accommodate field use. The fluidic chip consisted of a 160-nL PCR chamber, integrated Ti/Pt heaters with gold leads fabricated on the reverse side of the chip, a Ti/Pt four-point RTD fabricated within the PCR chamber, and a 7-cm μCE separation channel. The instrument integrated the following processing steps: PCR thermal cycling, electrophoretic separation, pneumatic valve fluidic control, and four-color laser-induced fluorescence detection. The multiplexing capability was demonstrated using a quadruplex Y-chromosome STR typing system consisting of the amelogenin gene and three Y STR loci (DYS390, DYS393, and DYS439). In this work, 35 PCR cycles for the amplification of all four loci, followed by CE separation and four-color fluorescence detection, was completed in 1.5 h. [10]. Liu and coworkers[258] also demonstrated the capabilities of this portable microsystem for crime scene investigation processes by performing real-time nine-plex autosomal STR analyses at a mock crime scene to secure male and female STR profiles (see Fig. 20) with 100-copy sensitivity. In this demonstration, all alleles were sized correctly in 2.5 h with a standard deviation ≤0.8 bp.

Fig. 16 (Continued) (**e**) PCR chamber with exit channel tapering before intersecting with the MR inlet channel (*Scale bar*: 1 mm). (**f**) Cross-T intersection (*Scale bar*: 1 mm). The relative sizes of the BR, SW, and BW channels create the difference in volume displacement during the pumping injection and affect how the resistance is dropped under an applied separation voltage. Reproduced from [10] with permission

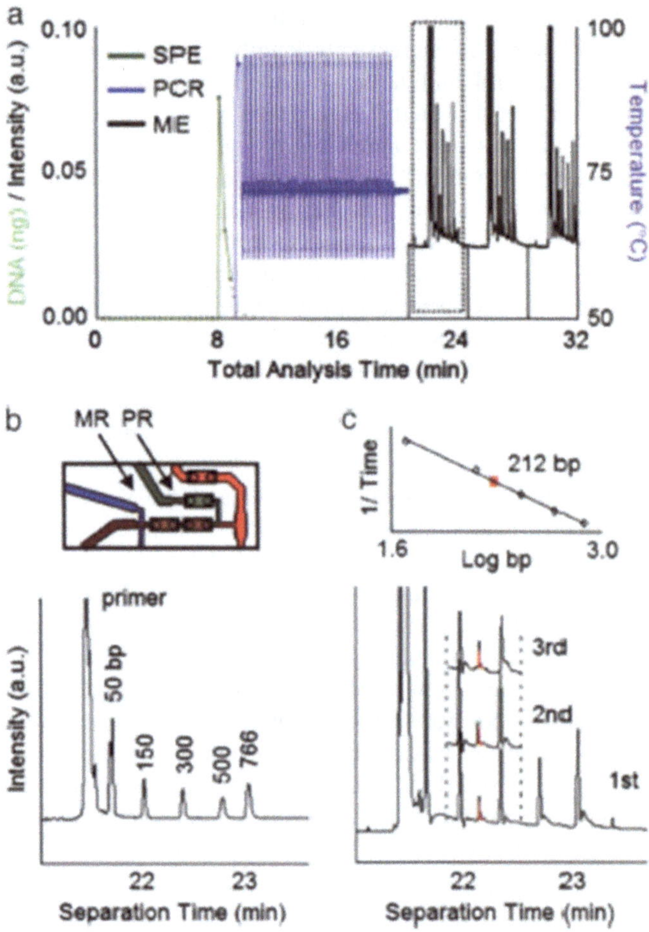

Fig. 17 Detection of *B. anthracis* from murine blood. (**a**) Detector responses during all three stages of sample processing and analysis are portrayed in terms of total analysis time. The SPE trace (*green*) was taken from off-line DNA extraction of the same murine sample and is representative of the total DNA concentration observed in a typical extraction. The temperature (*blue*) and fluorescence intensity (*black*) represent on-line data, with a total analysis time <24 min. Three sequential injections and separations were carried out to ensure the presence of amplified product. (**b**) Fluorescence data from an integrated analysis of a blank sample (no DNA) control with marker peaks labeled. The *inset* represents valve actuation during co-injection, with the PR and MR pumping inlets indicated by the *arrows*. (**c**) Zoomed view of the first separation shown in (**a**), with the product peak marked. The second and third runs are overlaid with the time axis cropped. *Inset* shows the sizing curve of inverse migration time vs. log(base pairs) with both the sizing standard peaks (*open diamonds*) and product (*square*) plotted for all three runs shown in (**a**). From these data, the product was 211 ± 2 bp. Reproduced from [10] with permission

Mathies and his group [54] reported a nanoliter-scale fluidic bioprocessor that integrated three processing steps of Sanger sequencing (thermal cycling, sample purification, and CE) into a single monolithic system. The fluidic system

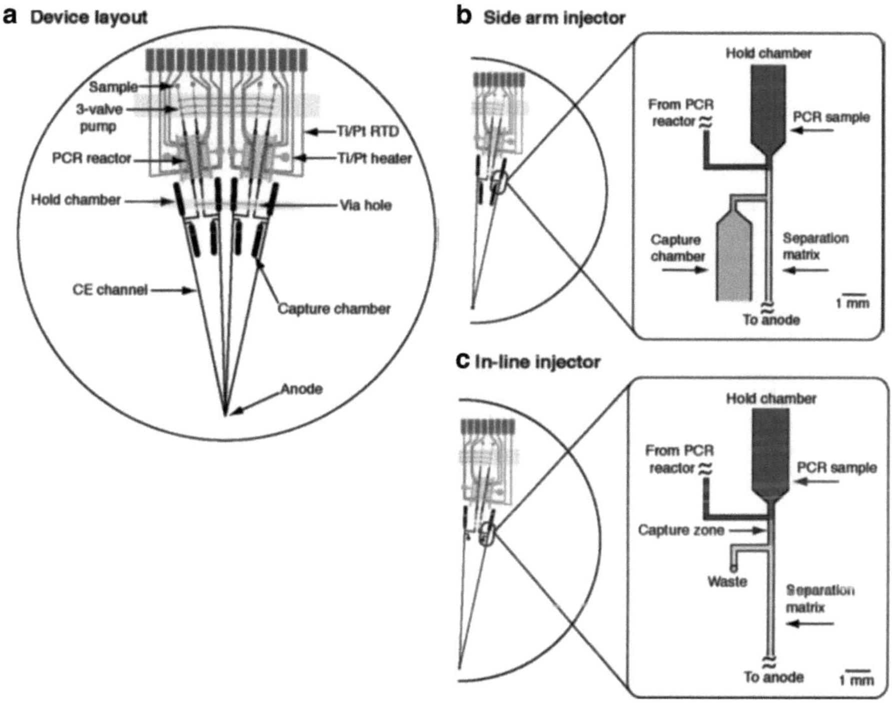

Fig. 18 (**a**) Layout of the PCR-CE microsystem with integrated sample cleanup. Each of the four analyzers is composed of a three-valve pump for fluidic control, an RTD temperature sensor and integrated heater for thermal cycling, a 250-nL PCR reaction chamber, etched chambers for analyte capture and purification, and μCE separation channels 5 cm in length. (**b**) Expanded view of the sidearm capture structure. A capture matrix made of 6.5% linear polyacrylamide (LPA) gel copolymerized with an oligonucleotide capture probe is manually loaded from the sidearm into the capture chamber. (**c**) Expanded view of the in-column capture structure. An in-situ photopolymerized capture matrix made of 5% acrylamide/bis gel copolymerized with an oligonucleotide capture probe is polymerized in a capture zone (40 mm deep × 120 mm wide × 1 mm long) directly in-line with the separation channel. Reproduced from [252] with permission

(see Fig. 21a) was microfabricated using a hybrid glass/PDMS wafer with the following three devices: (1) thermal cycling (TC) device, which consisted of a 250-nL reaction chamber with integrated resistive temperature probes, microvalves, and surface heaters; (2) affinity-capture purification chambers for cleaning up the reaction products prior to electrophoretic sorting; and (3) μCE channels 30 cm in length that was folded via turns 65 μm in width. The pneumatic valves and pumps were also included in the microsystem. Overall, the system was comprised of four-layers, three of which were glass wafers 100 mm in diameter that contained the fluidic channels and a final layer consisting of a PDMS membrane. The reaction chambers and μCE channels were etched between two thermally bonded glass wafers, while channel interconnections and microvalves were formed

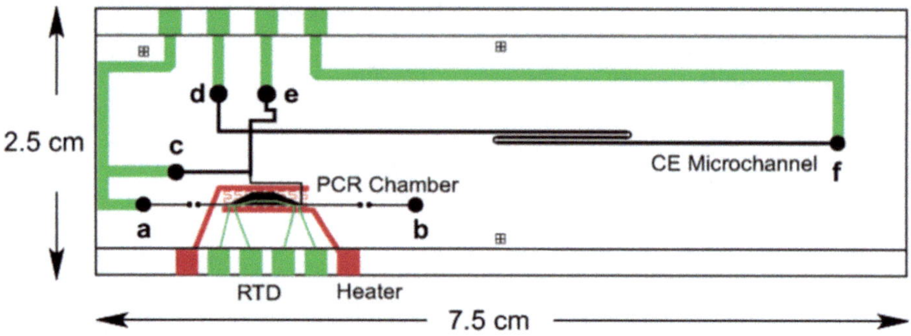

Fig. 19 Mask design for the portable PCR-CE system. The glass microchannels are indicated in *black*, the RTD and microfabricated electrodes are in *green*, and the heater (located on the back of the device) is shown in *red*. The PCR chamber is loaded through reservoirs *a* and *b*. Reservoir *c* is the co-inject reservoir, *d* is the cathode, *e* is the waste, and *f* is the anode. Reproduced from [111] with permission

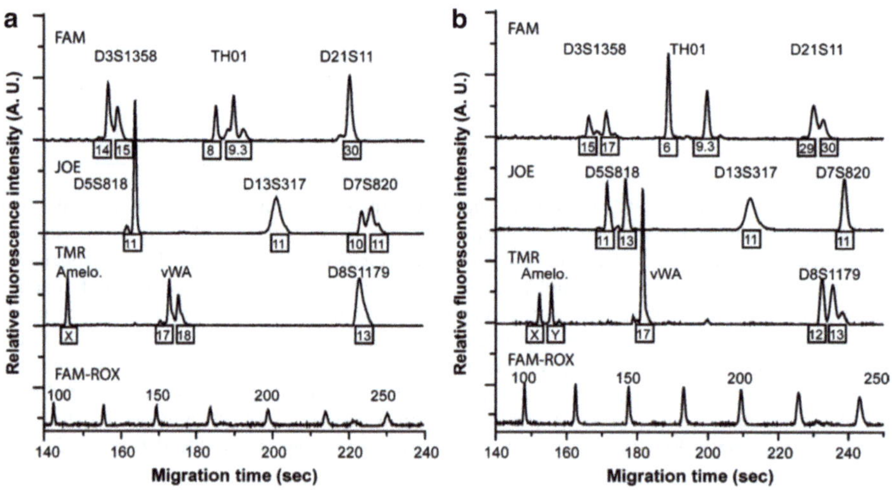

Fig. 20 Representative nine-plex STR profiles of (**a**) 9,947A female and (**b**) 9,948 male standard DNA obtained with 100 copies of DNA template in the PCR chamber of an integrated microfluidic system. Reproduced from [258] with permission

by the PDMS and a bottom manifold glass wafer. The system was demonstrated for complete Sanger sequencing from only 1 fmol of DNA template, with 556 continuous bases sequenced at 99% accuracy (see Fig. 21g), demonstrating read lengths appropriate for de novo sequencing of human and other complex genomes. The base-call accuracy estimates were accumulated using PHRED 0.020425.C. [259].

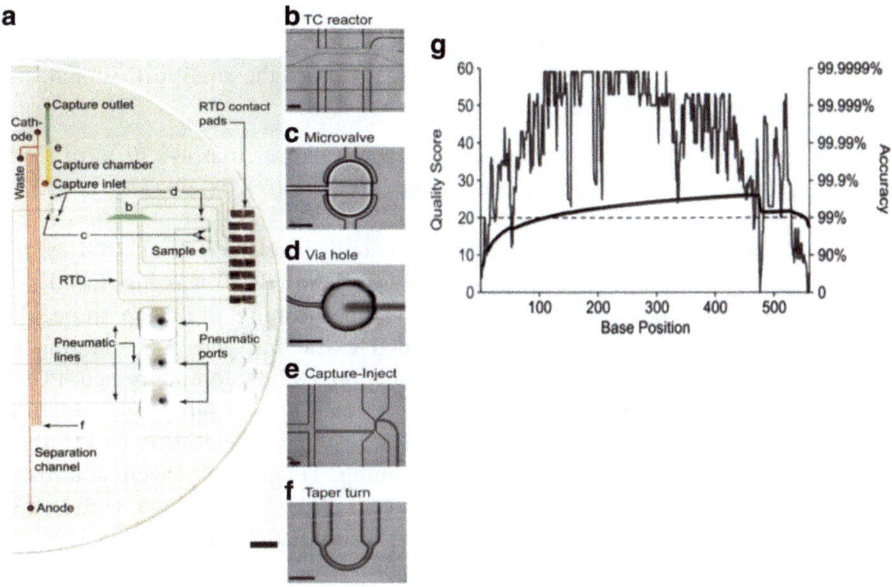

Fig. 21 (**a**) Photograph of the system bioprocessor components showing a DNA processing system. Colors indicate the location of sequencing reagent (*green*), capture gel (*yellow*), separation gel (*red*), and pneumatic channels (*blue*) (*Scale bar*: 5 mm). Component microphotographs b (*Scale bar*: 1 mm), c-f (*Scale bars*: 300 μm): (**b**) 250-nL thermal cycling reactor with RTDs; (**c**) 5-nL displacement volume microvalve; (**d**) via hole 500 g in diameter; (**e**) capture chamber and cross-injector; (**f**) taper 65 μm in width that folded the 30-cm channel to minimize turn-induced dispersion. (**g**) Base-call accuracies and sequence read length as predicted by PHRED. Percentage accuracy is related to the PHRED quality score by $100(1 - P_e)$, where P_e is the probability that the base call is incorrect and is equal to $1/10^{Q/10}$. A one-in-a-hundred error rate is indicated by the *dashed line*. PHRED quality scores are plotted at each base position. The *thick black line* charts predicted read accuracy at each base position: $100(\mathrm{Base}_i - \Sigma\ Pe_i)/\mathrm{Base}_i$, where Base_i is base position. Reproduced from [54] with permission

4.2 Integrated Systems with Microarray Readout

Unitizing the advantages of microchip-based DNA microarrays such as the highly parallel nature of the readout and the ability to screen DNA sequences with high specificity, numerous attempts have been made to incorporate front-end sample processing strategies with DNA microarrays used as the terminal readout step onto a single microfluidic platform. The front-end processing strategies that are needed prior to microarray readout are similar in nature to those required for μCE and are delineated in Fig. 1. The operational differences, in terms of microsystems using arrays versus μCE include:

1. No need for high voltage power supplies
2. Imaging over relatively larger areas when fluorescence detection is employed (imaging optics versus point detection of μCE)

3. On-chip heaters to control the temperature of the array, depending on the level of hybridization stringency required
4. Reduced processing times compared to μCE due to the small diffusional distances associated with microfluidics

Several examples of front-end process integration and microarrays to form functional microsystems have been reported in the literature [69, 78, 98, 114, 208, 210, 260–262].

One of the initial microsystems reported was by Anderson et al. [69] over a decade ago. A monolithic biochemical processing unit (BPU) was interfaced to a GeneChip commercialized by Affymetrix for performing multistep molecular processing of genomic samples and included extracting and concentrating of nucleic acids from a serum lysate, amplification (RT-PCR and nested PCR), enzymatic reactions (fragmentation, dephosphorylation, and labeling), metering, mixing, and hybridization to the GeneChip. The system was fabricated in PC using conventional computer-controlled micromachining. Temperatures were controlled by pressing thermal elements against the thin wall of the PC cartridge, with Peltier junctions used for heating and cooling. Fluidic manipulation was achieved through the use of a fluid barrier or hydrophobic membrane in conjunction with a pneumatically controlled diaphragm valve and hydrophobic vent. The system performance was evaluated using serum samples loaded with HIV virus. Analysis of the GeneChip results yielded an average accuracy of 99.7%, as determined by independent sequencing.

Liu et al. [98] presented a disposable, monolithic device that integrated PCR and a DNA microarray. The system was also fabricated in PC using CO_2 laser machining (see Fig. 22a). This system was assembled using a two-step process: First, thermal fusion bonding of PC was performed at 139 °C and under 2 tons of pressure for 45 min with a square window for the DNA array left open. Second, following surface activation and oligonucleotide probe immobilization using a Motorola proprietary attachment chemistry through the access window, the window was closed with another piece of properly sized PC using double-sided tape. The bonding was enforced by applying 2 tons of pressure for 2 min, and then the edges were sealed with epoxy. During PCR thermal cycling, the PCR device of the monolithic chip was sandwiched between two Peltier elements to allow thermal processing of the sample, with temperatures being monitored using thermocouples. Microfluidic control was accomplished through the use of three external syringe pumps docked to the system in combination with four on-chip Pluronics polymer valves and one hydrophobic valve. Asymmetrical PCR amplification and subsequent hybridization analysis of both *E. coli* and *Enterococcus faecalis* was demonstrated. However, the use of PC as the microarray platform generated a significant amount of autofluorescence, which degraded the detection limits for fluorescence readout (see Fig. 22b).

To overcome this problem, Hashimoto et al. [208] coupled CFPCR and continuous flow LDR (CFLDR) devices, both fabricated using a PC substrate, with a universal microarray fabricated using a PMMA substrate, which possesses better

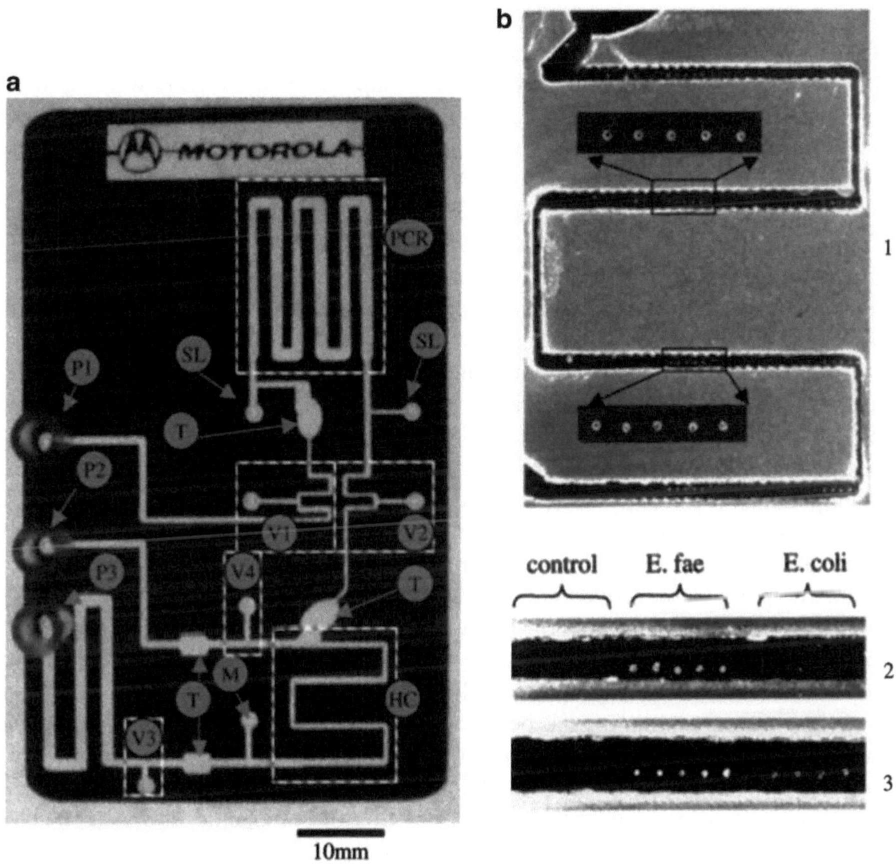

Fig. 22 (**a**) Monolithic integrated polycarbonate DNA analysis system. The system contained a serpentine PCR channel (*PCR*), a hybridization channel (*HC*), a syringe coupled to a hybridization wash solution channel, a waste channel coupled to a waste syringe, Pluronic traps (*T*), one hydrophobic air-permeable membrane valve (*M*), four Pluronics valves (*V1–V4*), two PCR reagent loading holes (*SL*), and three external syringe pumps interfaced to reservoirs: sample driving syringe pump (*P1*), waste-withdrawing syringe pump (*P2*), and wash syringe pump (*P3*). The dimensions of the system were 5.4 cm × 8.6 cm × 0.75 mm. (**b**) PCR hybridization results from the monolithic integrated system. *1 E. coli* 221-bp hybridization after amplification. Portions of the channel are enlarged for better viewing. *2* Fluorescent image of portion of the channel after *E. faecalis* amplification and hybridization. *3* Fluorescent image of portion of the channel after multiplex (*E. faecalis* and *E. coli*) amplification and hybridization. Reproduced from [98] with permission

optical properties with lower autofluorescence levels compared to PC (see Fig. 15). The chip was generated via microreplication (hot embossing) from metal molding tools fabricated using high-precision micromilling. The CFPCR/CFLDR chip was directly attached to thin film heaters for providing the set temperatures for the isothermal zones, with thermocouples embedded between the cover plate and the film heaters for monitoring the set temperatures required for both PCR and LDR.

In this work, low density universal microarrays were produced on the bottom floor of a UV-photoactivated PMMA microchannel, with the DNA zipcode probes attached to UV-generated carboxylic acid groups. PCR amplicons were used as templates for the allele-specific CFLDR, which produced single-stranded targets that were uniformly flowed over the universal array to reduce incubation times. Using a mixed population of genomic DNA as starting materials, one mutant in 80 wild-type sequences could be successfully discriminated in a total reaction time of 50 min, including 18.7 min for PCR, 8.1 min for LDR, 5 min for hybridization, 10 min for washing, and 2.6 min for fluorescence imaging of the low-density array. The authors also showed the ability to reduce reagent consumption by one order of magnitude compared to similar benchtop assays.

A self-contained biochip that integrated cell isolation and lysis with PCR amplification and electrochemical microarray-based detection was described by Liu et al. [262] The chip was machined in a PC substrate using a conventional computer-controlled milling machine and included a mixing unit for cell capture using immunomagnetic beads, a cell preconcentration/purification/lysis/PCR unit, and a DNA microarray chamber. In this work, fluidic components (e.g., paraffin-based microvalves, cavitation microstreaming mixers, and electrochemical or thermo-pneumatic pumps), embedded resistive heaters, and DNA microarray sensors were coupled to the system to perform DNA analysis of biological samples. Electrical power, PCR thermal cycling, DNA electrochemical signal readout, and magnetic elements for bead arrest were controlled by an off-chip instrument. Implementation of cavitation microstreaming has been shown to achieve cell capture efficiencies on the order of 73% using immunomagnetic beads and up to a fivefold reduction in hybridization time compared to passive incubation of the array with solution targets, as well as improved signal uniformity. Detection of pathogenic *E. coli* K12 cells seeded into rabbit blood and single-nucleotide polymorphism analysis from diluted blood samples were completed in 3.5 and 2.7 h, respectively.

Soper et al. [210] designed a polymer-based modular microsystem that could accept a crude sample and automatically carry out the entire molecular processing pipeline in an enclosed fluidic cartridge (see Fig. 23). The multistep assay included bacterial cell lysis, SPE of genomic DNA from the lysate, PCR amplification, LDR, and universal DNA array readout. The fluidic cartridge was generated via micro-replication from the appropriate metal molding tools, which were used to create structures on both sides of the polymer substrate (i.e., double-sided hot embossing). The integrated fluidic cartridge was comprised of a fluidic motherboard and two modules. One module was made from PC and used for SPE, while the other module was made from PMMA and contained DNA probes patterned on a planar wave-guide for evanescent excitation. These modules were interconnected to a fluidic motherboard fabricated in PC and were used for processing steps for thermal cell lysis, PCR, and LDR. Fluid handling, thermal management, and optical detection were controlled by off-chip supporting peripherals, which could be packaged into a small footprint instrument (1 ft^3). Identification of multidrug-resistant tuberculosis (MDR-TB) resulting from *Mycobacterium tuberculosis* (Mtb) strains in clinical sputum samples were demonstrated with a detection limit of ~50 bacterial cells

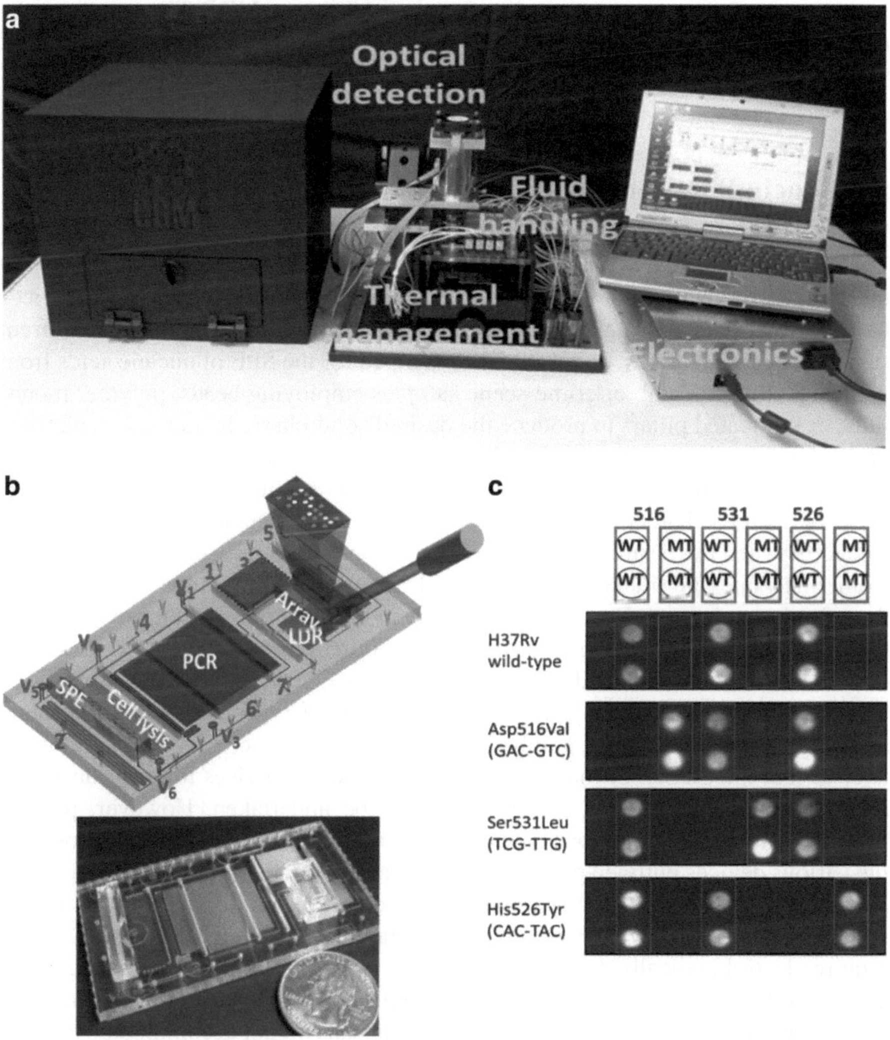

Fig. 23 Integrated, modular microfluidic chip for TB analysis. (**a**) Overview of the integrated system. The system had dimensions 12″ (length) × 12″ (width) × 12″ (height), and all fluid handling, thermal management, and optical detection were controlled by off-chip supporting peripherals and assembled into a small form factor instrument. (**b**) Schematic and photograph of the fluidic cartridge. The fluidic cartridge was composed of two modules and a fluidic motherboard. The fluidic motherboard was made from PC and consisted of processing steps for cell lysis, PCR, and LDR. One module was made from PC and used for SPE of genomic DNA, while the other module was made from PMMA and contained an air-embedded planar waveguide and the DNA array. Sample inlet (*1*), PCR mixture inlet (*2*), LDR mixture inlet (*3*), ethanol and air inlet (*4*), array wash inlet (*5*), vacuum connection (*6*), and waste outlet (*7*). *V1–V6* were on-chip membrane valves. *V2* is positioned next to the SPE module on the cell lysis microchannel and is not visible in current view. (**c**) Molecular assay results from drug-susceptible *Mycobacterium tuberculosis* (Mtb) strains and drug resistant Mtb strains. *516WT*, *531WT* and *526WT* are probes targeting drug-susceptible Mtb strains. *516MT*, *531MT*, and *526MT* are probes targeting drug-resistant Mtb strains. Reproduced from [210] with permission

from sputum (processing time <40 min). In addition to MDR-TB detection, the modular fluidic cartridge could be reconfigured for use with other assay formats, such as PCR-μCE.

5 Concluding Remarks

There have been extensive reports on devices designed to perform a single-step in the analysis of a variety of nucleic acids, such as DNAs and RNAs. These devices have been fabricated using a variety of micromanufacturing techniques in different substrate materials. Devices have been developed for the SPE of nucleic acids from clinical, environmental, or crime-scene samples employing beads, polymer monoliths, or fabricated pillars to produce the desired solid phase. In addition, a plethora of devices focused on thermal cycling (such as that required for PCR, cycle sequencing, or allele-specific LDR) have been detailed in the literature and typically use either a chamber-type approach, in which the chamber and its contents are cycled between the desired temperatures, or a continuous flow operation, in which isothermal zones are situated on the chip and the reaction fluid is transported through these isothermal zones. Devices have also been reported that perform μCE separation of DNAs using sieving matrices with various channel lengths to reach the desired resolution demanded of the separation. While these represent innovative concepts, the complete analysis of different sample types required for genetic analyses typically requires a number of processing steps (see Fig. 1). Therefore, it is clear that integration of many of the aforementioned devices to form functional and autonomously operated systems needs to be undertaken. However, process integration to form autonomous systems is not simply a matter of "hooking" together the various devices outlined previously. For example, many of the upstream processing steps and the reagents they require may be detrimental to those poised downstream. Also, some steps demand hydrodynamically driven flow, while others require electrokinetically driven flow. Another concern is unswept volumes, which can generate sample carryover artifacts or sample loss, especially when dealing with ultrasmall sample volumes. Some substrate materials do not accommodate particular processing steps, and high optical quality materials must be used for assays employing fluorescence as the detection mode. Finally, some process steps require thermal control, such as PCR, and these thermally actuated units must be isolated from those that are sensitive to temperature, such as microarrays.

Several examples of integrated microfluidic systems have been presented herein, most of which are proof-of-concept demonstrations with only a few examples that have actually dealt with clinical or "real-world" samples. Unfortunately, many of these demonstrators of integrated microfluidic systems, while attractive in terms of their ability to reduce sample processing time and reagent consumption, have only been utilized in research settings. Some of the more compelling applications for integrated microfluidic systems, such as in vitro diagnostics, homeland security, or forensics, will demand systems that can accommodate field analysis and/or one-time use operation. For example, in the case of in vitro diagnostics, it will be necessary to use the entire microfluidic system for a single patient sample, demanding that the cost of the chip be low. In addition, field analysis applications will require not only that the chip possesses a small footprint, but also that the support peripherals must have the same characteristic. These support peripherals include pumps, valves, reagent reservoirs, electronics, and optomechanics if some type of optical readout is used. The chip and support peripherals must all be packaged into a small form factor instrument and must consume minimal amounts of power to enable battery operation for extended periods of time. All this must be engineered without sacrificing assay performance in terms of reproducibility, LOD, sensitivity, and specificity.

Another interesting aspect is related to the manufacturability of the system. For wide spread commercialization, the fluidic system must be produced in high volume and at low cost. Chip production not only includes the microfabrication of the channel networks, but also chip assembly, integration of various components such as electrodes, optical elements, and valves and, finally, the surface attachment of necessary biologics to effect the desired process step. Although microfluidics has shifted from the use of silicon, glass, and other similar materials that require extensive microfabrication procedures to the use of polymer substrates that can use microreplication processes (similar to those used to produce CDs and DVDs) to produce the desired fluidic networks in a high production mode with good fidelity, the challenge still remains in chip finishing following production of the fluidic network.

The driving force behind the increasing development of integrated microfluidic systems is certainly due to their potential commercialization, but also in their diverse applications in such areas as biology, chemistry, and other disciplines that strongly demand the emergence of new analysis platforms to achieve higher performance and throughput. Because direct integration of PCR with other sample preparation protocols, including μCE, fluorescence, and microarrays has been demonstrated for a wide range of applications (pathogen detection, DNA typing, and DNA sequencing), these success stories will demand higher functionality at lower cost and with higher throughput. Such systems offer compelling advantages such as short assay turnaround times, automated operation, improved operator protection, lower cross-contamination, reduced human error, and lower overall assay cost. Minimization of potential carryover contamination from run-to-run is a key consideration in providing accurate and reliable results and the use of disposable fluidic cartridges will effectively minimize this risk.

References

1. Collins FS, Morgan M, Patrinos A (2003) Science 300:286
2. Aboud MJ, Gassmann M, McCord BR (2010) Electrophoresis 31:2672
3. Hopwood AJ, Hurth C, Yang J, Cai Z, Moran N, Lee-Edghill JG, Nordquist A, Lenigk R, Estes MD, Haley JP, McAlister CR, Chen X, Brooks C, Smith S, Elliott K, Koumi P, Zenhausern F, Tully G (2010) Anal Chem 82:6991

4. Horsman KM, Bienvenue JM, Blasier KR, Landers JP (2007) J Forensic Sci 52:784
5. Liu P, Seo TS, Beyor N, Shin KJ, Scherer JR, Mathies RA (2007) Anal Chem 79:1881
6. Chovan T, Guttman A (2002) Trends Biotechnol 20:116
7. Verpoorte E (2002) Electrophoresis 23:677
8. Khandurina J, McKnight TE, Jacobson SC, Waters LC, Foote RS, Ramsey JM (2000) Anal Chem 72:2995
9. Koh CG, Tan W, Zhao MQ, Ricco AJ, Fan ZH (2003) Anal Chem 75:4591
10. Easley CJ, Karlinsey JM, Bienvenue JM, Legendre LA, Roper MG, Feldman SH, Hughes MA, Hewlett EL, Merkel TJ, Ferrance JP, Landers JP (2006) Proc Natl Acad Sci USA 103:19272
11. Pal R, Yang M, Lin R, Johnson BN, Srivastava N, Razzacki SZ, Chomistek KJ, Heldsinger DC, Haque RM, Ugaz VM, Thwar PK, Chen Z, Alfano K, Yim MB, Krishnan M, Fuller AO, Larson RG, Burke DT, Burns MA (2005) Lab Chip 5:1024
12. deMello AJ (2006) Nature 442:394
13. Beebe DJ, Mensing GA, Walker GM (2002) Annu Rev Biomed Eng 4:261
14. Manz A, Graber N, Widmer HM (1990) Sens Actuators B Chem 1:244
15. Kan CW, Fredlake CP, Doherty EAS, Barron AE (2004) Electrophoresis 25:3564
16. Woolley AT, Mathies RA (1995) Anal Chem 67:3676
17. Schmalzing D, Adourian A, Koutny L, Ziaugra L, Matsudaira P, Ehrlich D (1998) Anal Chem 70:2303
18. Di Carlo D, Ionescu-Zanetti C, Zhang Y, Hung P, Lee LP (2005) Lab Chip 5:171
19. Irimia D, Tompkins RG, Toner M (2004) Anal Chem 76:6137
20. Waters LC, Jacobson SC, Kroutchinina N, Khandurina J, Foote RS, Ramsey JM (1998) Anal Chem 70:158
21. Belgrader P, Hansford D, Kovacs GTA, Venkateswaran K, Mariella R, Milanovich F, Nasarabadi S, Okuzumi M, Pourahmadi F, Northrup MA (1999) Anal Chem 71:4232
22. Taylor MT, Belgrader P, Furman BJ, Pourahmadi F, Kovacs GTA, Northrup MA (2001) Anal Chem 73:492
23. Lu H, Schmidt MA, Jensen KF (2005) Lab Chip 5:23
24. Lee SW, Tai YC (1999) Sens Actuators A Phys 73:74
25. Wang HY, Bhunia AK, Lu C (2006) Biosens Bioelectron 22:582
26. Nevill JT, Cooper R, Dueck M, Breslauer DN, Lee LP (2007) Lab Chip 7:1689
27. Abolmaaty A, El-Shemy MG, Khallaf MF, Levin RE (1998) J Microbiol Methods 34:133
28. Belgrader P, Benett W, Hadley D, Richards J, Stratton P, Mariella R, Milanovich F (1999) Science 284:449
29. Yang JM, Bell J, Huang Y, Tirado M, Thomas D, Forster AH, Haigis RW, Swanson PD, Wallace RB, Martinsons B, Krihak M (2002) Biosens Bioelectron 17:605
30. Fox MB, Esveld DC, Valero A, Luttge R, Mastwijk HC, Bartels PV, van den Berg A, Boom RM (2006) Anal Bioanal Chem 385:474
31. Wolfe KA, Breadmore MC, Ferrance JP, Power ME, Conroy JF, Norris PM, Landers JP (2002) Electrophoresis 23:727
32. Tian HJ, Huhmer AFR, Landers JP (2000) Anal Biochem 283:175
33. Breadmore MC, Wolfe KA, Arcibal IG, Leung WK, Dickson D, Giordano BC, Power ME, Ferrance JP, Feldman SH, Norris PM, Landers JP (2003) Anal Chem 75:1880
34. Legendre LA, Bienvenue JM, Roper MG, Ferrance JP, Landers JP (2006) Anal Chem 78:1444
35. Wu QR, Bienvenue JM, Hassan BJ, Kwok YC, Giordano BC, Norris PM, Landers JP, Ferrance JP (2006) Anal Chem 78:5704
36. Wen J, Guillo C, Ferrance JP, Landers JP (2007) Anal Chem 79:6135
37. Bhattacharyya A, Klapperich CM (2006) Anal Chem 78:788
38. Mahalanabis M, Al-Muayad H, Kulinski MD, Altman D, Klapperich CM (2009) Lab Chip 9:2811

39. Kulinski MD, Mahalanabis M, Gillers S, Zhang JY, Singh S, Klapperich CM (2009) Biomed Microdevices 11:671
40. Sauer-Budge AF, Mirer P, Chatterjee A, Klapperich CM, Chargin D, Sharon A (2009) Lab Chip 9:2803
41. Chung YC, Jan MS, Lin YC, Lin JH, Cheng WC, Fan CY (2004) Lab Chip 4:141
42. Wen J, Guillo C, Ferrance JP, Landers JP (2006) Anal Chem 78:1673
43. Legendre LA, Morris CJ, Bienvenue JM, Barron A, McClure R, Landers JP (2008) J Assoc Lab Auto 13:351
44. Christel LA, Petersen K, McMillan W, Northrup MA (1999) J Biomech Eng Trans ASME 121:22
45. Cady NC, Stelick S, Batt CA (2003) Biosens Bioelectron 19:59
46. Witek MA, Llopis SD, Wheatley A, McCarley RL, Soper SA (2006) Nucleic Acids Res 34:9
47. Witek MA, Hupert ML, Park DSW, Fears K, Murphy MC, Soper SA (2008) Anal Chem 80:3483
48. Park DSW, Hupert ML, Witek MA, You BH, Datta P, Guy J, Lee JB, Soper SA, Nikitopoulos DE, Murphy MC (2008) Biomed Microdevices 10:21
49. Kim J, Gale BK (2008) Lab Chip 8:1516
50. Kim J, Mauk M, Chen DF, Qiu XB, Gale B, Bau HH (2010) Analyst 135:2408
51. Xu YC, Vaidya B, Patel AB, Ford SM, McCarley RL, Soper SA (2003) Anal Chem 75:2975
52. Wang H, Chen JF, Zhu L, Shadpour H, Hupert ML, Soper SA (2006) Anal Chem 78:6223
53. Paegel BM, Yeung SHI, Mathies RA (2002) Anal Chem 74:5092
54. Blazej RG, Kumaresan P, Mathies RA (2006) Proc Natl Acad Sci USA 103:7240
55. Oleschuk RD, Shultz-Lockyear LL, Ning YB, Harrison DJ (2000) Anal Chem 72:585
56. Nakagawa T, Tanaka T, Niwa D, Osaka T, Takeyama H, Matsunaga T (2005) J Biotechnol 116:105
57. Cao WD, Easley CJ, Ferrance JP, Landers JP (2006) Anal Chem 78:7222
58. Saiki RK, Bugawan TL, Horn GT, Mullis KB, Erlich HA (1986) Nature 324:163
59. DeMello AJ (2003) Nature 422:28
60. Higuchi R, Fockler C, Dollinger G, Watson R (1993) Biotechnology 11:1026
61. Wiesner RJ (1992) Nucleic Acids Res 20:5863
62. Schönbrunner NJ, Fiss EH, Budker O, Stoffel S, Sigua CL, Gelfand DH, Myers TW (2006) Biochemistry 45:12786
63. Innis MA, Myambo KB, Gelfand DH, Brow MAD (1988) Proc Natl Acad Sci USA 85:9436
64. Northrup MA, Ching MT, White RM, Watson RT (1993) DNA amplification in a microfabricated reaction chamber. In: Proceedings seventh international conference on solid state sensors and actuators, Transducers'93, Yokohama, Japan, 7–10 June 1993, pp 924
65. Wilding P, Pfahler J, Bau HH, Zemel JN, Kricka LJ (1994) Clin Chem 40:43
66. Wilding P, Shoffner MA, Kricka LJ (1994) Clin Chem 40:1815
67. Wilding P, Shoffner MA, Cheng J, Hvichia G, Kricka LJ (1995) Clin Chem 41:1367
68. Soper SA, Ford SM, Xu YC, Qi SZ, McWhorter S, Lassiter S, Patterson D, Bruch RC (1999) J Chromatogr A 853:107
69. Anderson RC, Su X, Bogdan GJ, Fenton J (2000) Nucleic Acids Res 28:E60
70. Burns MA, Mastrangelo CH, Sammarco TS, Man FP, Webster JR, Johnson BN, Foerster B, Jones D, Fields Y, Kaiser AR, Burke DT (1996) Proc Natl Acad Sci USA 93:5556
71. Huhmer AFR, Landers JP (2000) Anal Chem 72:5507
72. Ferrance JP, Wu QR, Giordano B, Hernandez C, Kwok Y, Snow K, Thibodeau S, Landers JP (2003) Anal Chim Acta 500:223
73. Giordano BC, Ferrance J, Swedberg S, Huhmer AFR, Landers JP (2001) Anal Biochem 291:124
74. Giordano BC, Copeland ER, Landers JP (2001) Electrophoresis 22:334
75. Wilding P, Kricka LJ, Cheng J, Hvichia G, Shoffner MA, Fortina P (1998) Anal Biochem 257:95
76. Zhao Z, Cui Z, Cui DF, Xia SH (2003) Sens Actuators A Phys 108:162

77. Woolley AT, Hadley D, Landre P, deMello AJ, Mathies RA, Northrup MA (1996) Anal Chem 68:4081
78. Trau D, Lee TMH, Lao AIK, Lenigk R, Hsing IM, Ip NY, Carles MC, Sucher NJ (2002) Anal Chem 74:3168
79. Lagally ET, Emrich CA, Mathies RA (2001) Lab Chip 1:102
80. Liu J, Hansen C, Quake SR (2003) Anal Chem 75:4718
81. Zou QB, Miao YB, Chen Y, Sridhar U, Chong CS, Chai TC, Tie Y, Teh CHL, Lim TM, Heng C (2002) Sens Actuators A Phys 102:114
82. Zou QB, Sridhar U, Chen Y, Singh J (2003) IEEE Sens J 3:774
83. Dunn WC, Jacobson SC, Waters LC, Kroutchinina N, Khandurina J, Foote RS, Justice MJ, Stubbs LJ, Ramsey JM (2000) Anal Biochem 27:157
84. Matsubara Y, Kerman K, Kobayashi M, Yamamura S, Morita Y, Tamiya E (2005) Biosens Bioelectron 20:1482
85. Koh CG, Tan W, Zhao MQ, Ricco AJ, Fan ZH (2003) Anal Chem 75:6379
86. Easley CJ, Karlinsey JM, Landers JP (2006) Lab Chip 6:601
87. Kopp MU, de Mello AJ, Manz A (1998) Science 280:1046
88. Chiem N, Harrison DJ (1997) Anal Chem 69:373
89. Hashimoto M, Chen PC, Mitchell MW, Nikitopoulos DE, Soper SA, Murphy MC (2004) Lab Chip 4:638
90. Chen PC, Nikitopoulos DE, Soper SA, Murphy MC (2008) Biomed Microdevices 10:141
91. Chen PC, Park DS, You BH, Kim N, Park T, Soper SA, Nikitopoulos DE, Murphy MC (2010) Sens Actuators B Chem 149:291
92. Park DSW, Chen PC, You BH, Kim N, Park T, Lee TY, Datta P, Desta Y, Soper SA, Nikitopoulos DE, Murphy MC (2010) J Micromech Microeng 20:085011
93. Park DS, Wang H, Chen P-C, Park T, Kim N, You BH, Nikitopoulos DE, Soper SA, Murphy MC (2010) Passive micro-assembly of a fluidic control chip and a multi-well continuous flow PCR chip for high throughput applications. In: Proceedings 14th international conference on miniaturized systems for chemistry and life sciences, μTas2010, 3–7 October 2010, Groningen, Netherlands, pp 1126
94. Yeung SHI, Liu P, Del Bueno N, Greenspoon SA, Mathies RA (2009) Anal Chem 81:210
95. West J, Karamata B, Lillis B, Gleeson JP, Alderman J, Collins JK, Lane W, Mathewson A, Berney H (2002) Lab Chip 2:224
96. Roper MG, Easley CJ, Landers JP (2005) Anal Chem 77:3887
97. Chen JF, Wabuyele M, Chen HW, Patterson D, Hupert M, Shadpour H, Nikitopoulos D, Soper SA (2005) Anal Chem 77:658
98. Liu YJ, Rauch CB, Stevens RL, Lenigk R, Yang J, Rhine DB, Grodzinski P (2002) Anal Chem 74:3063
99. Curcio M, Roeraade J (2002) Anal Chem 75:1
100. Park N, Kim S, Hahn JH (2003) Anal Chem 75:6029
101. Liu J, Enzelberger M, Quake S (2002) Electrophoresis 23:1531
102. Sadler DJ, Changrani R, Roberts P, Chou CF, Zenhausern F (2003) IEEE Trans Compon Pack Technol 26:309
103. Chow AW (2002) AlChE J 48:1590
104. Zheng W, Chen S (2001) Proc SPIE 4560:256
105. Schneegass I, Kohler JM (2001) J Biotechnol 82:101
106. Hupert ML, Witek MA, Wang Y, Mitchell MW, Liu X, Bejat Y, Nikitopoulos DE, Goettert J, Murphy MC, Soper SA (2003) Proc SPIE 4982:52
107. Mitchell MW, Liu X, Bejat Y, Nikitopoulos DE, Soper SA, Murphy MC (2003) Proc SPIE 4982:83
108. Chen Z, Qian S, Abrams WR, Malamud D, Bau HH (2004) Anal Chem 76:3707
109. Zhang CS, Xu JL, Ma WL, Zheng WL (2006) Biotechnol Adv 24:243
110. Bu MQ, Melvin T, Ensell G, Wilkinson JS, Evans AGR (2003) J Micromech Microeng 13: S125

111. Lagally ET, Scherer JR, Blazej RG, Toriello NM, Diep BA, Ramchandani M, Sensabaugh GF, Riley LW, Mathies RA (2004) Anal Chem 76:3162
112. Lee DS, Park SH, Yang HS, Chung KH, Yoon TH, Kim SJ, Kim K, Kim YT (2004) Lab Chip 4:401
113. Lee DS, Wu MH, Ramesh U, Lin CW, Lee TM, Chen PH (2004) Sens Actuators B Chem 100:401
114. Lee TMH, Carles MC, Hsing IM (2003) Lab Chip 3:100
115. Rodriguez I, Lesaicherre M, Tie Y, Zou QB, Yu C, Singh J, Meng LT, Uppili S, Li SFY, Gopalakrishnakone P, Selvanayagam ZE (2003) Electrophoresis 24:172
116. Chou CF, Changrani R, Roberts P, Sadler D, Burdon J, Zenhausern F, Lin S, Mulholland A, Swami N, Terbrueggen R (2002) Microelectron Eng 61–2:921
117. Poser S, Schulz T, Dillner U, Baier V, Kohler JM, Schimkat D, Mayer G, Siebert A (1997) Sens Actuators A Phys 62:672
118. Belgrader P, Young S, Yuan B, Primeau M, Christel LA, Pourahmadi F, Northrup MA (2001) Anal Chem 73:286
119. Sun K, Yamaguchi A, Ishida Y, Matsuo S, Misawa H (2002) Sens Actuators B Chem 84:283
120. Fukuba T, Yamamoto T, Naganuma T, Fujii T (2004) Chem Eng J 101:151
121. Friedman NA, Meldrum DR (1998) Anal Chem 70:2997
122. Yang H, Choi CA, Chung KH, Jun CH, Kim YT (2004) Anal Chem 76:1537
123. Bruckner-Lea CJ, Tsukuda T, Dockendorff B, Follansbee JC, Kingsley MT, Ocampo C, Stults JR, Chandler DP (2002) Anal Chim Acta 469:129
124. Cady NC, Stelick S, Kunnavakkam MV, Batt CA (2005) Sens Actuators B Chem 107:332
125. Chaudhari AM, Woudenberg TM, Albin M, Goodson KE (1998) J Microelectromech Syst 7:345
126. Erill I, Campoy S, Rus J, Fonseca L, Ivorra A, Navarro Z, Plaza JA, Aguilo J, Barbe J (2004) J Micromech Microeng 14:1558
127. Grodzinski P, Liu RH, Chen B, Blackwell J, Liu Y, Rhine D, Smekal T, Ganser D, Romero C, Yu H, Chan T, Kroutchinina N (2001) Biomed Microdevices 3:275
128. Obeid PJ, Christopoulos TK, Crabtree HJ, Backhouse CJ (2003) Anal Chem 75:288
129. Zhou X, Liu D, Zhong R, Dai Z, Wu D, Wang H, Du Y, Xia Z, Zhang L, Mei X, Lin B (2004) Electrophoresis 25:3032
130. Gulliksen A, Solli L, Karlsen F, Rogne H, Hovig E, Nordstrøm T, Sirevåg R (2003) Anal Chem 76:9
131. Sethu P, Mastrangelo CH (2004) Sens Actuators B Chem 98:337
132. Shen K, Chen X, Guo M, Cheng J (2005) Sens Actuators B Chem 105:251
133. Slyadnev MN, Tanaka Y, Tokeshi M, Kitamori T (2001) Anal Chem 73:4037
134. Oda RP, Strausbauch MA, Huhmer AFR, Borson N, Jurrens SR, Craighead J, Wettstein PJ, Eckloff B, Kline B, Landers JP (1998) Anal Chem 70:4361
135. Wittwer CT, Fillmore GC, Hillyard DR (1989) Nucleic Acids Res 17:4353
136. Wittwer CT, Fillmore GC, Garling DJ (1990) Anal Biochem 186:328
137. Soper SA, Ford SM, Xu Y, Qi S, McWhorter S, Lassiter S, Patterson D, Bruch RC (1999) Proc SPIE 3602:392
138. Swerdlow H, Jones BJ, Wittwer CT (1997) Anal Chem 69:848
139. Wang Q, Gong H (2003) Proc SPIE 5119:77
140. Tanaka Y, Slyadnev MN, Hibara A, Tokeshi M, Kitamori T (2000) J Chromatogr A 894:45
141. Fermér C, Nilsson P, Larhed M (2003) Eur J Pharm Sci 18:129
142. Orrling K, Nilsson P, Gullberg M, Larhed M (2004) Chem Commun (7):790
143. Dietrich J, Schmitt P, Zieger M, Preve B, Rolland J-L, Chaabihi H, Gueguen Y (2002) FEMS Microbiol Lett 217:89
144. Ke C, Berney H, Mathewson A, Sheehan MM (2004) Sens Actuators B Chem 102:308
145. Pal D, Venkataraman V (2002) Sens Actuators A phys 102:151
146. Krishnan M, Ugaz VM, Burns MA (2002) Science 298:793
147. Chandrasekhar S (1961) Hydrodynamic and hydromagnetic stability. Clarendon, Oxford

148. Issaq HJ, Janini GM, Atamna IZ, Muschik GM (1991) J Liq Chromatogr 14:817
149. Jorgenson JW, Lukacs KD (1983) Science 222:266
150. Jorgenson JW, Lukacs KD (1981) Anal Chem 53:1298
151. Issaq HJ, Xu H, Chan KC (2001) J Liq Chromatogr Relat Technol 24:2381
152. Harrison DJ, Fluri K, Seiler K, Fan ZH, Effenhauser CS, Manz A (1993) Science 261:895
153. Woolley AT, Mathies RA (1994) Proc Natl Acad Sci USA 91:11348
154. Landers JP (1995) Clin Chem 41:495
155. Giddings JC (1969) Separ Sci 4:181
156. Boček P, Vespalec R, Giese RW (2000) Anal Chem 72:586A
157. Luckey JA, Norris TB, Smith LM (1993) J Phys Chem 97:3067
158. Lagally ET, Mathies RA (2004) J Phys D Appl Phys 37:R245
159. Meagher RJ, Won JI, McCormick LC, Nedelcu S, Bertrand MM, Bertram JL, Drouin G, Barron AE, Slater GW (2005) Electrophoresis 26:331
160. Stellwagen NC, Gelfi C, Righetti PG (1997) Biopolymers 42:687
161. Olivera BM, Baine P, Davidson N (1964) Biopolymers 2:245
162. Fawcett JS, Morris CJOR (1966) Separ Sci 1:9
163. Ogston AG (1958) Trans Faraday Soc 54:1754
164. Rodbard D, Chrambach A (1970) Proc Natl Acad Sci USA 65:970
165. Chrambac A, Rodbard D (1971) Science 172:440
166. Ferguson KA (1964) Metabolism 13:985
167. Stellwagen NC, Stellwagen E (2009) J Chromatogr A 1216:1917
168. Chiari M, Cretich M, Consonni R (2002) Electrophoresis 23:536
169. Barbier V, Buchholz BA, Barron AE, Viovy J-L (2002) Electrophoresis 23:1441
170. Madabhushi RS, Vainer M, Dolnik V, Enad S, Barker DL, Harris DW, Mansfield ES (1997) Electrophoresis 18:104
171. Chiari M, Riva S, Gelain A, Vitale A, Turati E (1997) J Chromatogr A 781:347
172. Heller C (1999) Electrophoresis 20:1962
173. Xu F, Baba Y (2004) Electrophoresis 25:2332
174. Sunada WM, Blanch HW (1997) Electrophoresis 18:2243
175. Cottet H, Gareil P (2002) Electrophoresis 23:2788
176. Albarghouthi MN, Buchholz BA, Doherty EAS, Bogdan FM, Zhou H, Barron AE (2001) Electrophoresis 22:737
177. Buchholz BA, Barron AE (2001) Electrophoresis 22:4118
178. Njoroge SK, Witek MA, Hupert ML, Soper SA (2010) Electrophoresis 31:981
179. Doherty EAS, Kan CW, Barron AE (2003) Electrophoresis 24:4170
180. Doherty EAS, Kan CW, Paegel BM, Yeung SHI, Cao ST, Mathies RA, Barron AE (2004) Anal Chem 76:5249
181. Ekani-Nkodo A, Tinland B (2003) Phys Rev E 67:051920
182. Shi X, Hammond RW, Morris MD (1995) Anal Chem 67:1132
183. Viovy J-L, Duke T (1993) Electrophoresis 14:322
184. Klein J (1978) Macromolecules 11:852
185. Mayer P, Slater GW, Drouin G (1994) Anal Chem 66:1777
186. Netter H (1969) Theoretical biochemistry. Wiley, New York, 87
187. Heller C, Slater GW, Mayer P, Dovichi N, Pinto D, Viovy JL, Drouin G (1998) J Chromatogr A 806:113
188. Desruisseaux C, Drouin G, Slater GW (2001) Macromolecules 34:5280
189. Desruisseaux C, Slater GW, Drouin G (1998) Macromolecules 31:6499
190. Ulanovsky L, Drouin G, Gilbert W (1990) Nature 343:190
191. Ren H, Karger AE, Oaks F, Menchen S, Slater GW, Drouin G (1999) Electrophoresis 20:2501
192. Vreeland WN, Meagher RJ, Barron AE (2002) Anal Chem 74:4328
193. Sinville R, Coyne J, Meagher RJ, Cheng Y-W, Barany F, Barron A, Soper SA (2008) Electrophoresis 29:4751

194. Fodor SPA, Read JL, Pirrung MC, Stryer L, Lu AT, Solas D (1991) Science 251:767
195. Fodor SPA, Rava RP, Huang XHC, Pease AC, Holmes CP, Adams CL (1993) Nature 364:555
196. Schena M, Shalon D, Davis RW, Brown PO (1995) Science 270:467
197. Churchill GA (2002) Nat Genet 32:490
198. Heller MJ (2002) Annu Rev Biomed Eng 4:129
199. Venkatasubbarao S (2004) Trends Biotechnol 22:630
200. Syvanen AC (2005) Nat Genet 37:S5
201. Hoheisel JD (2006) Nat Rev Genet 7:200
202. Guo Z, Guilfoyle RA, Thiel AJ, Wang RF, Smith LM (1994) Nucleic Acids Res 22:5456
203. Joos B, Kuster H, Cone R (1997) Anal Biochem 247:96
204. Erickson D, Liu XZ, Krull U, Li DQ (2004) Anal Chem 76:7269
205. Yuen PK, Li GS, Bao YJ, Muller UR (2003) Lab Chip 3:46
206. Lenigk R, Liu RH, Athavale M, Chen ZJ, Ganser D, Yang JN, Rauch C, Liu YJ, Chan B, Yu HN, Ray M, Marrero R, Grodzinski P (2002) Anal Biochem 311:40
207. Wang Y, Vaidya B, Farquar HD, Stryjewski W, Hammer RP, McCarley RL, Soper SA, Cheng YW, Barany F (2003) Anal Chem 75:1130
208. Hashimoto M, Barany F, Soper SA (2006) Biosens Bioelectron 21:1915
209. Xu F, Datta P, Wang H, Gurung S, Hashimoto M, Wei S, Goettert J, McCarley RL, Soper SA (2007) Anal Chem 79:9007
210. Hupert ML, Wang H, Chen H-W, Chen PC, Stryjewski W, Patterson D, Witek MA, Datta P, Goettert J, Murphy MC, Soper SA (2008) A field-deployable system for automated molecular testing using modular microfluidics. In: 12th international conference on miniaturized systems for chemistry and life sciences, μTAS2008, 12-16 October 2008, San Diego, California, USA, pp 1946
211. Situma C, Wang Y, Hupert M, Barany F, McCarley RL, Soper SA (2005) Anal Biochem 340:123
212. Rasmussen SR, Larsen MR, Rasmussen SE (1991) Anal Biochem 198:138
213. Csaki A, Moller R, Straube W, Kohler JM, Fritzsche W (2001) Nucleic Acids Res 29:E81
214. Ahn S, Walt DR (2005) Anal Chem 77:5041
215. Fan ZH, Mangru S, Granzow R, Heaney P, Ho W, Dong QP, Kumar R (1999) Anal Chem 71:4851
216. Ali MF, Kirby R, Goodey AP, Rodriguez MD, Ellington AD, Neikirk DP, McDevitt JT (2003) Anal Chem 75:4732
217. Schena M, Shalon D, Heller R, Chai A, Brown PO, Davis RW (1996) Proc Natl Acad Sci USA 93:10614
218. Cheung VG, Morley M, Aguilar F, Massimi A, Kucherlapati R, Childs G (1999) Nat Genet 21:15
219. McCarley RL, Vaidya B, Wei SY, Smith AF, Patel AB, Feng J, Murphy MC, Soper SA (2005) J Am Chem Soc 127:842
220. Wei SY, Vaidya B, Patel AB, Soper SA, McCarley RL (2005) J Phys Chem B 109:16988
221. Liu YJ, Rauch CB (2003) Anal Biochem 317:76
222. Wei CW, Cheng JY, Huang CT, Yen MH, Young TH (2005) Nucleic Acids Res 33:e78
223. Yaralioglu GG, Wygant IO, Marentis TC, Khuri-Yakub BT (2004) Anal Chem 76:3694
224. Vanderhoeven J, Pappaert K, Dutta B, Vanhummelen P, Baron GV, Desmet G (2004) Electrophoresis 25:3677
225. Sternberg JC (1966) Adv Chromatogr 2:205
226. Harvey MC, Stearns SD (1983) J Chromatogr Sci 21:473
227. Claemsens HA, Burcinova A, Cramers CA, Mussche P, van Tilburg CCE (1990) J Microcolumn Sep 2:132
228. Blas M, Delaunay N, Rocca J-L (2008) Electrophoresis 29:20
229. Fan ZH, Harrison DJ (1994) Anal Chem 66:177
230. Harrison DJ, Fan ZH, Seiler K, Manz A, Widmer HM (1993) Anal Chim Acta 283:361

231. Effenhauser CS, Manz A, Widmer HM (1993) Anal Chem 65:2637
232. Effenhauser CS, Paulus A, Manz A, Widmer HM (1994) Anal Chem 66:2949
233. Wenclawiak BW, Puschl R (2006) Anal Lett 39:3
234. Figeys D, Ahmadzedeh H, Arriaga E, Dovichi NJ (1996) J Chromatogr A 744:325
235. Lin CC, Chen CC, Lin CE, Chen SH (2004) J Chromatogr A 1051:69
236. Kim DK, Kang SH (2005) J Chromatogr A 1064:121
237. Jacobson SC, Koutny LB, Hergenroder R, Moore AW, Ramsey JM (1994) Anal Chem 66:3472
238. Chien RL, Burgi DS (1991) J Chromatogr 559:141
239. Bharadwaj R, Santiago JG, Mohammadi B (2002) Electrophoresis 23:2729
240. Khandurina J, Jacobson SC, Waters LC, Foote RS, Ramsey JM (1999) Anal Chem 71:1815
241. Kelly RT, Li Y, Woolley AT (2006) Anal Chem 78:2565
242. Strausbauch MA, Landers JP, Wettstein PJ (1996) Anal Chem 68:306
243. Pedersen-Bjergaard S, Rasmussen KE, Halvorsen TG (2000) J Chromatogr A 902:91
244. Muscate A, Natt F, Paulus A, Ehrat M (1998) Anal Chem 70:1419
245. Burns MA, Johnson BN, Brahmasandra SN, Handique K, Webster JR, Krishnan M, Sammarco TS, Man PM, Jones D, Heldsinger D, Mastrangelo CH, Burke DT (1998) Science 282:484
246. Walker GT, Fraiser MS, Schram JL, Little MC, Nadeau JG, Malinowski DP (1992) Nucleic Acids Res 20:1691
247. Hashimoto M, Barany F, Xu F, Soper SA (2007) Analyst 132:913
248. Simpson PC, Woolley AT, Mathies RA (1998) Biomed Microdevices 1:7
249. Belgrader P, Smith JK, Weedn VW, Northrup MA (1998) J Forensic Sci 43:315
250. Northrup MA, Benett B, Hadley D, Landre P, Lehew S, Richards J, Stratton P (1998) Anal Chem 70:918
251. Zhang N, Tan H, Yeung ES (1999) Anal Chem 71:1138
252. Thaitrong N, Toriello NM, Del Bueno N, Mathies RA (2009) Anal Chem 81:1371
253. Olsen KG, Ross DJ, Tarlov MJ (2002) Anal Chem 74:1436
254. Toriello NM, Douglas ES, Thaitrong N, Hsiao SC, Francis MB, Bertozzi CR, Mathies RA (2008) Proc Natl Acad Sci USA 105:20173
255. Toriello NM, Liu CN, Blazej RG, Thaitrong N, Mathies RA (2007) Anal Chem 79:8549
256. Budowle B, van Daal A (2009) Biotechniques 46:339
257. Grover WH, Skelley AM, Liu CN, Lagally ET, Mathies RA (2003) Sens Actuators B 89:315
258. Liu P, Yeung SHI, Crenshaw KA, Crouse CA, Scherer JR, Mathies RA (2008) Forensic Sci Int Genet 2:301
259. Ewing B, Hillier L, Wendl MC, Green P (1998) Genome Res 8:175
260. Yeung SW, Lee TMH, Cai H, Hsing IM (2006) Nucleic Acids Res 34:e118
261. Yeung SSW, Lee TMH, Hsing IM (2008) Anal Chem 80:363
262. Liu RH, Yang JN, Lenigk R, Bonanno J, Grodzinski P (2004) Anal Chem 76:1824

Top Curr Chem (2011) 304: 261–294
DOI: 10.1007/128_2011_152

Published online: 16 June 2011

Integrated Multifunctional Microfluidics for Automated Proteome Analyses

John K. Osiri, Hamed Shadpour, Małgorzata A. Witek, and Steven A. Soper

Abstract Proteomics is a challenging field for realizing totally integrated microfluidic systems for complete proteome processing due to several considerations, including the sheer number of different protein types that exist within most proteomes, the large dynamic range associated with these various protein types, and the diverse chemical nature of the proteins comprising a typical proteome. For example, the human proteome is estimated to have $>10^6$ different components with a dynamic range of $>10^{10}$. The typical processing pipeline for proteomics involves the following steps: (1) selection and/or extraction of the particular proteins to be analyzed; (2) multidimensional separation; (3) proteolytic digestion of the protein sample; and (4) mass spectral identification of either intact proteins (top-down proteomics) or peptide fragments generated from proteolytic digestions (bottom-up proteomics). Although a number of intriguing microfluidic devices have been designed, fabricated and evaluated for carrying out the individual processing steps listed above, work toward building fully integrated microfluidic systems for protein analysis has yet to be realized. In this chapter, information will be provided on the nature of proteomic analysis in terms of the challenges associated with the sample type and the microfluidic devices that have been tested to carry out individual processing steps. These include devices such as those for multidimensional electrophoretic separations, solid-phase enzymatic digestions, and solid-phase extractions, all of which have used microfluidics as the functional platform for their implementation. This

J.K. Osiri, H. Shadpour, M.A. Witek, and S.A. Soper (✉)
Department of Chemistry, Louisiana State University, Baton Rouge, LA 70817, USA
and
Department of Mechanical Engineering, Louisiana State University, Baton Rouge, LA 70817, USA
e-mail: chsope@lsu.edu

will be followed by an in-depth review of microfluidic systems, which are defined as units possessing two or more devices assembled into autonomous systems for proteome processing. In addition, information will be provided on the challenges involved in integrating processing steps into a functional system and the approaches adopted for device integration. In this chapter, we will focus exclusively on the front-end processing microfluidic devices and systems for proteome processing, and not on the interface technology of these platforms to mass spectrometry due to the extensive reviews that already exist on these types of interfaces.

Keywords Integrated microsystems · Isoelectric focusing · Microfluidics · Orthogonality · Proteome · Top-down and bottom-up analysis · Two-dimensional electrophoresis

Contents

1 Introduction 263
1.1 The Proteome, Proteomics and Protein Analyses 263
1.2 Challenges in Proteomics 263
2 Approaches to Protein Analysis 265
2.1 Top-Down Strategy 265
2.2 Bottom-Up Strategy 266
3 Proteomics: Analytical Challenges and Microfluidics 267
3.1 Sample Preparation 267
3.2 Protein Separations 270
3.3 Protein Digestion 274
3.4 Peptide Separation and Mass Spectrometry Analysis 275
4 Integrated Microsystems for Protein Analysis 276
4.1 Integrated Systems with Two Devices 278
4.2 Integrated Systems with Three Devices 282
4.3 Integrated Systems with More Than three Devices 285
5 Concluding Remarks 287
References 289

Abbreviations

BMA	Butyl methacrylate
BSA	Bovine serum albumin
CGE-PP	Chip gel electrophoresis protein profiling
EDMA	Ethylene dimethylacrylate
EGDMA	Ethylene glycol dimethacrylate
ESI	Electrospray ionization
GBP	Gold binding peptide
GMA	Glycidyl methacrylate
IEF	Isoelectric focusing

IMAC	Immobilized metal affinity chromatography
LMA	Lauryl methacrylate
MEKC	Micellar electrokinetic capillary electrophoresis
MudPIT	Multidimensional protein identification technology
NDA	Naphthalene-2,3-dicarboxaldehyde
PCR	Polymerase chain reaction
PDMS	Poly(dimethylsiloxane)
PMMA	Poly(methyl methacrylate)
PVDF	Poly(vinylidene difluoride)
SDS-PAGE	Sodium dodecylsulfate–poly(acrylamide) gel electrophoresis
SPE	Solid phase extraction

1 Introduction

1.1 The Proteome, Proteomics and Protein Analyses

The word "proteome" was first used in 1994 at the "Two-dimensional electrophoresis" meeting in Siena (Italy) in reference to the total protein complement of a cell, tissue, or body fluid [1]. The original definition did not account for the numerous post-translational modifications that occur, significantly increasing the total number of different protein types in any entity [2, 3]. It is understood now that the evolutionary complexity of an organism is governed by the number of gene product modifications, for example *Escherichia coli* has ~1.5 times more proteins than genes, while in humans 10–40 proteins per gene have been estimated [4]. The proteome of *E. coli* K12 W110 consists of 4,226 proteins, [5, 6], which is significantly less than that expected within one average human cell [7, 8].

Proteomics refers to the study of the proteome in any given organism and is concerned with the determination of structure, expression, interaction, function, activity, and localization of all proteins comprising the organism. Most of the discoveries and studies on proteins have centered on expression because it is approachable in general, whereas protein structure and function are far from being understood at a systemic level [9].

1.2 Challenges in Proteomics

Although most recognize the importance of proteomics, securing the extensive information required on the entire complement of proteins comprising the proteome is fraught with challenges. These challenges can be categorized according to: (1) the nature of the sample; (2) current techniques of analysis; and (3) data analysis.

1.2.1 The Nature of the Sample

The complexity of a proteome is simply overwhelming. The human genome is composed of approximately 3×10^4 genes, [10, 11], which expresses an estimated $>10^6$ different proteins in the human proteome [4]. Furthermore, proteins are a very diverse class of molecules: size can range from a few tens of amino acid residues (e.g., toxins) to several thousands (e.g., human titin is composed of 26,926 amino acids); the isoelectric point (pI) can vary from 3 to 11; and there is a wide range of hydrophilic or hydrophobic characteristics [12]. For example, cell membrane-associated proteins contain hydrophobic domains making them extremely difficult to analyze using conventional methods such as isoelectric focusing (IEF). The complexity of the proteome is further compounded by the large dynamic range of many constituents. For example, the dynamic range of protein concentration in human serum is $\sim 10^{10}$ and the dynamic range of proteins within a single cell can span nearly six orders of magnitude [13]. To further highlight the challenge posed by sample complexity, let us assume that a particular protein has a single copy within a cell and that ~10 fmol of a peptide generated from a proteolytic digestion of a protein is needed to produce a viable mass spectrum for finger printing, therefore approximately 6×10^{10} cells would be required for the analysis [14].

1.2.2 Current Techniques for Analysis

As noted above, proteomics attempts to determine the identities and quantities of constituents of the proteome. Unfortunately, typical analytical tools can only quantitatively analyze less than 50 constituents simultaneously, which pales compared to the number of different proteins present in a single sample [15]. Sodium dodecyl sulfate–polyacrylamide gel electrophoresis (SDS-PAGE) coupled to IEF is the typical strategy employed to form 2D electrophoresis platforms, which can afford dynamic ranges of 10^2–10^4 and can routinely resolve >2,000 components per gel. In this method, the IEF is run in the first dimension and sorts components based on their pI. SDS-PAGE is used in the second dimension to sort components on the basis of their molecular weights (MWs). However, while the peak capacities of these systems are impressive, they are well below the 300,000 to 1×10^6 proteins present in a human proteome. Moreover, IEF/SDS-PAGE requires well-trained personnel, is time-consuming, and has limited automation potential [16]. Mass spectrometry (MS), which is the common readout modality used for proteomics, provides a limit-of-detection (LOD) of $\sim 10^{-15}$ moles for peptides and possesses high specificity, but has a limited dynamic range [17]. Furthermore, there is no amplification technique for proteins, such as PCR in genomics, and thus, proteins must be analyzed in their native state.

1.2.3 Data Analysis

The number of measurable parameters in proteomics is enormous, requiring a large number of biological and methodological replicates, which is difficult to attain due to the limited number of test subjects, and problems pertaining to statistics and bias [15]. Bias in protein data analysis is so pervasive that it is viewed as a threat to the validity of protein biomarkers for cancer diagnostics, where results have been disputed or poorly reproduced [18, 19]. This issue is apparent when dealing with large data sets, such as those presented in proteomic analyses. It is inherently easier to find a correlation irrespective of real cause and effect, due to false positives probably outnumbering true positives [15]. According to Lay et al. [15], "the failure of many proteomic studies probably correlates with the failure to consider the analytical need to define quality standards, including method validation and standardization". Unfortunately, due to the lack of quality standards across the board, it is difficult to compare results generated from various laboratories.

2 Approaches to Protein Analysis

Ideally, a proteomic platform should quantitatively analyze the entire proteome in a high-throughput fashion and at high sensitivity [20]. Two approaches toward protein characterization are commonly recognized, namely bottom-up and top-down strategies (see Fig. 1) [21]. In bottom-up strategies, a protein mixture is sorted into individual components and the isolated components are enzymatically digested into peptides with the mass of the individual fragments used to "piece" the protein back together again. This strategy is also called the "gel approach" because it relies heavily on the use of 2D IEF/SDS-PAGE to separate the proteins prior to proteolytic digestion. In the top-down approach, the intact protein mixture is fractionated into individual components, and then the intact proteins are submitted for exact mass analysis. Typically, the top-down and bottom-up strategies are complementary to each other in terms of proteomic coverage [9]. A more concise description of these techniques follows in Sects. 2.1 and 2.2.

2.1 Top-Down Strategy

In top-down methods, as noted above, intact proteins are subjected to mass analysis following some type of fractionation step. Several advantages are associated with this strategy, including the ability to identify translational start/stop sites, mRNA splice variants, and post-translational modifications because intact proteins are analyzed directly via MS. In terms of the MS ionization source, electrospray ionization (ESI) is commonly used for top-down strategies. Also, liquid chromatography (LC), sometimes used for the separation/fractionation of the proteins, couples naturally to ESI due to the ability to generate continuous sample infusion into the ESI source from the chromatography column and due to compatibility of the flow rates [22].

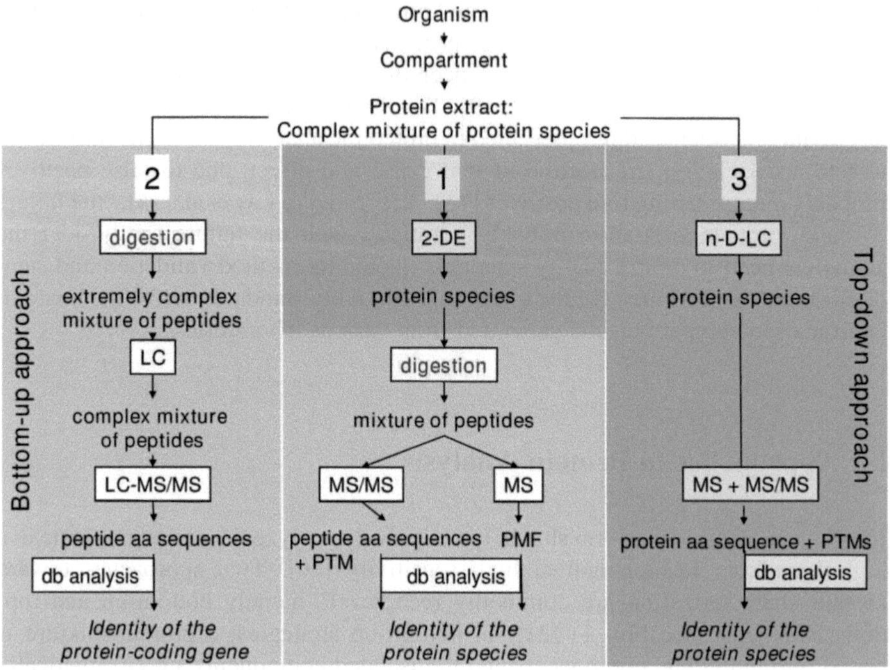

Fig. 1 Bottom-up and top-down strategies for analyzing proteomic samples. A description of these two processes can be found in Sect. 2. *2-DE* two dimensional electrophoresis, *n-D-LC* multidimensional liquid chromatography, *PMF* Peptide mass fingerprinting, *PTM* Post-translational modification, *db* database. Adapted with permission from Schlüter et al. [21]

2.2 Bottom-Up Strategy

Bottom-up proteomics involves the digestion of proteins proteolytically into peptides prior to MS analysis. There are several advantages for using a bottom-up strategy, most notably it provides the ability to analyze a large number of proteins in a single analysis. For bottom-up strategies, one can employ a strategy in which the proteins are exhaustively sorted using, for example, IEF/SDS-PAGE. Conversely, the proteins can be digested without subjecting the sample to a sorting step and, instead, digesting the entire protein sample at once, which is sometimes called "shotgun proteomics" or multidimensional protein identification technology (MudPIT). In this technique peptides are separated using two chromatography steps: a strong cationic exchange and reversed-phase high performance liquid chromatography. The advantages of a high efficiency sorting step prior to proteolytic digestion is that the proteins can be visualized immediately after SDS-PAGE and protein profiles can be generated to provide expression information. Also, resolved protein spots are digested, which results in fewer peptides being fed into the MS and thus producing simpler peptide mass spectra compared to the MS obtained from a shotgun analysis.

3 Proteomics: Analytical Challenges and Microfluidics

The term "microfluidics" refers to analytical tools where fluids are driven, either hydrodynamically or electrokinetically, through microstructured channels. Key attributes of microfluidics are the high degree of parallelization they offer to provide high-throughput processing and the integration of different functions required to analyze a particular sample, providing high levels of process automation. There are additional advantages: First, microsystems can provide the ability to analyze ultrasmall sample volumes. Second, the high surface-to-volume ratio afforded by miniaturized platforms is preferable for solid-phase extractions or solid-phase microreactions, where analyte–wall interactions are required and the interaction is diffusion-controlled. Third, short analysis times can be achieved due to reduced length scales without sacrificing efficiency. For example, short electrophoresis development times can be realized that still provide separation efficiencies comparable to their macroscale counterparts. Finally, microsystems provide the ability to seamlessly integrate processing steps into a single or multiple chips, without consequences arising from unswept volumes and, at the same time, improve process reliability and reproducibility as well as process automation. In the Sects. 3.1–3.4, a brief introduction to microfluidic platforms that can be used for the various processing steps of protein analysis will be discussed, including sample preparation, separation, digestion and mass spectral analysis of the digests.

3.1 Sample Preparation

Before pursuing the appropriate strategy for analyzing a large number of proteins, the targets must be extracted from the biological matrix to exclude sample impurities that may affect downstream processing steps. In addition, it may be advisable to select a subset of the proteome to make downstream processing more tractable by reducing the complexity of the sample input in terms of reducing the number of components. For example, one may wish to selectively isolate cytosolic proteins from a group of cells and submit them for processing. Conversely, one can selectively analyze the membrane-associated proteins, free from the cytosolic components. Sample preparation can involve, but is not limited to, cell lysis and target protein solid-phase extraction/selection, preconcentration, denaturing, and labeling.

To release proteins encased in cells for analysis, the cells, including the cellular and nuclear membranes, must be destroyed through lysis. Cell lysis can be carried out in an appropriate solubilization solution [23]. Lysis methods includes two categories, gentle lysis or vigorous lysis.

Gentle lysis methods are generally employed when the sample of interest consists of easily lysed cells (such as tissue culture cells, blood cells and some microorganisms). Gentle lysis methods can also be employed when only one particular subcellular fraction is to be analyzed. The most commonly used gentle lysis methods are osmotic lysis [24], freeze–thaw lysis [23, 25, 26], detergent lysis [27, 28], and enzymatic lysis [29, 30]. Sometimes these techniques are combined (e.g., osmotic lysis following enzymatic treatment or freeze–thaw in the presence of detergent) to provide a more selective and complete lysis process.

Vigorous lysis methods are employed when cells are less easily disrupted, for example, cells in solid tissues. These methods include sonication [31–33], French pressure [29, 30, 34], grinding [32, 35–37], mechanical homogenization [24, 25, 38], and glass bead homogenization [29, 30]. Vigorous lysis methods usually result in complete disruption of the cell membranes and some organelles.

In terms of cell lysis on-chip, Li and Harrison [39] were the first to report cell lysis through combination of a chemical method using SDS and electrical techniques. Since then, other on-chip lysis methods have appeared, including mechanical [40], thermal [41], ultrasonic [42], and electrophoretic [43, 44].

When cells are lysed, proteases (enzymes that break peptide bonds in proteins) are often activated. Degradation of proteins through protease action greatly complicates the analysis by 2D electrophoresis, so action should be taken to avoid this problem. If possible, it is advisable to inhibit proteases by disrupting the sample directly into strong denaturants such as 8 M urea, 10% trichloroacetic acid (TCA), or 2% SDS [45–47]. Proteases are less active at lower temperatures, so sample preparation at low temperature is recommended. In addition, proteolysis can often be inhibited by preparing the sample in the presence of Tris base, sodium carbonate or basic carrier ampholyte mixtures [48, 49].

Because many components of biological samples can interfere with downstream analysis, it is necessary to remove them. Insoluble substances can be removed by centrifugation. For 2D electrophoresis followed by MS, it is necessary to remove salts before analysis. This can be achieved by dialysis, size-exclusion filtering, protein precipitation, or reverse-phase chromatography [50, 51]. Frequently, abundant proteins, such as albumin or immunoglobulins, need to be removed as well [13]. Complex samples can also be fractionated before analysis to obtain simpler subfractions and to decrease the dynamic range of components [13]. Affinity purification is another powerful approach for reduction of the complexity of a sample by isolating specific classes of proteins or protein complexes [52]. These preparation steps are often more time-consuming than the subsequent analysis steps and influence the sensitivity and discriminative power of MS-based protein identification [51, 53].

In the case of solid-phase extraction, targets are retained on a stationary phase either through hydrophobic, electrostatic, or affinity interactions to affect both preconcentration of the targets and removal of potential interfering components. In terms of on-chip solid-phase extractions, Kutter et al. [54] modified a glass surface to create a C_{18} phase, which was subsequently used to extract appropriate components that had high retention for this phase. They estimated an enrichment factor of 80. Besides on-chip wall modifications to generate the appropriate extraction phase, integration of polymeric membranes or monoliths are alternative methods of producing a high surface-area, solid-phase extraction phase. Through simple modifications of only the surface of a wall, a surface-to-volume ratio of 230–781 mm^{-1} [55] can result and 1,500 mm^{-1} for three-dimensional (3D) polymeric networks occupying the open channel, which can result in increased loading of the target material. These 3D networks can consist of such materials as polyvinylidene difluoride (PVDF), which has been placed at the inlet of a nanospray tip for protein desalting prior to ESI-MS [56].

As an example of monoliths used for the solid-phase extraction of certain targets, Mao et al. utilized lectin affinity chromatography to extract glycoproteins [57]. In this example (see Fig. 2), the monolith consisted of ethylene dimethylacrylate (EDMA) and glycidyl methacrylate (GMA) monomers UV-polymerized in the presence of a porogen. Following polymerization, *Pisum sativum* glutinin was immobilized onto the monolith to target the glycoproteins. The use of this monolith was coined "lectin affinity chromatography". The device was successfully demonstrated to retain three different glycoproteins, namely turkey ovalbumin, chicken ovalbumin, and ovomucoid. Electrophoregrams of turkey ovalbumin in the presence

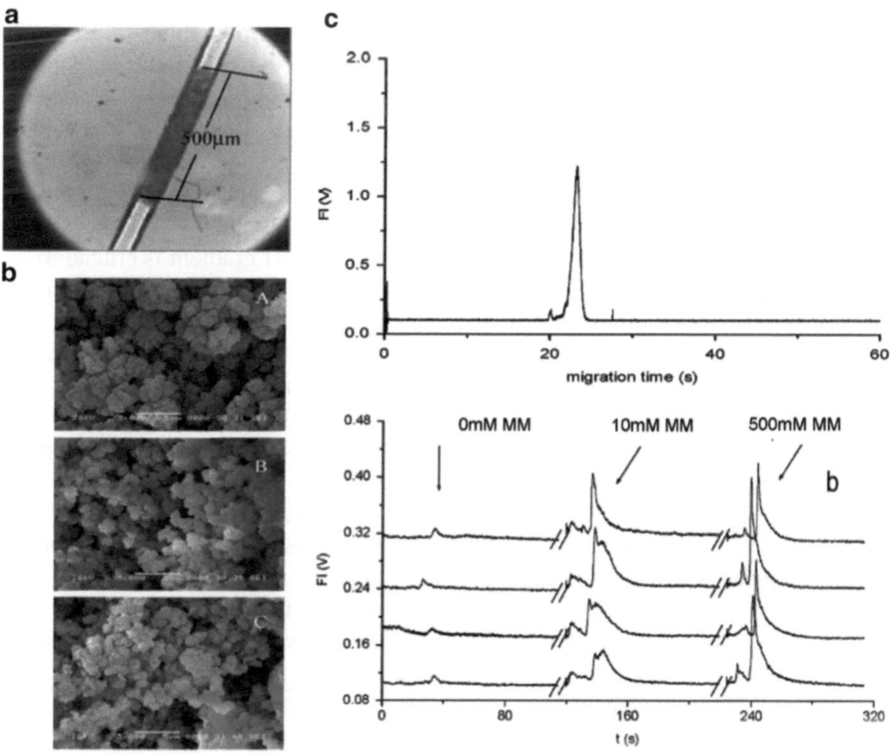

Fig. 2 (*a*) Monolith of ethylene dimethylacrylate and glycidyl methacrylate monomers UV-polymerized in the presence of a porogen into a fluidic microchannel. (*b*) Three high-resolution SEM images of the monolith used for glycolated protein extraction. (*c*) Electrophoresis of turkey ovalbumin analyzed by capillary zone electrophoresis (*top graph*) and by the polymer monolith (*bottom graph*). The separation buffer was comprised of 20 mM HEPES (pH 7.49) and 0.1% Triton X-100, 1 mM $MnCl_2$ and 1 mM $CaCl_2$. Adapted with permission from Mao et al. [57]

and absence of the polymer monolith are shown in Fig. 2. As can be seen, the turkey ovalbumin protein was retained on this extraction phase.

3.2 Protein Separations

Protein separation techniques sort complex protein mixtures into nearly purified components. The goal is to obtain individual components comprised of a single protein so that each protein can be easily identified without interference from others. Separating protein mixtures into individual components is usually impractical if not impossible, especially when dealing with a complex proteome that contains numerous proteins. Conventional 2D separations involve a charge-based separation, where proteins are separated on the basis of their pIs in IEF, followed by a size-based separation, where proteins are separated according to their MW using SDS-PAGE. This method has the advantage of routinely resolving up to 2,000 proteins, with reports of up to 5,000 [58, 59] and 10,000 when large-sized gels are used [60]. But, the process is time-consuming, laborious, and not very reproducible [61, 62].

IEF separates proteins according to their pI and is based on the amphoteric nature of these molecules. Proteins carry either positive, negative, or zero net charge depending on the pH of their environment. The net charge of a protein is the sum of all the negative and positive charges of its amino and carboxylic acid side chains. The pI represents the specific pH at which the net charge of the protein is zero. Proteins are positively charged at pH values below their pI and negatively charged at pH values above their pI. The presence of a pH gradient is crucial to the IEF technique. In a pH gradient under the influence of an electric field, a protein will move to the position in the gradient where its net charge is zero. A protein with a net positive charge will migrate toward the cathode, becoming progressively less positively charged as it moves through the pH gradient until it reaches its pI, and vice versa for proteins with negative net charges. This then represents the focusing effect of IEF, which can concentrate proteins at their respective pIs. IEF performed under denaturing conditions typically gives the highest resolution.

Commercial carrier ampholyte mixtures are comprised of hundreds of individual polymeric species with pIs spanning a specific pH range. When a voltage is applied across a carrier ampholyte mixture, the carrier ampholytes with the highest pI (and the most negative charge) move toward the anode and the carrier ampholytes with the lowest pI (and the most positive charge) move toward the cathode. The other carrier ampholytes align themselves between the extremes, according to their pIs and buffer their environment to the corresponding pHs. The result is a continuous pH gradient.

SDS-PAGE is performed in polyacrylamide gels containing SDS. SDS and proteins form complexes with structures composed of protein-decorated micelles connected by short flexible polypeptide segments [63]. The result of this structure is that large amounts of SDS are incorporated into the SDS–protein complex in a ratio of approximately 1.4 g SDS/g protein. SDS masks the charge of the proteins themselves, and the formed anionic complexes have a roughly constant net negative charge per unit mass. Besides SDS, a reducing agent (e.g., 2-mercaptoethanol) is typically added to break any disulfide bonds present in the proteins. When proteins are treated with both SDS and a reducing agent, the degree of electrophoretic separation within a polyacrylamide gel depends largely on the MW of the proteins. In fact, there is an approximately linear relationship between the logarithm of the MW and the rate of migration of the SDS–polypeptide complex. This linear relationship is only valid for a certain MW range, which is determined by the polyacrylamide gel composition. The logarithm of the mobility versus the percent monomer composition (%*T*) of the gel is also linear. Tris-containing buffers are the most commonly used buffer systems for SDS-PAGE [64, 65]. This buffer system separates proteins at high pH, which confers the advantage of minimal protein aggregation and clean separation, even at relatively heavy protein loads.

The need for utilizing multidimensional separations is a direct consequence of the sheer complexity of the samples associated with proteomics. Because of this fact, the peak capacity of any separation platform must be high. The peak capacity (*P*) describes the number of components that can be resolved in any given separation [66]. Unfortunately, *P* is typically not adequate using a single LC or capillary electrophoresis (CE) procedure by itself [67]. Therefore, attempts have been made using different LC or CE modes and a combination of these techniques (multidimensional separations) to generate the required *P* for proteome analysis [68–71]. A requirement of any successful multidimensional procedure is orthogonality, which means that the selected dimensions possess different, but compatible, separation mechanisms. Furthermore, the subsequent dimension in any multidimensional separation should not destroy the resolution achieved by the previous separation [68, 69].

According to Giddings [69], the *P* of a multidimensional separation is the product of the peak capacities of its constituent one-dimensional (1D) methods ($P_1 \ldots P_n$). A 2D separation actually generates this theoretically available peak capacity only if the constituent 1D dimensions are completely orthogonal. A high degree of retention correlation between the dimensions can reduce a 2D separation to what is, in fact, a 1D separation with peaks distributed along the diagonal of a plot of retention times between the constituent dimensions [72]. The information content of any multidimensional system is the sum of the mean information content of each individual dimension minus the cross-information [73]. Minimizing cross-information is therefore important in any multidimensional separation [72].

Changes in migration order and migration time can serve as indicators of different separation mechanisms responsible for the migration behavior in each dimension [74]. To evaluate the orthogonality of various dimensions [75–77], protein migration maps can be constructed. To perform this, the migration times ($\mathrm{MT_i}$) for each protein comprising the separation is acquired in each separation mode. Then, the normalized migration times ($\mathrm{MT_{i,norm}}$) for both dimensions are calculated using the following equation:

$$\mathrm{MT_{i,norm}} = [\mathrm{MT_i} - \mathrm{MT_{min}}]/[\mathrm{MT_{max}} - \mathrm{MT_{min}}], \qquad (1)$$

where MT_{max} and MT_{min} represent the migration times of the most and least retained proteins, respectively, for each mode. The orthogonality (O) can then be calculated using the following equation:

$$O = \left[\sum \text{bins} - \sqrt{P_{\max}}\right] / [0.63 P_{\max}], \tag{2}$$

where $\sum$bins is the number of bins in the 2D plot containing data points, and P_{max} is the total peak capacity obtained as a sum of all bins [75]. The practical peak capacity (N_p) is usually lower than the theoretical peak capacity because only a fraction of the surface is utilized for separation [75]. The practical P can be evaluated using the following equation [75]:

$$N_p = P_{2D}\left[\sum \text{bins}\right] / P_{\max}. \tag{3}$$

Microchip separations compared to macroscale separations can offer reduced analysis times, require minimal sample volumes, and can be automated. In spite of scaling issues (i.e., the column lengths in microchips are typically shorter than conventional separation systems), microchip separations can produce high peak capacities in short time periods. Numerous groups have reported different microchip 2D systems for protein separations [78]. For example, Li et al. reported a chip for IEF and SDS-PAGE analysis in a plastic microfluidic network [79]. The chip was made from polycarbonate (see Fig. 3) and could complete the 2D separation in <10 min using a 2 × 2 cm footprint with a peak capacity of ~1,700. Resolution and focusing effects generated in the IEF phase of the analysis were maintained into the second dimension by using an electrokinetic transfer process. In addition, the chip configuration included a staggered geometry of the upper and lower second dimension channels (see Fig. 3), which allowed for transfer of the entire contents of the IEF dimension. The peak capacity of the system could be increased by increasing

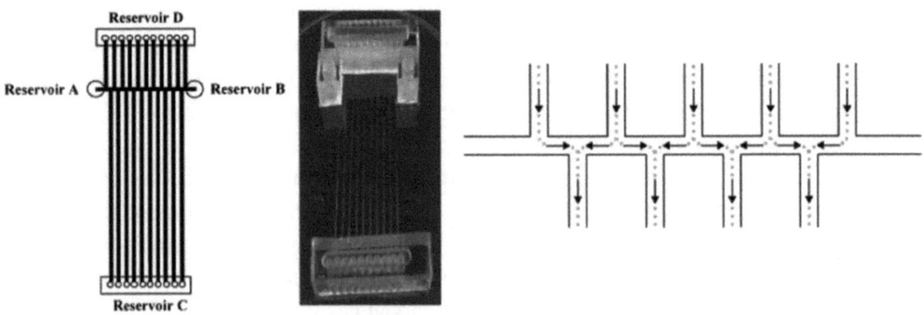

Fig. 3 *Left*: Microfluidic chip for IEF/SDS-PAGE. *Center*: Chip used for the 2D electrophoretic separation. The chip was made from a polycarbonate substrate from a Si template via hot embossing. *Right*: Paths for the electrokinetic transfer of focused protein bands from the first (IEF) into the second (SDS-PAGE) dimension. Adapted with permission from Li et al. [79]

the density of channels in the array. Yang and coworkers [80] reported a 2D IEF/PAGE separation of *E. coli* proteins using a PMMA microchip.

Shadpour et al. [81] and Osiri et al. [82] employed SDS micro-capillary gel electrophoresis (SDS μ-CGE) and micellar electrokinetic capillary (MEKC) electrophoresis in the first and second dimensions, respectively, to sort intact proteins using a poly(methylmethacrylate), PMMA, microchip. A diagram of the microchip is shown in Fig. 4. The electrophoresis commenced in the first dimension for a prescribed amount of time and, then, the bands from the first dimension were sequentially injected into the second dimension for development. The 2D electrophoresis system could generate a peak capacity of ~2,600 for proteins isolated from fetal calf serum (see Fig. 4).

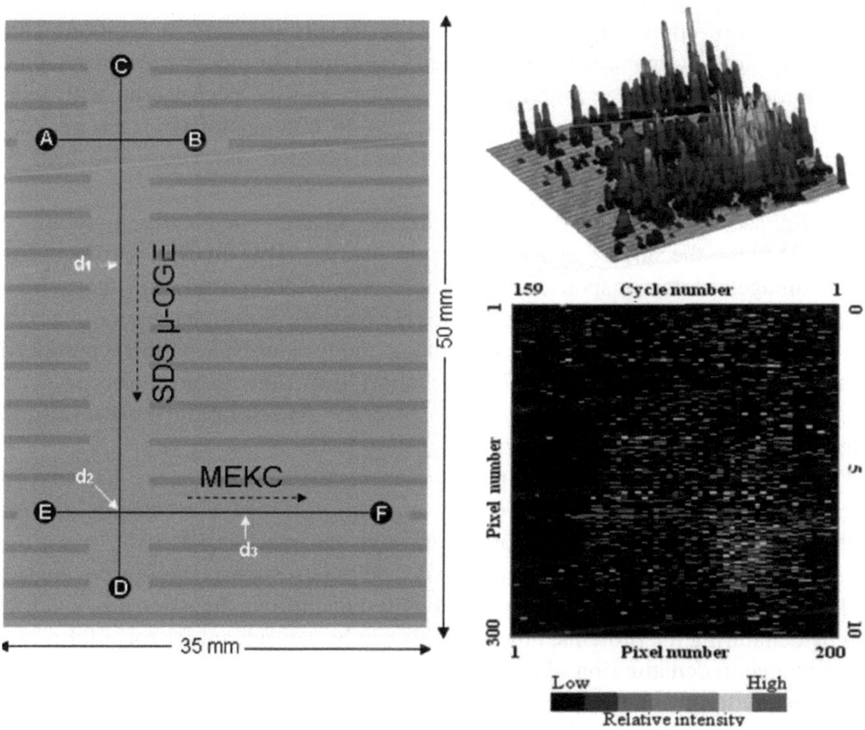

Fig. 4 *Left*: PMMA-based 2D electrophoresis chip made for the multidimensional separation of proteins using SDS μ-CGE and MEKC. The solution reservoirs were: (*a*) sample reservoir, (*b*) sample waste reservoir, (*c*) SDS μ-CGE buffer reservoir, (*d*) SDS μ-CGE buffer waste reservoir, (*e*) MEKC buffer reservoir, (*f*) MEKC buffer waste reservoir. SDS μ-CGE channel: Injection length 10 mm, separation length 40 mm, effective separation length = 30 mm; MEKC channel: separation length 25 mm, effective separation length 10 mm. *Right*: 2D separation of a fecal calf serum proteome using SDS μ-CGE/MEKC. The 2D SDS μ-CGE × MEKC were performed at 300 V/cm and 400 V/cm, respectively. A 10 s separation time was utilized in the first dimension prior to performing the serial 10 s MEKC cycles. A total of 159 MEKC cycles was used with a 1 s transfer time from the first to second dimension. All of the proteins were labeled with a fluorescent dye prior to the separation. Reprinted with permission from Shadpour and Soper [81] and Osiri et al. [82]

3.3 Protein Digestion

An integral part of most proteomic processing pipelines is the use of proteolytic digestion to produce a series of peptides that can be used for identifying the protein using bottom-up processing strategies [83, 84]. Three different approaches can be used for proteolytic digestion: in-gel [85], in-solution [86], and solid phase [87]. In-gel digestion is accomplished by cutting spots from gel electrophoretic bands that contain the proteins of interest, which are then subjected to in situ digestion [88]. Drawbacks of this method include limited accessibility to the proteins inside the gel [89] and gel destaining that can cause poor digestion yields due to residual destaining solvents [90]. Furthermore, the process cannot easily be transitioned to microfluidic chip formats to realize process automation.

In the second approach, proteins are enzymatically digested in solution. This approach requires long incubation times due to the need for low proteolytic enzyme concentrations to minimize autodigestion artifacts and the need to run with relatively high temperatures to achieve high digestion efficiencies [91]. Excessively high temperatures, enzyme concentrations, and reaction times can also lead to nonspecific cleavage and deamidation [91–93].

Solid-phase digestion uses proteolytic enzymes that are chemically immobilized or adsorbed onto the surface of a solid support [94, 95]. This digestion protocol has the advantages of fast response, low sample consumption, and high throughput [96], and is easily adaptable to microfluidic chip formats. Solid-phase microreactors also minimize sample loss during treatment and reduce autolysis products produced from the proteolytic enzymes [97]. The short diffusion distance for properly designed reactors and a high enzyme-to-substrate concentration ratio results in faster digestions than using in-solution digestion [98].

The digestion efficiency of solid-phase microreactors depends on the geometry of the reactor, the digestion temperature, the composition of digestion solvents, and the transport velocity of the target proteins through the reactor [98–101]. Also, the digestion efficiency can be enhanced by physical means such as microwave energy [102] or ultrasound [103, 104]. Organic solvents can improve digestion efficiency as well by denaturing the proteins; however, this can be a disadvantage for solid-phase digestion due to denaturation of the immobilized enzyme by the solvent [105].

The geometry of the microreactor can be an open channel or a 3D structure. An open channel format is the simplest configuration, but digestion is limited by the relatively long diffusional distances, which can limit the digestion rate. Diffusion rates of proteins also depend on their concentrations [106]. Lower diffusion rates are expected at higher concentrations of proteins, thereby lowering digestion efficiency due to limited encounter numbers between the surface-immobilized proteolytic enzymes and proteins. A 3D solid phase reactor format can be configured from a monolithic porous network [6, 107] or from a packed channel [108]. The high surface-to-volume ratio compared to open channels reduces the digestion time due to smaller diffusional lengths, which allows for more encounters between the substrate and immobilized enzyme [109]. A microfluidic chip packed with trypsin-derivatized beads in a fluidic channel has been reported [108]. The bead-packed chip provided faster protein digestion and fewer trypsin autolysis products compared to a homogeneous digestion.

Solid-phase microreactors can be formed in situ by immobilization of enzymes through covalent attachment to supports or encapsulation within gel matrices [6, 87, 107, 110]. This avoids the difficulties arising from packing beads into microchannels. Monolithic supports for immobilizing proteolytic enzymes can be generated through a polymerization reaction of a monomer solution into a microfluidic channel [111]. For example, a porous organic polymer monolithic microreactor was developed in which trypsin was immobilized within the monolith using azlactone functional groups for covalent attachment of the enzyme [107, 110]. The sequence coverage of tryptic peptides from myoglobin was determined by off-line (matrix-assisted laser desorption/ionization (MALDI) and found to be 67% for a 12 s residence time.

A silica sol–gel monolith containing zeolite nanoparticles has been reported [87]. This device had a high surface area for the immobilized enzyme, allowing for a high load within the microreactor. A 0.5 μL volume containing 0.2 μg/μL of the proteins (cytochrome *c* and BSA) was digested within 5 s in the microreactor, as indicated by off-line MALDI-TOF (time-of-flight) MS. The microreactor could be used repeatedly and the enzyme remained active for more than a month when it was stored at temperatures below 4 °C.

A pepsin microreactor was developed using a sol–gel monolithic column photopolymerized within a fused silica capillary [6]. The column was used for on-line ESI CE/MS. Although monolithic microreactors are fast and efficient, the process of their preparation may require more than 24 h and can be difficult to reproduce.

A 3D microreactor can be created using microstructures, such as pillars, within the reactor channel, which bypasses problems associated with bead packing or monolith formation in the channel [112, 113]. The channel footprint as well as the pillars occupying the reactor bed can be produced in polymeric materials, such as PMMA, using hot embossing or injection molding. Therefore, the entire high surface-area microreactor can be created in a single step, simplifying device preparation. Furthermore, the support structures are placed in a desired location within the device, with a fixed distance between the packing material providing unrestricted substrate access to the immobilized enzyme.

3.4 Peptide Separation and Mass Spectrometry Analysis

In order to identify the protein, the resulting peptides or intact protein must be fed into a mass spectrometer. Great progress has been made towards interfacing various microchips to a mass spectrometer via ionization-based interfaces for ESI and transfer interfaces to MALDI plates. For example, Musyimi et al. [114] electrophoretically separated cyctochrome *c* digests on PMMA microchips coupled to online MALDI-MS using a rotating ball inlet. Work involving interfacing microchips to

MS has received significant attention and has spawned extensive research efforts resulting in numerous publications. Detailed information on this aspect of the proteome processing pipeline is beyond the scope of this chapter. For more information on this topical area, the reader is referred to a recent review [115].

4 Integrated Microsystems for Protein Analysis

According to the theoretical physicist and accomplished author, Freeman J. Dyson, "new directions in science are launched by new tools, much more often than by new concepts" [116]. It is believed that integrating various proteomic processes onto a chip will improve process automation and make the arduous task of proteome analyses more available to non-trained users. This will improve fundamental discoveries in protein biology, especially in clinical settings where the user needs to input the sample into the system and get a readout. Various devices performing a single process in the proteomic pipeline are currently used in isolation, as discussed earlier. However, several groups have attempted to integrate processing devices onto a single common microfluidic platform to improve process automation. As a matter of definition, we define a "device" as a chip that performs a single processing step, for example cell lysis. Devices can then be integrated using the appropriate interconnection technology into a single wafer or combined to a control board as separate devices to form the microsystem, which is composed of two or more devices.

For a conventional analysis of complex protein samples, the analysis time can be on the order of several days, the sample volumes required in the microliter–milliliter range, the process strategy labor-intensive, and the transfer of sample from one processing step to the next a source of material loss and/or contamination. In this context, it is advantageous to consider the use of a fully integrated system, whereby the sample is moved through processing steps without requiring operator intervention. This would allow for the reduction of material loss, reduce the analysis time, improve result reproducibility, and eliminate most sample handling by the operator. These advantages have been highlighted and reviewed by a number of groups [117–119]. In this section, we will introduce work that has been initiated to realize the goal of creating fully integrated microsystems for protein analysis.

The ideal system would integrate all of the processing steps into a single platform to provide efficient and fast analysis of the sample with minimum need for operator intervention and sample handling. A view of such a system is shown in Fig. 5. As depicted in Fig. 5, an integrated microsystem would be comprised of several different processing steps (i.e., devices) including:

1. A device that can lyse the cells once a particular cell type has been selected from a heterogeneous population if required.
2. Solid-phase extraction device used to select the target molecules from a mixed-population for preconcentration and purification. This can be used to extract

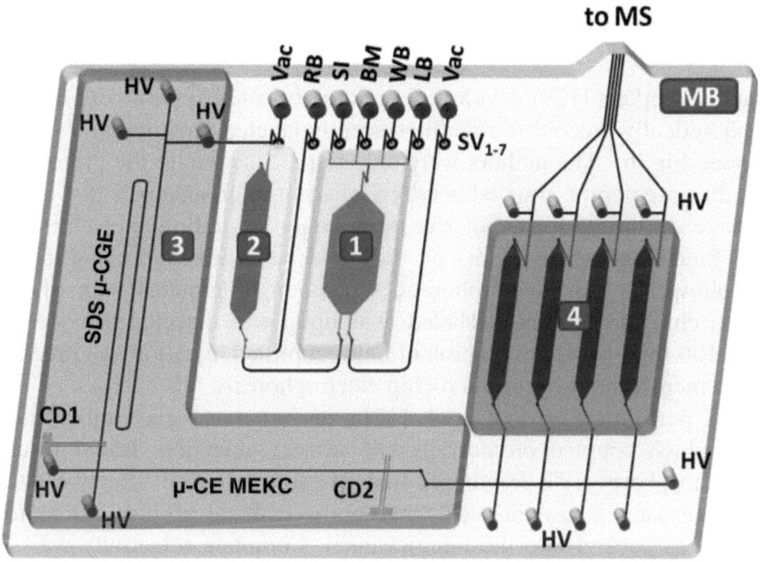

Fig. 5 Layout of an integrated microchip for full protein process analysis. The system consists of a cell lysis unit (*1*), a unit for the solid phase extraction of the entire proteome or a subpopulation of the proteome (*2*), 2D electrophoresis unit to sort the protein mixture into discrete components (*3*), solid-phase proteolytic reactors (*4*), and an on-line interface to MS, either ESI or MALDI

all of the proteins comprising the proteome or a specific subclass from the proteome.

3. A sorting device. The protein mixture can be sorted into discrete components using a multidimensional separation process, such as 2D electrophoresis, prior to the proteolytic digestion. Although this can be done using a variety of different electrophoresis modalities, those depicted in Fig. 5 use SDS µ-CGE in the first dimension and MEKC to sort co-migrating proteins by their hydrophobicity in the second dimension.
4. Proteolytic digestion device to generate peptides.
5. Introduction of peptides or intact proteins into a mass spectrometer for protein identification.

We considered platforms wherein two or more protein processing devices were combined to build the microsystem. Our discussions are primarily focused on the integration of various processing steps prior to MS analysis. Therefore, the interface to the MS is not specifically outlined in this discussion nor is it counted as a device in the system being discussed. This survey will provide several unique perspectives: (1) processing strategies that can be transitioned to the microscale and challenges associated with such a transition; (2) challenges with building integrated microsystems for protein analysis arising from the complexities of the sample; and (3) the benefits of automation using microsystems in terms of processing time, limit-of-detection and system reliability.

4.1 Integrated Systems with Two Devices

Foote and coworkers [120] developed a microfabricated system with the ability to electrophoretically preconcentrate fluorescently labeled proteins prior to their separation (see Fig. 6). The authors were able to preconcentrate the proteins using a porous silica membrane situated between adjacent microchannels that allowed for the passage of buffer ions, but excluded larger migrating molecules, such as proteins. Preconcentration factors of ~600-fold were achieved using this on-chip format followed by an electrophoretic separation of proteins with SDS-PAGE. Using this chip, fluorescently labeled ovalbumin was detected at concentrations as low as 100 fmol by a combination of field-amplified injection and preconcentration at the membrane prior to microchip electrophoresis.

In work performed by Yue et al. [121], an integrated glass microfluidic was reported, which coupled proteolysis with affinity selection. Initial results with standard phosphorpeptide fragments from β-casein showed selective capture of the phosphorylated fragments using immobilized metal affinity chromatography (IMAC) beads packed into the microchannel. Complete selectivity was seen for angiotensin used as a model, with capture of only the phosphorylated forms.

An integrated poly(dimethylsiloxane), PDMS, microchip for SPE and CE followed by ESI/TOF MS has been developed and evaluated by Dahlin and coworkers [122]. The microchip (see Fig. 7) was fabricated in a two-level cross design with PDMS cast over steel wires. Following PDMS polymerization and removal of the wires, 50 μm

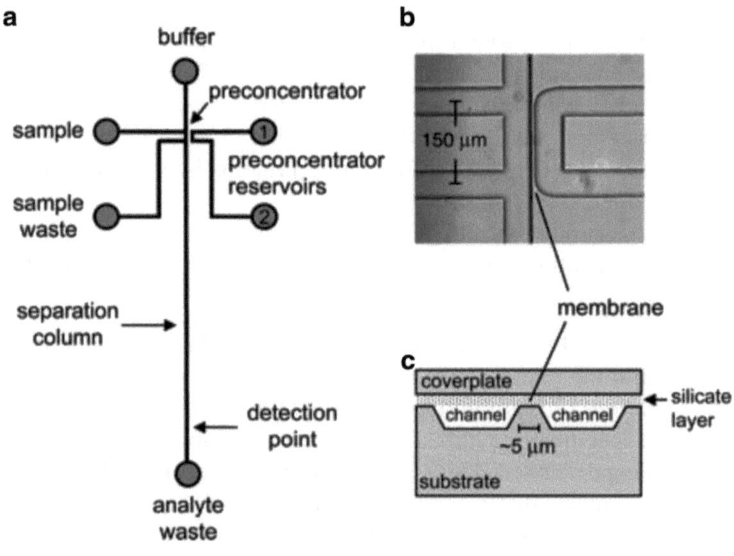

Fig. 6 (**a**) Microchip used for preconcentration and the subsequent electrophoretic separation of the preconcentrated proteins using SDS-PAGE. (**b**) Microscopic image of preconcentrator-injector channels. (**c**) Cross-section through injector and preconcentrator channels. Reproduced, with permission, from [120]

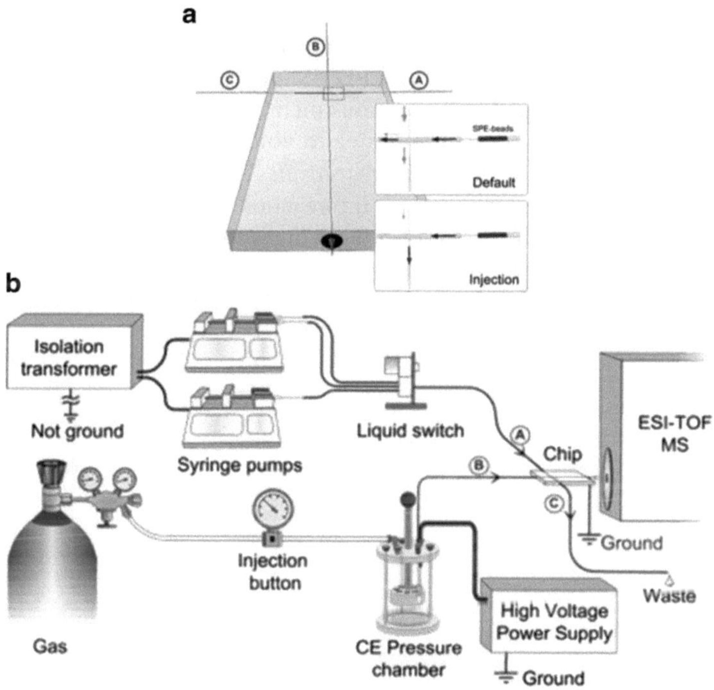

Fig. 7 (**a**) PDMS microchip device: *A* sample inlet channel, *B* CE channel, *C* waste channel. (**b**) Instrument setup and the connection of the microchip to the ESI/TOF MS. Reproduced from [122]

cylindrical channels were formed and fused-silica capillaries were then inserted into the structure to produce an interconnection. The inner walls of the capillaries and the PDMS microchip channels were modified with a positively charged polymer. The channel at the lower level was packed with 5 μm hyper-crosslinked polystyrene beads acting as a SPE medium used for desalting. In the upper level channel, microchip CE was performed. The device was equipped with an emitter tip coated with conducting graphite to facilitate electrical contact for ESI. Six standard peptide mixtures were dissolved in physiological salt solution, injected, desalted, separated, and sprayed into the MS for analysis, with an LOD in the femtomole regime.

In a similar effort to combine preconcentration with electrophoretic separations, Fortier et al. [123] investigated the analytical performances of a microfluidic system comprised of an enrichment column, a reversed phase separation channel, and a nanoelectrospray emitter embedded together in polyimide layers. The authors demonstrated that the configuration minimized transfer lines and connections and reduced peak broadening and dead volumes, resulting in good reproducibility of retention time and peak intensity. The microchip was interfaced to both ion trap and TOF MS. Measurements were performed for a dilution series of protein digests spiked into rat plasma samples and provided an LOD of 1–5 fmol.

In another study, Gao and coworkers [124] developed an integrated microchip system for rapid and sensitive protein identification generated by on-line protein digestion and analysis of the digested proteins using transient capillary isotachophoresis/capillary zone electrophoresis (CZE) with MS. A miniaturized membrane reactor was constructed by fabricating a PDMS microchip and coupling the microfluidic to a poly(vinylidene fluoride) porous membrane with adsorbed trypsin (see Fig. 8), which produced a large surface area-to-volume ratio through the use of the porous membrane media with adsorbed trypsin. The residence time of proteins inside the trypsin-adsorbed membrane, the reaction temperature, and the protein concentration controlled the extent of protein digestion.

A microfluidic system was described by Wang and coworkers [108] in which an electrospray interface to MS was integrated to a CE channel and a protein digestion

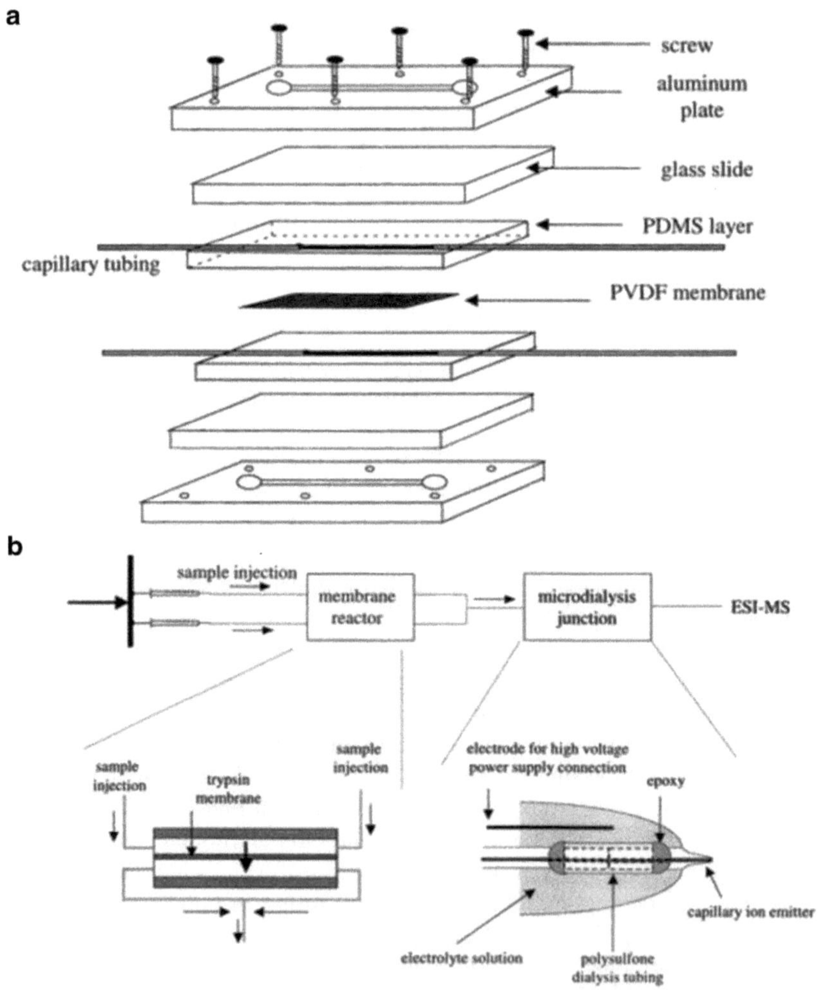

Fig. 8 (**a**) Membrane reactor assembly. (**b**) Setup for performing ESI MS analysis of peptide mixtures from the trypsin membrane reactor. Reproduced from [124]

bed situated on a monolithic substrate (see Fig. 9). To perform the analysis, an 800 μm wide, 150 μm deep and 15 mm long channel served as a reactor bed for trypsin, which was immobilized on 40–60 μm diameter beads. Separation was performed in channels feeding a capillary attached to the chip with a low dead-volume coupler. Then, the sample including melittin, cytochrome *c* and bovine serum albumin (BSA), was pumped through the reactor bed at flow rates between 0.5 and 60 μL/min and the application of the system for rapid digestion, separation, and identification of proteins was demonstrated. A flow rate of 1 or 0.5 μL/min was found to be adequate for complete digestion of cytochrome c or BSA, respectively, corresponding to a digestion time of 3–6 min at room temperature.

Hardouin et al. [125] reported a nano-HPLC–chip–MS system that integrated a sample enrichment column and a nanoscale C_{18} reversed phase separation column with a nanoelectrospray tip. The chip was automatically loaded and positioned into an MS nanospray chamber. To enhance protein identification, the method combined *m*/*z* data generated from the MS with highly reproducible peptide retention times (relative standard deviation, RSD = 1.2%) from the nano-LC. Proteins were separated and digested off-chip prior to the nano-HPLC–chip–MS analysis using conventional 2D IEF/SDS-PAGE, protein spot excision from the 2D gels after separation, and digestion of protein spots with trypsin. The nano-HPLC–chip–MS technique was able to analyze and detect as low as 10 fmol of a BSA digest.

Long et al. [126] coupled a preconcentration SPE system to microchip electrophoresis in a multilayer PDMS configuration with the various layers separated by a nanoporous thin membrane. A representation of the system is shown in Fig. 9. The SPE, carried out in the upper unit, could be performed either by pressure or electroosmosis pumping. The SPE device was made by loading the upper channel with a suspension of C_{18}-coated 10 μm silica beads. Sample was delivered to the bead bed from R1 to R2 (see Fig. 10), followed by washing with 20/80 (v/v) acetonitrile–borate buffer and, then, elution of concentrated analytes from the SPE column with 60/40 (v/v) acetonitrile–borate buffer. To introduce an injection plug into the lower CE device, a 2 s pulse voltage was applied between R2 and R4. CE detection was monitored by laser-induced fluorescence (LIF). Although the system was tested with a dye-labeled ephedrine solution, its application could be

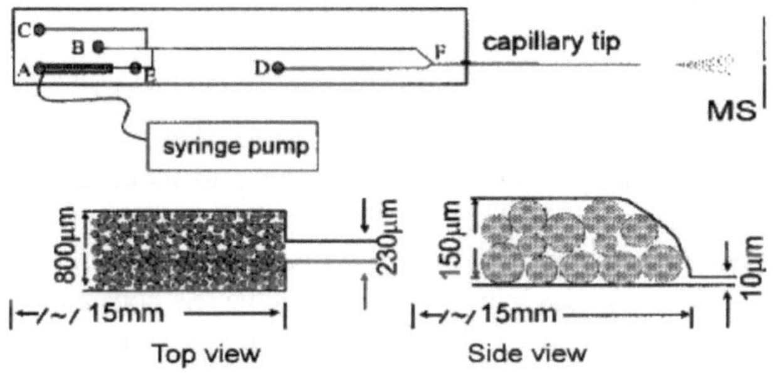

Fig. 9 Integrated enzyme reaction bed and CE microchip. *Top* and *side views* show a blow up of the packed trypsin bead. Reproduced from [108]

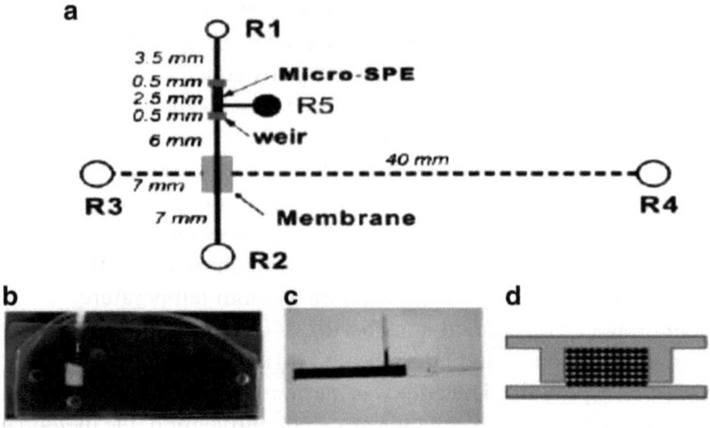

Fig. 10 (**a**) Layout of an integrated SPE-CE multilayer system consisting of a small piece of nanoporous membrane sandwiched between the upper (*continuous line from R1 to R2; R stands for Reservoir*) and lower (*broken line* from *R3* to *R4*) PDMS layers. (**b**) Photograph of the multilayer system. (**c**) Micrograph and (**d**) diagram of the packed μ-SPE column between two shallow weirs. Reproduced from [126]

extended to proteins as well. Similar systems in the sense of integrating a preconcentrator to a separation device have been reported [127–132].

An integrated microfabricated system composed of a proteolytic reactor and chromatographic column with direct interface to ESI-MS was reported by Carlier et al. [133] The system is represented in Fig. 11 and was fabricated from SU-8. The chromatographic end of the chip was terminated with a nano-ESI interface. The digestion module was composed of trypsin covalently attached to a monolithic polymer, which was also used to prepare a hydrophobic stationary phase for the separation of peptides prior to MS analysis. Monoliths were made in situ by photopolymerizing ethylene glycoldimethacrylate (EDMA) monomers in the presence of lauryl methacrylate (LMA) or butyl methacrylate (BMA) crosslinkers.

4.2 *Integrated Systems with Three Devices*

Gottschlich et al. [134] developed a microfluidic system that integrated enzymatic reactions, electrophoretic separation of the reactants from the products, and post-separation labeling of the proteins and peptides prior to fluorescence detection (see Fig. 12). Tryptic digestion of oxidized insulin β-chain was performed in 15 min under stopped flow conditions in a heated channel serving as the reactor, and the separation was completed in 60 s. Localized thermal control of the reaction channel was achieved using a resistive heating element. The separated reaction products were then labeled with naphthalene-2,3-dicarboxaldehyde (NDA) and detected by fluorescence detection.

A PDMS microfluidic system has been reported by Dodge and coworkers [135] that combined on-line protein electrophoretic separation, selection, and digestion of

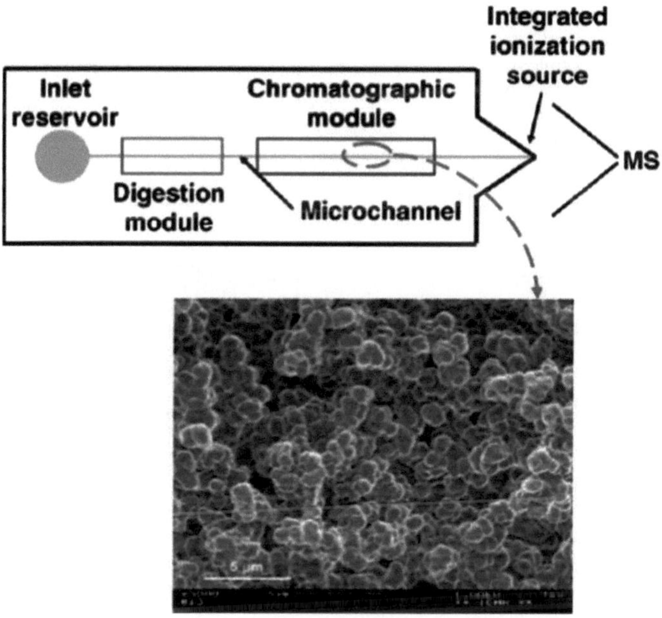

Fig. 11 *Top*: SU-8 based microfluidic system, which includes an enzymatic microreactor, a chromatographic device, and an integrated ionization emitter tip. *Bottom*: SEM photograph of a section of a monolithic phase prepared from LMA/EDMA. Reproduced from [133]

the protein of interest for subsequent identification by MS (see Fig. 13). The microfluidic system contained eight valves and a micropump to control fluid flow. To evaluate the system performance, myoglobin was successfully isolated from serum, digested, and the proteolytic peptides recovered from the micromixer for protein identification. Total analysis time from sample injection to protein identification was ~30 min, with sample volumes in the range of tens of nanoliters.

Huh et al. [136] developed the integrated microfluidic system shown in Fig. 14, which was used to successfully lyse cells by mixing cells with a lysing buffer at 500 rpm for 10 min. Mixing was made possible by embedding a microdisk in a microchamber and applying an external rotating magnetic force. Elution buffer was then introduced into the chamber to shuttle the lysed sample into a fritless SPE device that was made by in situ photopolymerization of ethylene glycol dimethacrylate (EGDMA) and acrylamide monomer solutions. Most cellular debris and negatively charged proteins fused to a recombinant gold binding peptide (GBP) from *E. coli* bound to the SPE surface, allowing up to 60% of the total proteome to be shuttled to a detection device where GBP-fusion proteins were electrochemically detected using three gold electrodes integrated into the microchannels.

Herr et al. [137] developed a microfluidic chip for saliva-based analysis of proteins (see Fig. 15) [137], which allowed for automated measurement of total salivary metalloproteinase-8 (MMP-8) concentration. The chip performed three tasks: (1) enrichment of targets from saliva via a size-exclusion membrane inside

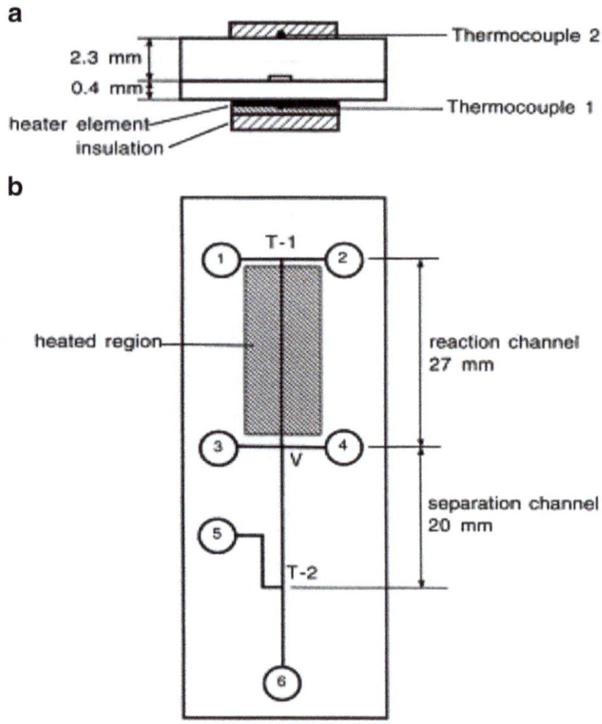

Fig. 12 (**a**) Cross-sectional view of a microchip, heating element, and thermocouples. (**b**) Diagram of the microchip used for on-chip proteolytic reactions, separations, and post-column labeling for generating fluorescent moieties. The fluid reservoirs are: (*1*) substrate, (*2*) enzyme, (*3*) buffer, (*4*) sample waste, (*5*) NDA, and (*6*) waste. Reproduced from [134]

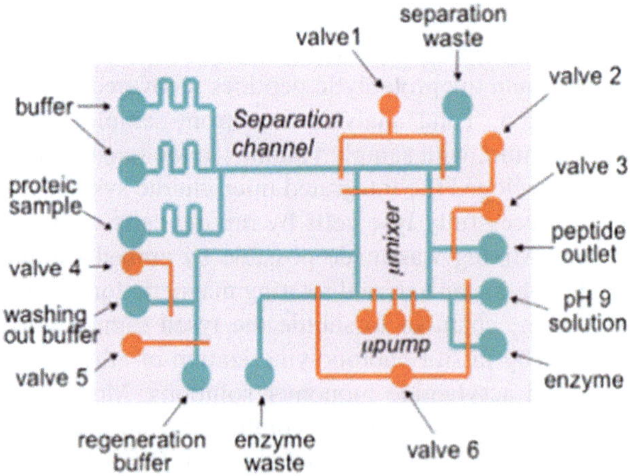

Fig. 13 Integrated PDMS microfluidic system comprising four modules: an injection/separation module in which pumping was entirely supported by electroosmotics; a protein trapping module; a circular micromixer where pumping was mechanically achieved; and an enzyme reaction module. Fluidic channels are in *red*, actuation channels are in *blue-green*. Valve actuation channels were filled with water in order to avoid air entering the fluidic channels through the PDMS membranes. Integrated valves are numbered from *1* to *6*. Reproduced from [135]

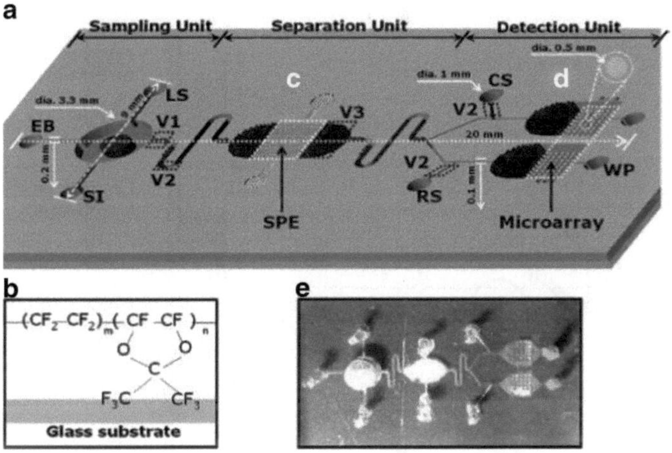

Fig. 14 Integrated microfluidic system for on-chip analysis of biomolecules. Fully integrated microfluidic system showing (**a**) sample mixing zone for disruption of cells, (**c**) the sample purification zone with acrylamide-functionalized SPE packed by in situ polymerization, and (**d**) the gold microcircle patterned analysis zone for detection of infectious pathogens. Reservoirs in the mixing chamber are labeled *SI*, *EB*, and *LS* for sample inlet, elution buffer, and lysis solution, respectively. Detection reservoirs are labeled *RS*, *CS*, and *WP* for reaction sample, control sample, and waste port, respectively. (**b**) Chemical structure of Teflon AF 1600 used for the hydrophobic film valve (*V2*). (**e**) Photograph of the integrated system coupled with a magnetic stirrer. The cylinder-type micropillar for packing SPE is for removing debris (*V3*). Device dimension are 10.0 (width) × 18.0 (length) × 5 (height) mm. Reproduced from [136]

a channel; (2) immunobinding of fluorescently labeled MMP-8 antibodies; and (3) discrimination of MMP-8 proteins from MMP-8–antibody complexes using a size-based electrophoretic separation in an on-chip photopolymerized gel. The authors reported a lower limit of quantitation of 130 ng/mL, which made it possible to detect MMP-8 in clinical patients; average MMP-8 concentrations in the saliva of aperiodontally diseased patients was determined to be 623.8 ± 204.0 ng/mL ($p < 0.05$). The chip was made from glass and was part of a portable analytical reader with a menu-based user interface, high-voltage power supply connections, software control, data acquisition and communication components, which were all integrated into a single platform.

4.3 Integrated Systems with More Than three Devices

Stachowiak et al. [138] coupled an automated sample preparation (ASP) system to a chip gel electrophoresis protein profiling (CGE-PP) system to execute an autonomous microfluidic sample preparation and detection of aerosolized bacterial cells and spores on the basis of protein profiling. The combined system, which was field-deployable, was capable of differentiating between bacterial organisms. It operated by collecting

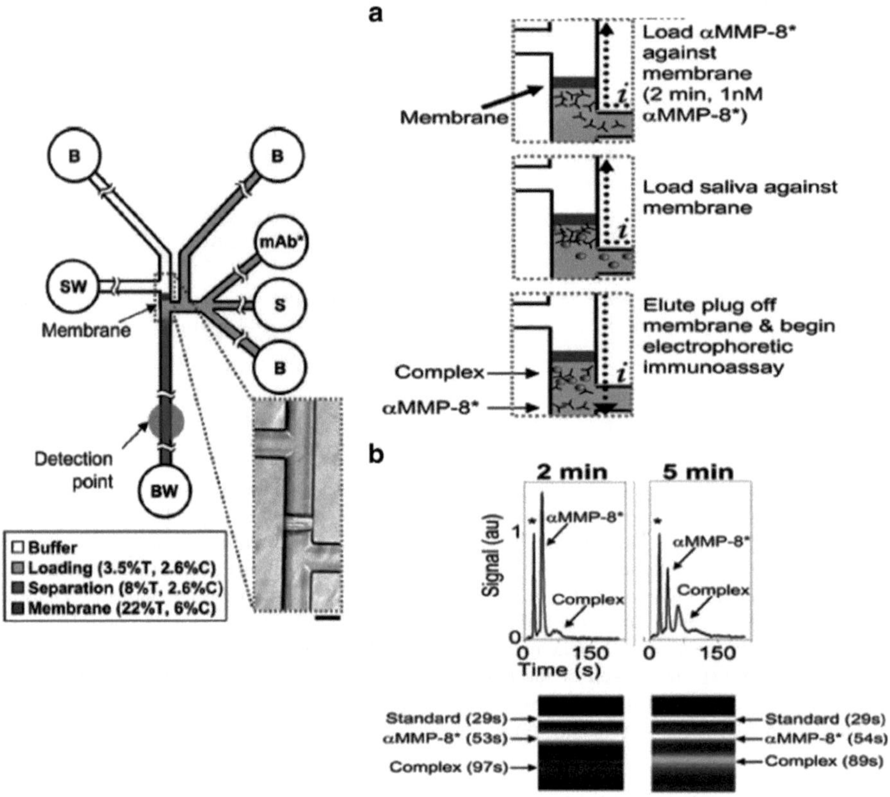

Fig. 15 *Left*: Device layout of immunoseparation chip. Reservoirs are as follows: *S* sample, *B* buffer, *SW* sample waste, *BW* buffer waste, *mAb** fluorescently labeled monoclonal antibody to MMP-8. Polyacrylamide gel composition is indicated by *grayscale shading* (%T and %C are percentage of total acrylamide and bis-acrylamide cross-linker, respectively). *Inset* shows a 40× bright-field image of the size-exclusion membrane. *Right*: On-chip sample enrichment. (**a**) Chip operation scheme. Saliva sample and labeled antibody result in co-enrichment at the size-exclusion membrane. An electric potential is applied across the membrane, causing the enriched species to elute into the separation channel, thus initiating the electrophoretic immunoassay followed by a potential switch to omit the membrane from the current path. Current flow is indicated by *i*. (**b**) Electropherograms and gel-like plots resulting after a 5-min enrichment. Protein internal standard is marked with an *asterisk* on the electropherograms. Reproduced with permission from [137]

aerosol samples using a collector, preconcentrating the organisms, thermochemically lysing the bacterial cells or spores, subjecting the resulting proteins to size-exclusion chromatography prior to fluorescent labeling the proteins with dyes and, finally, separating the labeled proteins using gel electrophoresis to generate a protein profile. The entire process strategy was completed in approximately 10 min with a sensitivity of 16 agent-containing particles per liter of air for *Bacillus subtilis* spores. By profiling the proteins of pathogenic organisms, the technique could detect any pathogen, unlike PCR-based or immunoassay-based techniques, which requires the pathogen's sequence to be known or monoclonal antibodies available for affinity selection.

5 Concluding Remarks

Developing fully integrated multifunctional microfluidic systems for the automated analysis of complex proteomes using either a top-down or bottom-up strategy will have significant ramifications in a number of important application areas such as biomarker discovery, protein biology, and in vitro diagnostics to name a few. The challenges in designing and fabricating such systems are numerous and include accommodating the diverse array of different protein types, the large dynamic range, and the low copy number of some types of proteins. This is just from a chemical perspective; the engineering requirements of such a system are also immense and include material selection to optimize the particular processing step, creating interconnects that have low dead volumes, providing flexibility to accommodate different process strategies (i.e., top-down or bottom up or some hybrid strategy) and a proper interface to the mass spectrometer required on the back-end of the analysis pipeline.

Each processing strategy adopted for proteomics has its own merits and challenges when contemplating transitioning it to a microchip platform. For example, a bottom-up approach using a shotgun strategy requires no initial multidimensional electrophoresis step because all proteins in the sample mixture are digested in one proteolytic step, but it does require sorting of the generated peptide fragments using a variety of LC-based methods, which can be difficult to transition to a chip format. On the other hand, the other bottom-up format requires the digestion of individual proteins that have been sorted electrophoretically using multidimensional formats and some success has been generated in developing such devices. The challenge is in interfacing the 2D separation with the subsequent proteolytic digestion step when employing a bottom-up strategy, but this is not required in top-down approaches. The coupling of 2D electrophoresis with proteolytic digestion for a bottom-up approach would require either the microreactor to be configured such that the digestion time was short enough to accommodate the spot production rate of the 2D electrophoresis or to generate on-chip multiple solid-phase microreactors to accommodate the rate of elution of components from the multidimensional separation. For example, if a 2D separation generates 131 spots per minute [139], a single solid-phase microreactor would have to produce complete digestion in $<$7.6 ms to prevent bottlenecking of the protein processing. Of course, the use of multiple solid-phase microreactors could provide longer times for digestion, with the time allocated to this processing step directly related to the number of reactors included in the system. For a top-down approach, no proteolytic digestion is required following a fractionation step of the entire protein sample with the intact proteins fed directly into the mass spectrometer and this would eliminate this potential bottleneck.

There are also other challenges facing the development of fully integrated systems for protein analysis. For example, different devices may require different buffer conditions for their optimized performance; the separation buffer conditions required to isolate proteins prior to protein digestion are different and might be incompatible. The integrated chip presented by Yue et al. [121] required different pH buffers for trypsin digestion and IMAC enrichment. Thus, when proteins were transported through the bead-based trypsin reactor prior to transport through the IMAC beads, 4% acetic acid was pumped through a side inlet to adjust the pH of the digested protein solution (pH 8.0 to pH 4–5 adjustment). Peterson et al. [140] used acetonitrile to release accumulated proteins from a SPE bed that was directly coupled to a proteolytic digestion reactor and reported high protein recovery. However, the use of pure organic solvents to release the proteins can be impractical where a separation step follows the extraction because organic solvents may not be compatible with the subsequent chromatographic process. Also, in many cases, SDS additives are used to sort proteins by their molecular weight and SDS can denature proteins, especially proteolytic enzymes used in the subsequent digestion step. However, it has been reported that trypsin digestion in the presence of 0.1% SDS results in better sequence coverage compared to digestions obtained without the use of SDS, but concentrations of 1% SDS completely disrupted the digestion [141]. Therefore, careful control of the SDS concentration prior to proteolytic digestion must be undertaken.

There is also an issue of incomplete digestion due to the limited amount of immobilized proteolytic enzyme associated with solid-phase microreactors. One way to approach this problem is to provide high surface area microreactors and this can be achieved by replacing large diameter beads onto which the proteolytic enzyme has been attached with smaller diameter particles or by immobilization of the enzyme onto a sol–gel matrix that is formed within the channel [121]. One can also create solid supports with a high surface area using a pillared architecture fabricated by microreplication in the same step used to create the fluidic network [142, 143]. This has the advantages of not requiring additional processing steps once the fluidic channels have been produced.

Preconcentrating protein samples after extraction and before separation in an integrated system for proteomic analysis is very important; enough material is needed for the subsequent MS analysis. For example, an electrophoretic peak volume of 10 pL is not uncommon in multidimensional electrophoresis platforms using microchips [81]. If the MS phase of the analysis has a mass detection limit of 1 fmol, the concentration of each protein required to accommodate the MS would be 0.1 mM in the 10 pL electrophoretic band. One way to preconcentrate the sample is in the separation injector [120]. However, peak distortions and loss of resolution during microchip CE can be observed following long preconcentration times [120]. Solid-phase extractions can also assist in providing not only preselection of the target material, but also preconcentration of the target material to accommodate the MS.

Finally, the dead volumes developed when capillary connections and other junctions to the fluidic chip to provide off-chip processing, such as the interface to MS, can significantly degrade separation performance and introduce sample carryover or sample loss. For example, Gao and coworkers [124] estimated ~4 μL for the dead volume associated with their miniaturized trypsin membrane reactor. As noted above, the typical peak volume associated with microchip electrophoresis is on the order of 10 pL, which demands that nearly all protein processing post-separation be done directly on-chip so as not to degrade separation performance, because these connections can create extra column effects that result in significant degradation in peak capacity through loss of plate numbers.

To develop a fully integrated chip for complete protein analysis on–chip, a combination of both bottom-up approaches (dubbed a bottom "combo" approach) has been considered. This processing pipeline would include the following: (1) cell lysis and protein solid-phase extraction; (2) moderate peak capacity separation, where the electrophoretic bands are actually mixtures of components instead of individual components; (3) digestion of the protein band mixture, similar in nature to the shotgun approach except that a smaller subpopulation of the proteome is analyzed; and (4) multidimensional peptide separations prior to introduction into the mass spectrometer. If one were to implement the bottom combo approach on a mammalian cell proteome, which contains ~10,000 proteins per cell [7, 8], it would first separate the 10,000 proteins after cell lysis using multidimensional electrophoresis. If we assume that we were able to resolve ~2,500 proteins using the multidimensional electrophoresis system [82], each band would contain four different proteins on average. As a typical example, the tryptic digestion of cytochrome *c* (104 amino acids) results in about ten peptides [114], and an average protein contains ~360 amino acids, so we would expect to generate roughly 140 peptides for four proteins (for the bottom combo approach). In addition, having sufficient signal-to-noise ratio is needed for the mass spectral readout. About 1.4×10^{-15} mole is expected from the 10 pL protein peak volume [81] considering an average concentration of 3 nM for each protein and an average molecular mass of 46 kDa [81].

References

1. Wilkins MR, Sanchez JC, Gooley AA, Appel RD, HumpherySmith I, Hochstrasser DF, Williams KL (1996) Biotechnol Genet Eng Rev 13:19
2. de Hoog CL, Mann M (2004) Annu Rev Genomics Hum Genet 5:267
3. Wilkins MR, Pasquali C, Appel RD, Ou K, Golaz O, Sanchez JC, Yan JX, Gooley AA, Hughes G, HumpherySmith I, Williams KL, Hochstrasser DF (1996) Biotechnology 14:61
4. Jackson LO, Borkman S, Liyanage R, Wilkens CL (2006) Trends Anal Chem 25:1046
5. Ishihama Y, Schmidt T, Rappsilber J, Mann M, Hartl FU, Kerner MJ, Frishman D (2008) Bmc Genomics 9
6. Kato M, Sakai-Kato K, Jin H, Kubota K, Miyano H, Toyo'oka T, Dulay MT, Zare RN (2004) Anal Chem 76:1896
7. Anderson NL, Anderson NG (1998) Electrophoresis 19:1853
8. Hu S, Michels DA, Fazal MA, Ratisoontorn C, Cunningham ML, Dovichi NJ (2004) Anal Chem 76:4044
9. Kussmann M, Affolter M, Fay LB (2005) Comb Chem High Throughput Screen 8:679
10. Venter JC, Adams MD, Myers EW, Li PW, Mural RJ, Sutton GG, Smith HO, Yandell M, Evans CA, Holt RA, Gocayne JD, Amanatides P, Ballew RM, Huson DH, Wortman JR, Zhang Q, Kodira CD, Zheng XQH, Chen L, Skupski M, Subramanian G, Thomas PD, Zhang JH, Miklos GLG, Nelson C, Broder S, Clark AG, Nadeau C, McKusick VA, Zinder N, Levine AJ, Roberts RJ, Simon M, Slayman C, Hunkapiller M, Bolanos R, Delcher A, Dew I, Fasulo D, Flanigan M, Florea L, Halpern A, Hannenhalli S, Kravitz S, Levy S, Mobarry C, Reinert K, Remington K, Abu-Threideh J, Beasley E, Biddick K, Bonazzi V, Brandon R,

Cargill M, Chandramouliswaran I, Charlab R, Chaturvedi K, Deng ZM, Di Francesco V, Dunn P, Eilbeck K, Evangelista C, Gabrielian AE, Gan W, Ge WM, Gong FC, Gu ZP, Guan P, Heiman TJ, Higgins ME, Ji RR, Ke ZX, Ketchum KA, Lai ZW, Lei YD, Li ZY, Li JY, Liang Y, Lin XY, Lu F, Merkulov GV, Milshina N, Moore HM, Naik AK, Narayan VA, Neelam B, Nusskern D, Rusch DB, Salzberg S, Shao W, Shue BX, Sun JT, Wang ZY, Wang AH, Wang X, Wang J, Wei MH, Wides R, Xiao CL, Yan CH, Yao A, Ye J, Zhan M, Zhang WQ, Zhang HY, Zhao Q, Zheng LS, Zhong F, Zhong WY, Zhu SPC, Zhao SY, Gilbert D, Baumhueter S, Spier G, Carter C, Cravchik A, Woodage T, Ali F, An HJ, Awe A, Baldwin D, Baden H, Barnstead M, Barrow I, Beeson K, Busam D, Carver A, Center A, Cheng ML, Curry L, Danaher S, Davenport L, Desilets R, Dietz S, Dodson K, Doup L, Ferriera S, Garg N, Gluecksmann A, Hart B, Haynes J, Haynes C, Heiner C, Hladun S, Hostin D, Houck J, Howland T, Ibegwam C, Johnson J, Kalush F, Kline L, Koduru S, Love A, Mann F, May D, McCawley S, McIntosh T, McMullen I, Moy M, Moy L, Murphy B, Nelson K, Pfannkoch C, Pratts E, Puri V, Qureshi H, Reardon M, Rodriguez R, Rogers YH, Romblad D, Ruhfel B, Scott R, Sitter C, Smallwood M, Stewart E, Strong R, Suh E, Thomas R, Tint NN, Tse S, Vech C, Wang G, Wetter J, Williams S, Williams M, Windsor S, Winn-Deen E, Wolfe K, Zaveri J, Zaveri K, Abril JF, Guigo R, Campbell MJ, Sjolander KV, Karlak B, Kejariwal A, Mi HY, Lazareva B, Hatton T, Narechania A, Diemer K, Muruganujan A, Guo N, Sato S, Bafna V, Istrail S, Lippert R, Schwartz R, Walenz B, Yooseph S, Allen D, Basu A, Baxendale J, Blick L, Caminha M, Carnes-Stine J, Caulk P, Chiang YH, Coyne M, Dahlke C, Mays AD, Dombroski M, Donnelly M, Ely D, Esparham S, Fosler C, Gire H, Glanowski S, Glasser K, Glodek A, Gorokhov M, Graham K, Gropman B, Harris M, Heil J, Henderson S, Hoover J, Jennings D, Jordan C, Jordan J, Kasha J, Kagan L, Kraft C, Levitsky A, Lewis M, Liu XJ, Lopez J, Ma D, Majoros W, McDaniel J, Murphy S, Newman M, Nguyen T, Nguyen N, Nodell M, Pan S, Peck J, Peterson M, Rowe W, Sanders R, Scott J, Simpson M, Smith T, Sprague A, Stockwell T, Turner R, Venter E, Wang M, Wen MY, Wu D, Wu M, Xia A, Zandieh A, Zhu XH (2001) Science 291:1304

11. Lander ES, Linton LM, Birren B, Nusbaum C, Zody MC, Baldwin J, Devon K, Dewar K, Doyle M, FitzHugh W, Funke R, Gage D, Harris K, Heaford A, Howland J, Kann L, Lehoczky J, LeVine R, McEwan P, McKernan K, Meldrim J, Mesirov JP, Miranda C, Morris W, Naylor J, Raymond C, Rosetti M, Santos R, Sheridan A, Sougnez C, Stange-Thomann N, Stojanovic N, Subramanian A, Wyman D, Rogers J, Sulston J, Ainscough R, Beck S, Bentley D, Burton J, Clee C, Carter N, Coulson A, Deadman R, Deloukas P, Dunham A, Dunham I, Durbin R, French L, Grafham D, Gregory S, Hubbard T, Humphray S, Hunt A, Jones M, Lloyd C, McMurray A, Matthews L, Mercer S, Milne S, Mullikin JC, Mungall A, Plumb R, Ross M, Shownkeen R, Sims S, Waterston RH, Wilson RK, Hillier LW, McPherson JD, Marra MA, Mardis ER, Fulton LA, Chinwalla AT, Pepin KH, Gish WR, Chissoe SL, Wendl MC, Delehaunty KD, Miner TL, Delehaunty A, Kramer JB, Cook LL, Fulton RS, Johnson DL, Minx PJ, Clifton SW, Hawkins T, Branscomb E, Predki P, Richardson P, Wenning S, Slezak T, Doggett N, Cheng JF, Olsen A, Lucas S, Elkin C, Uberbacher E, Frazier M, Gibbs RA, Muzny DM, Scherer SE, Bouck JB, Sodergren EJ, Worley KC, Rives CM, Gorrell JH, Metzker ML, Naylor SL, Kucherlapati RS, Nelson DL, Weinstock GM, Sakaki Y, Fujiyama A, Hattori M, Yada T, Toyoda A, Itoh T, Kawagoe C, Watanabe H, Totoki Y, Taylor T, Weissenbach J, Heilig R, Saurin W, Artiguenave F, Brottier P, Bruls T, Pelletier E, Robert C, Wincker P, Rosenthal A, Platzer M, Nyakatura G, Taudien S, Rump A, Yang HM, Yu J, Wang J, Huang GY, Gu J, Hood L, Rowen L, Madan A, Qin SZ, Davis RW, Federspiel NA, Abola AP, Proctor MJ, Myers RM, Schmutz J, Dickson M, Grimwood J, Cox DR, Olson MV, Kaul R, Shimizu N, Kawasaki K, Minoshima S, Evans GA, Athanasiou M, Schultz R, Roe BA, Chen F, Pan HQ, Ramser J, Lehrach H, Reinhardt R, McCombie WR, de la Bastide M, Dedhia N, Blocker H, Hornischer K, Nordsiek G, Agarwala R, Aravind L, Bailey JA, Bateman A, Batzoglou S, Birney E, Bork P, Brown DG, Burge CB, Cerutti L, Chen HC, Church D, Clamp M, Copley RR, Doerks T, Eddy SR, Eichler EE, Furey TS, Galagan J, Gilbert JGR, Harmon C, Hayashizaki Y, Haussler D, Hermjakob H, Hokamp K, Jang WH,

Johnson LS, Jones TA, Kasif S, Kaspryzk A, Kennedy S, Kent WJ, Kitts P, Koonin EV, Korf I, Kulp D, Lancet D, Lowe TM, McLysaght A, Mikkelsen T, Moran JV, Mulder N, Pollara VJ, Ponting CP, Schuler G, Schultz JR, Slater G, Smit AFA, Stupka E, Szustakowki J, Thierry-Mieg D, Thierry-Mieg J, Wagner L, Wallis J, Wheeler R, Williams A, Wolf YI, Wolfe KH, Yang SP, Yeh RF, Collins F, Guyer MS, Peterson J, Felsenfeld A, Wetterstrand KA, Patrinos A, Morgan MJ (2001) Nature 409:860
12. Lion N, Rohner TC, Dayon L, Arnaud IL, Damoc E, Youhnovski N, Wu ZY, Roussel C, Josserand J, Jensen H, Rossier JS, Przybylski M, Girault HH (2003) Electrophoresis 24:3533
13. Anderson NL, Anderson NG (2002) Mol Cell Proteomics 1:845
14. Kim H, Page GP, Barnes S (2004) Nutrition 20:155
15. Lay JO, Borgmann S, Liyanage R, Wilkins CL (2006) Trends Analyt Chem 25:1046
16. Ahmed N, Rice GE (2005) J Chromatogr B 815:39
17. Reinders J, Lewandrowski U, Moebius J, Wagner Y, Sickmann A (2004) Proteomics 4:3686
18. Ransohoff DF (2005) Nat Rev Cancer 5:142
19. Coombes KR, Morris JRS, Hu JH, Edmonson SR, Baggerly KA (2005) Nat Biotechnol 23:291
20. Haynes PA, Yates JR (2000) Yeast 17:81
21. Schluter H, Apweiler R, Holzhutter H-G, Jungblut PR (2009) Chem Cent J 3:1
22. Emmett MR, Caprioli RM (1994) J Am Soc Mass Spectrom 5:605
23. Bollag DM, Edelstein SJ (eds) (1991) Protein methods, Wiley-Liss, New York
24. Dignam JD (1990) Meth Enzymol 182:194
25. Lenstra JA, Bloemendal H (1983) Eur J Biochem 135:413
26. Toda T, Ishijima Y, Matsushita H, Yoshida M, Kimura N (1994) Electrophoresis 15:984
27. Sanchez J-C, Appel RD, Golaz O, Pasquali C, Ravier F, Bairoch A, Hochstrasser DF (1995) Electrophoresis 16:1131
28. Portig I, Pankuweit S, Lottspeich F, Maisch B (1996) Electrophoresis 17:803
29. Cull M, McHenry CS (1990) Meth Enzymol 182:147
30. Jazwinski SM (1990) Meth Enzymol 182:154
31. Kawaguchi S-i, Kuramitsu S (1995) Electrophoresis 16:1060
32. Goerg A, Postel W, Guenther S (1988) Electrophoresis 9:531
33. Teixeira-Gomes AP, Cloeckaert A, Bezard G, Dubray G, Zygmunt MS (1997) Electrophoresis 18:156
34. Ames GFL, Nikaido K (1976) Biochemistry 15:616
35. Goerg A, Boguth G, Obermaier C, Posch A, Weiss W (1995) Electrophoresis 16:1079
36. Goerg A, Postel W, Domscheit A, Guenther S (1988) Electrophoresis 9:681
37. Posch A, van den Berg BM, Berg HCJ, Goerg A (1995) Electrophoresis 16:1312
38. Gegenheimer P (1990) Meth Enzymol 182:174
39. Li PCH, Harrison DJ (1997) Anal Chem 69:1564
40. Di Carlo D, Jeong KH, Lee LP (2003) Lab Chip 3:287
41. Waters LC, Jacobson SC, Kroutchinina N, Khandurina J, Foote RS, Ramsey JM (1998) Anal Chem 70:158
42. Taylor MT, Belgrader P, Furman BJ, Pourahmadi F, Kovacs GTA, Northrup MA (2001) Anal Chem 73:492
43. Lee SW, Tai YC (1999) Sens Actuators A Phys 73:74
44. Lu H, Schmidt MA, Jensen KF (2005) Lab Chip 5:23
45. Damerval C, De Vienne D, Zivy M, Thiellement H (1986) Electrophoresis 7:52
46. Wu FS, Wang MY (1984) Anal Biochem 139:100
47. Harrison PA, Black CC (1982) Plant Physiol 70:1359
48. Rabilloud T (1996) Electrophoresis 17:813
49. Theillet C, Delpeyroux F, Fiszman M, Reigner P, Esnault R (1982) Planta 155:478
50. Badock V, Steinhusen U, Bommert K, Otto A (2001) Electrophoresis 22:2856
51. Issaq HJ, Conrads TP, Janini GM, Veenstra TD (2002) Electrophoresis 23:3048
52. Bauer A, Kuster B (2003) Eur J Biochem 270:570

53. Peng J, Gygi SP (2001) J Mass Spectrom 36:1083
54. Kutter JP, Jacobson SC, Ramsey JM (2000) J Microcolumn Sep. 12:93–97
55. Kutter JP, Jacobson SC, Matsubara N, Ramsey JM (1998) Anal Chem 70:3291
56. Lion N, Gellon JO, Jensen H, Girault HH (2003) J Chromatogr A 1003:11
57. Mao X, Luo Y, Dai Z, Wang K, Du Y, Lin BC (2004) Anal Chem 76:6941
58. Service RF (2001) Science 294:2074
59. Smith RD (2000) Nat Biotechnol 18:1041
60. Wang H, Hanash S (2003) J Chromatogr B 787:11
61. Chen XX, Wu HK, Mao CD, Whitesides GM (2002) Anal Chem 74:1772
62. Molloy MP, Brzezinski EE, Hang JQ, McDowell MT, VanBogelen RA (2003) Proteomics 3:1912
63. Ibel K, May RP, Kirschner K, Szadkowski H, Mascher E, Lundahl P (1990) Eur J Biochem 190:311
64. Laemmli UK (1970) Nature (London, United Kingdom) 227:680
65. Schaegger H, Von Jagow G (1987) Anal Biochem 166:368
66. Giddings JC (1967) Anal Chem 39:1027
67. Stroink T, Ortiz MC, Bult A, Lingeman H, de Jong GJ, Underberg WJM (2005) J Chromatogr B 817:49
68. Giddings JC (1984) Anal Chem 56:1258A
69. Giddings JC (1987) J High Resolut Chromatogr 10:319
70. Cortes HJ (1992) J Chromatogr 626:3
71. Cortes HJ (ed) (1990) Multidimensional chromatography: techniques and applications. In: Chromatographic science series, vol 50. Marcel Dekker, New York
72. Venkatramani CJ, Xu J, Phillips JB (1996) Anal Chem 68:1486
73. Erni F, Frei RW (1978) J Chromatogr 149:561
74. Rocklin RD, Ramsey RS, Ramsey JM (2000) Anal Chem 72:5244
75. Gilar M, Olivova P, Daly AE, Gebler JC (2005) Anal Chem 77:6426
76. Slonecker PJ, Li XD, Ridgway TH, Dorsey JG (1996) Anal Chem 68:682
77. Gray M, Dennis GR, Wormell P, Shalliker RA, Slonecker P (2002) J Chromatogr A 975:285
78. Chen H, Fan ZH (2009) Electrophoresis 30:758
79. Li Y, Buch JS, Rosenberger F, Devoe DL, Lee CS (2004) Anal Chem 76:742
80. Yang S, Liu JK, Lee CS, Devoe DL (2009) Lab Chip 9:592
81. Shadpour H, Soper SA (2006) Anal Chem 78:3519
82. Osiri JK, Shadpour H, Park S, Snowden BC, Chen ZY, Soper SA (2008) Electrophoresis 29:4984
83. Ethier M, Hou W, Duewel HS, Figeys D (2006) J Proteome Res 5:2754
84. Liu Y, Zhong W, Meng S, Kong J, Lu H, Yang P, Girault HH, Liu B (2006) Chem Eur J 12:6585
85. Rosenfeld J, Capdevielle J, Guillemot JC, Ferrara P (1992) Anal Biochem 203:173
86. Ru QC, Zhua LA, Katenhusena RA, Silbermana J, Brzeskia H, Liebmana M, Shriver CD (2006) J Chromatogr A 1111:175
87. Huang Y, Shan W, Liu B, Liu Y, Zhang Y, Zhao Y, Lu H, Tang Y, Yang P (2006) Lab Chip 6:534
88. Kumarathasan P, Mohottalage S, Goegan P, Vincent R (2005) Anal Biochem 346:85
89. Wilm M, Shevchenko A, Houthaeve T, Breit S, Schweigerer L, Fotsis T, Mann M (1996) Nature 379:466
90. Gharahdaghi F, Weinberg CR, Meagher DA, Imai BS, Mische SM (1999) Electrophoresis 20:601
91. Klammer AA, MacCoss MJ (2006) J Proteome Res 5:695
92. Lundell N, Schreitmüller T (1999) Anal Biochem 266:31
93. Schaefer H, Chamrad DC, Marcus K, Reidegeld KA, Blüggel M, Meyer HE (2005) Proteomics 5:846
94. Křenková J, Foret F (2004) Electrophoresis 25:3550
95. Massolini G, Calleri E (2005) J Sep Sci 28:7

96. Duan J, Liang Z, Yang C, Zhang J, Zhang L, Zhang W, Zhang Y (2006) Proteomics 6:412
97. Slysz GW, Lewis DF, Schriemer DC (2006) J Proteome Res 5:1959
98. Duan J, Sun L, Liang Z, Zhang J, Wang H, Zhang L, Zhang W, Zhang Y (2006) J Chromatogr A 1106:165
99. Strader MB, Tabb DL, Hervey WJ, Pan C, Hurst GB (2006) Anal Chem 78:125
100. Lazar IM, Ramsey RS, Ramsey JM (2001) Anal Chem 73:1733
101. Sim TS, Kim E-M, Joo HS, Kim BG, Kim Y-K (2006) Lab Chip 6:1056
102. Pramanik BN, Mirza UA, Ing YH, Liu Y-H, Bartner PL, Weber PC, Bose AK (2002) Protein Sci 11:2676
103. López-Ferrer D, Capelo JL, Vázquez J (2005) J Proteome Res 4:1569
104. López-Ferrer D, Hixson KK, Smallwood H, Squier TC, Petritis K, Smith RD (2009) Anal Chem 81:6272
105. Liu J, Lin S, Qi D, Deng C, Yang P, Zhang X (2007) J Chromatogr A 1176:169
106. Nesmelova IV, Skirda VD, Fedotov VD (2002) Biopolymers 63:132
107. Peterson DS, Rohr T, Svec F, Fréchet JMJ (2002) Anal Chem 74:4081
108. Wang C, Oleschuk R, Ouchen F, Li J, Thibault P, Harrison DJ (2000) Rapid Commun Mass Spectrom 14:1377
109. Slovakova M, Minc N, Bilkova Z, Smadja C, Faigle W, Fütterer C, Taverna M, Viovy J-L (2005) Lab Chip 5:935
110. Xie S, Svec F, Fréchet JMJ (1999) Biotechnol Bioeng 62:30
111. Svec F (2006) Electrophoresis 27:947
112. McCarley RL, Vaidya B, Wei S, Smith AF, Patel AB, Feng J, Murphy MC, Soper SA (2005) J Am Chem Soc 127:842
113. Qi S, Liu X, Ford S, Barrows J, Thomas G, Kelly K, McCandless A, Lian K, Goettert J, Soper SA (2002) Lab Chip 2:88
114. Musyimi HK, Guy J, Narcisse DA, Soper SA, Murray KK (2005) Electrophoresis 26:4703
115. Lee J, Soper SA, Murray KK (2009) J Mass Spectrom 44:579
116. Hood L (2002) J Proteome Res 1:399
117. Laurell T, Marko-Varga G (2002) Proteomics 2:345
118. O'Connor CD, Pickard K (2003) In: Day INM (ed) Microarrays & microplates: applications in biomedical sciences. BIOS Scientific, Abingdon, UK, pp 61–88
119. Chang H-T, Huang Y-F, Chiou S-H, Chiu T-C, Hsieh M-M (2004) Curr Proteomics 1:325
120. Foote RS, Khandurina J, Jacobson SC, Ramsey JM (2005) Anal Chem 77:57
121. Yue GE, Roper MG, Balchunas C, Pulsipher A, Coon JJ, Shabanowitz J, Hunt DF, Landers JP, Ferrance JP (2006) Anal Chim Acta 564:116
122. Dahlin AP, Bergstroem SK, Andren PE, Markides KE, Bergquist J (2005) Anal Chem 77:5356
123. Fortier M-H, Bonneil E, Goodley P, Thibault P (2005) Anal Chem 77:1631
124. Gao J, Xu J, Locascio LE, Lee CS (2001) Anal Chem 73:2648
125. Hardouin J, Duchateau M, Joubert-Caron R, Caron M (2006) Rapid Commun Mass Spectrom 20:3236
126. Long ZC, Shen Z, Wu DP, Qin JH, Lin BC (2007) Lab Chip 7:1819
127. Broyles BS, Jacobson SC, Ramsey JM (2003) Anal Chem 75:2761
128. Ramsey JD, Collins GE (2005) Anal Chem 77:6664
129. Yang WC, Sun XH, Pan T, Woolley AT (2008) Electrophoresis 29:3429
130. Proczek G, Augustin V, Descroix S, Hennion MC (2009) Electrophoresis 30:515
131. Tuomikoski S, Virkkala N, Rovio S, Hokkanen A, Siren H, Franssila S (2006) J Chromatogr A 1111:258
132. Hatch AV, Herr AE, Throckmorton DJ, Brennan JS, Singh AK (2006) Anal Chem 78:4976
133. Carlier J, Arscott S, Thomy V, Camart JC, Cren-Olive C, Le Gac S (2005) J Chromatogr A 1071:213–222
134. Gottschlich N, Culbertson CT, McKnight TE, Jacobson SC, Ramsey JM (2000) J Chromatogr B 745:243

135. Dodge A, Brunet E, Chen S, Goulpeau J, Labas V, Vinh J, Tabeling P (2006) Analyst 131:1122
136. Huh YS, Park TJ, Lee EZ, Hong WH, Lee SY (2008) Electrophoresis 29:2960
137. Herr AE, Hatch AV, Throckmorton DJ, Tran HM, Brennan JS, Giannobile WV, Singh AK (2007) "Microfluidic immunoassays as rapid saliva-based clinical diagnostics." *Proc Natl Acad Sci USA*, 104(13), 5268–5273
138. Stachowiak JC, Shugard EE, Mosier BP, Renzi RF, Caton PF, Ferko SM, de Vreugde JLV, Yee DD, Haroldsen BL, VanderNoot VA (2007) Anal Chem 79:5763
139. Osiri JK, Shadpour H, Soper SA (2010) Anal Bioanal Chem 398:489
140. Peterson DS, Rohr T, Svec F, Frechet JMJ (2003) Anal Chem 75:5328
141. Zhang N, Li L (2004) Rapid Commun Mass Spectrom 18:889
142. Lee J, Musyimi HK, Soper SA, Murray KK (2008) J Am Soc Mass Spectrom 19:964
143. Xu YC, Vaidya B, Patel AB, Ford SM, McCarley RL, Soper SA (2003) Anal Chem 75:2975

Top Curr Chem (2011) 304: 295–321
DOI: 10.1007/128_2011_147

Published online: 20 May 2011

Cells in Microfluidics

Chi Zhang and Danny van Noort

Abstract Microfluidic devices offer a realistic environment for cell cultures as it is related to scales found in biological systems. The aim is to create more in vivo like systems, in comparison to 2D plate cultures. Creating 3D cell culture constructs increase the cell's functionality. By controlling the microenvironment (e.g., cell matrix, flow rate, temperature) cell functionality can be increased even more. As microfluidic devices allow for precise control of the microenvironment, they are a paramount tool to study stem cells and their differentiation caused by external factors. We will give an overview of the use of microfluidic devices for some biological problems, and especially as a cell culture platforms. We focus on 3D cell cultures and stem cells and their microenvironment.

Keywords 3D cell cultures, Microenvironment, Microfluidics, Stem cells

Contents

1 Introduction 297
2 Microfluidic Devices as Cell Culture Platforms 297
  2.1 Introduction to Microfluidic Systems 298
  2.2 Flow Rates 298
  2.3 Laminar Flows 299
3 Microfluidics for Biological Applications 300
  3.1 Chemotaxis 301
  3.2 Quorum Sensing 302
  3.3 Cell Sorting 305

C. Zhang
Division of Nanobiotechnology, AlbaNova University Center, Royal Institute of Technology, 106 91, Stockholm, Sweden

D. van Noort (✉)
MechanoBiology Institute, Singapore, National University of Singapore, T-Lab, 5A Engineering Drive 1, 117411 Singapore, Singapore
e-mail: drr.dvn@gmail.com

4 3D Cell Cultures ... 305
4.1 Importance of 3D Cell Culture ... 305
4.2 Controlling the Soluble Microenvironment in Microfluidics ... 306
4.3 3D Cell Culture Platforms in Fluidics ... 307
4.4 Drug Screening ... 308
4.5 Tumour Spheroids ... 308
5 Stem Cells ... 310
5.1 Stem Cell Niche ... 311
5.2 Protein Patterns ... 311
5.3 Soluble Factors ... 312
5.4 Shear Stress ... 313
6 High-Throughput Multi-array Chip ... 315
7 Conclusion ... 317
References ... 318

Abbreviations

μCCA	Microscale cell culture analogue
3D-μFCCS	Three-dimensional microfluidic cell culture system
ALP	Alkaline phosphatase
bFGF	Basic fibroblast growth factor
BMSC	Bone marrow stem cell
BSA	Bovine serum albumin
cDNA	Complementary deoxyribonucleic acid
CEPC	Circulating endothelial progenitor cell
CNS	Central nervous system
CTC	Circulating tumour cell
DC	Direct current
ECM	Extracellular matrix
EGF	Epidermal growth factor
EPC	Endothelial progenitor cell
ESC	Embryonic stem cell
FACS	Fluorescence-activated cell sorter
FBS	Fetal bovine serum
FITC	Fluorescein isothiocyanate
FSS	Flow shear stress
GFP	Green fluorescent protein
hESC	Human embryonic stem cell
hNSC	Human neural stem cell
HSC	Haematopoietic stem cell
IC_{50}	50% maximal inhibitory concentration
iPS	Induced pluripotent stem
mRNA	Messenger ribonucleic acid

MSC	Mesenchymal stem cells
PDGF	Platelet-derived growth factor
PDMS	Polydimethylsiloxane
PEG	Polyethylene glycol
PZT	Piezoelectric transducer
SMC	Smooth muscle cell
TGF	Transforming growth factor
UHTP	Ultrahigh-throughput

1 Introduction

Microfluidics provides a good basis for biological research, as micro (and nano) scale is compatible with biological structures and scales. Cells are in the micrometer range, typically around 10–15 μm. Compared to conventional cell culture techniques, microfluidic devices allow for a precise control of microenvironments, which is especially important when studying stem cell behaviours. Control of the microenvironment allows for changes (e.g., in flow rate, pH, O_2 levels) that influence mechanical and biochemical factors in a cell and, therefore, cell functionality. These controls are readily implementable in microfluidic devices, using pumps, valves and smart microchannel design.

In this chapter, we will discuss the advantages for microfluidic devices as cell culture platforms. Then we will introduce 3D cell cultures as a way to overcome the limitations of the standard 2D cell culture methods, especially in microfluidic devices. Furthermore, we will show how to control the microenvironment inside the devices and how it affects cell cultures. Finally, applications in stem cell differentiation highlight the need of culture techniques and control inside the microfluidic devices.

2 Microfluidic Devices as Cell Culture Platforms

Conventional cell culture techniques have been well established over the last several decades and essentially consist of immersing cells in a homogeneous culture medium. The cells are usually maintained in a static environment and have limited cell–cell interactions [1]. However, in vivo, cells do respond to spatially and temporally organized signals from the surrounding microenvironment, which consists of neighbouring cells and an extracellular matrix (ECM). In vivo cells exhibit a very distinct geometry, which is not observed when maintained in conventional cell cultures [2]. In order to gain more cellular insights to different biological problems, it is important to perform cellular experiments that more closely reflect the in vivo situation with extensive cell–cell, cell–matrix and cell–soluble factor interactions [3].

Different cell culture technologies have been developed to establish a better cellular microenvironment, thus increasing their functionality. For example, cells were seeded on surfaces coated with ECM proteins (e.g., laminin, collagen, fibronectin) to stimulate cell–matrix interactions [4, 5]. Also, different cell types were co-cultured to enhance growth and function [6, 7]. Among the newly emerging technologies, microfluidic systems allow for precise control of the cellular microenvironment in cultures [8–10]. Various microfluidic cell culture devices have been developed to create microenvironments of greater physiological relevance that are adaptable to high-throughput platforms for biological applications [10–12].

2.1 Introduction to Microfluidic Systems

Microfluidics usually refers to liquid flows generated in chips with microscale channels, including fabrication, handling and practical use of the chips. Streams of gases or fluidized solids/particles in microchips are also included in the definition of microfluidic [8, 13]. Microfluidic channels can be obtained after the process of replica moulding. Elastomeric materials such as polydimethylsiloxane (PDMS) are often used to mould against a rigid master, with microscale structures fabricated using standard photolithography [14]. After bonding the elastomeric material with a substrate surface, the device is connected to liquid reservoirs by tubing and pumps to deliver a flow to the microfluidic channels [15–17]. The development of clean room technology made mass production of microscale devices at low cost possible. The initial motivation of developing microfluidic devices was to significantly reduce reagent consumption and increase efficiency. Gradually, the low cost of mass producing the microfluidic devices and their capability to be integrated for large-scale automation made microfluidic devices ideal systems for commercialisation [18]. Tailor-made microfluidic devices are sold by companies such as Agilent Technologies, Dolomite and Fluidigm Technology. With the integration of on-chip sensors, detectors or even optical components, it is possible to perform cell culture, cell sorting and biological assays on one chip.

2.2 Flow Rates

Fluid flow in microfluidic devices mimics vasculature in vivo [19], therefore the incorporation of fluid flow to cell cultures allows for efficient exchange of nutrients and metabolites between cells and culture medium, maintaining a constant cell culture microenvironment [20, 21]. As such, cellular functions can be enhanced in microfluidic systems compared to conventional static cell cultures [17]. However, in microfluidic devices, fluid flow exerts shear stress (FSS), which can modulate cellular behaviour on the cells. FSS can induce the reorganization of the cytoskeleton. Bone marrow stem cells (BMSCs) tend to align in the direction of the shear vector [22]. Furthermore, primary rat hepatocytes are known to be very sensitive to

FSS, whereby a FSS higher than 5 dyns/cm^2 can impede their liver-specific functions [23]. FSS also plays a role in the extravasations of tumour cells as tumour–endothelial cell interactions are dependent on shear [24].

Flow rates influence the rate of cell growth inside a perfusion system. For example, 1.0 mL/min flow rate for 1 week resulted in substantial cell death of the MC3T3-E1 osteoblast-like cells, while a lower flow rate increased the viability [25]. Also, the viability of cell cultures is affected by the flow rate. By controlling the flow rate, the viability of cell cultures can be optimised [26] (Fig. 1). By using a curved main channel, a gradient velocity, and therefore a gradient, can be generated. Also, by having two inlets it is possible to create a chemical gradient [27] (Fig. 2).

L929 mouse fibroblast cells in this device changed their morphology due to the forces exerted on them. A gradient was created by introducing a standard culture medium with fetal bovine serum (FBS) at one inlet and a serum-free medium at the other. FBS contains proteins, such as serum albumin, that are needed for cell survival, division and growth. The cell density and attachment depended on a combination of sheer stress and FBS concentration [28].

2.3 Laminar Flows

The flows inside microfluidic channels are laminar, meaning that when two or more parallel flows are inside a single channel, mixing only takes place by diffusion at the interfaces [29]. This phenomenon gives the possibility to precisely pattern cells and

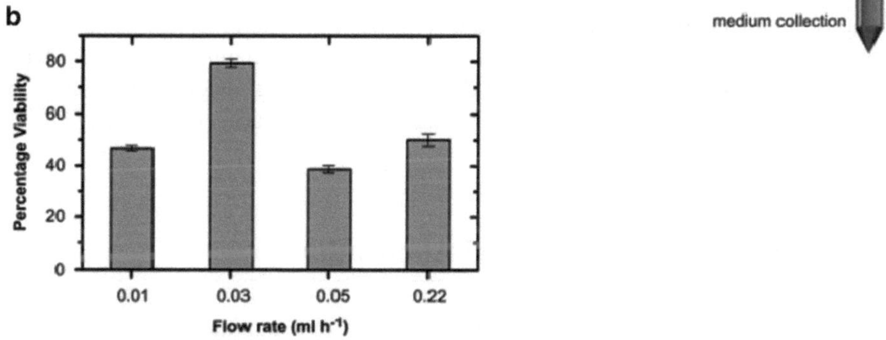

Fig. 1 Perfusion culture of the cells in a gel-free 3D microfluidic cell culture system (*3D-μFCCS*). (**a**) Representation of a one-pass perfusion culture system. (**b**) Optimization of perfusion culture flow rate for maximum cell viability [26]

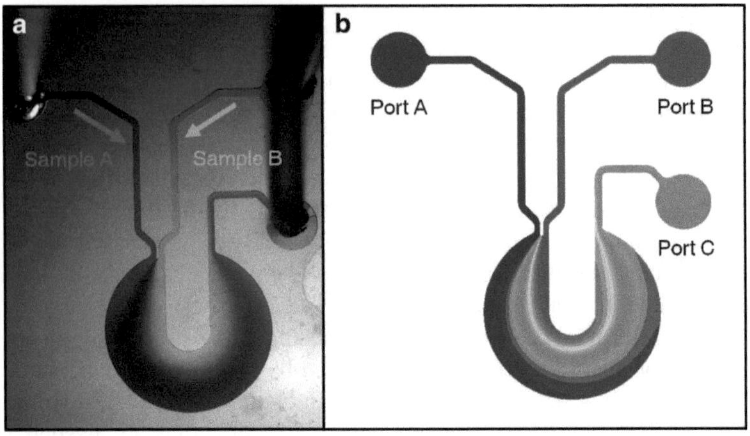

Fig. 2 Comparison of experimental and computational diffusion results in a curved microchannel using an osmotic pump: (**a**) experimental data and (**b**) computational simulation [27]

their environments using parallel flows of different solutions in microchannels [29, 30]. It allows spatial control of the culture medium itself in a highly controlled manner [31]. This method can also be used to study subcellular processes by positioning the interface between two adjacent flows over a single cell spreading across the floor of the microchannel [29]. Two different dyes or reactants can be directed to different regions of the same cell, resulting in different behaviour at either ends of the cell. This formation of networks by laminar streams is a novel approach for local stimulation of cells and can also be useful for drug testing applications. When laminar flows containing different compounds at different concentrations are applied to cell cultures, data on cellular responses can be rapidly acquired.

The laminar flows within microfluidic channels can also be used to control the positioning of soluble factors relative to cells [32]. With the incorporation of gradient generators, a range of concentrations of soluble factors can be generated by sequential merging, mixing and splitting of two or more inlet flows, each of which contain a particular environmental stimulus [33]. Therefore, different growth conditions can be created in parallel or in a combinatorial manner [33]. Such stable gradients have been used to study cell chemotaxis, yielding insights into neutrophil migration in response to various concentrations of interleukin (IL)-8 [34].

3 Microfluidics for Biological Applications

Microfluidic systems are promising platforms for fundamental biological studies such as hepatitis B viral replication [35]; axonal injury, regeneration and transport in the central nervous system (CNS) [36]; or the control of biochemical and biomechanical forces to evaluate and quantify capillary growth and endothelial cell migration [37]. Many microfluidic-based systems have been developed as cell culture platforms. In the rest of this section we give some general applications of microfluidic devices.

3.1 Chemotaxis

Cells respond to chemical gradients. Depending on the concentration of the chemical, motile cells will move at different speeds towards or away from the gradient, depending on whether the chemical is an attractant or repellent. Microfluidic devices can go beyond traditional chemotaxis assays and enable the quantification of the cell's transport parameters.

Ciliates are very motile and a good model system for motility experiments. *Escherichia coli* excretes amino acids [38], which can attract ciliates, as bacteria are their food source. Using a two-chamber system with a connecting channel that is open/close switchable, it was shown that ciliates move towards or away from excreted amino acids [39] and could detect 10 pM glycine–proline (Fig. 3) [40]. To simulate an in vivo gradient profile response for lymphocytes, another method was used that consisted of a Y-shaped device for easy and fast generation of single and superimposed gradients with diffusive profiles. Chemokines CCL19 and CXCL12 were selected as they have been shown to regulate lymphocyte homing in vivo. It was shown that human T cells chemotax toward the CCL19 gradient, but not the CXCL12 gradient [41] (Fig. 4).

Microfluidic experiments have shown that it is possible to quantify population-scale transport parameters (chemotactic sensitivity and random motility) of a population of bacteria. Traditionally, transport parameters have been theoretically derived from single-cell behaviour using probabilistic models. But, with a microfluidic capillary it is possible to generate and measure gradients of chemoattractant (α-methylaspartate) while simultaneously capturing the swimming

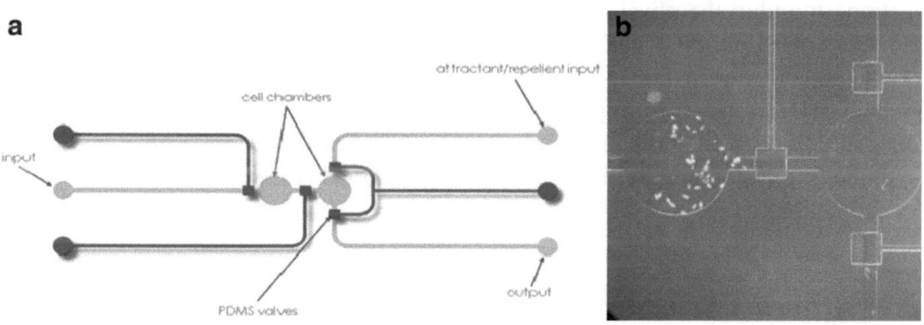

Fig. 3 (**a**) Overview of PDMS valve-operated microfluidic device consisting of cell chambers to load ciliates and chemoattractant or repellent. (**b**) Initially the ciliates were placed in the left chamber, while the attractant was loaded in the right chamber. Both chambers were separated from each other and the outside world by valves [39]

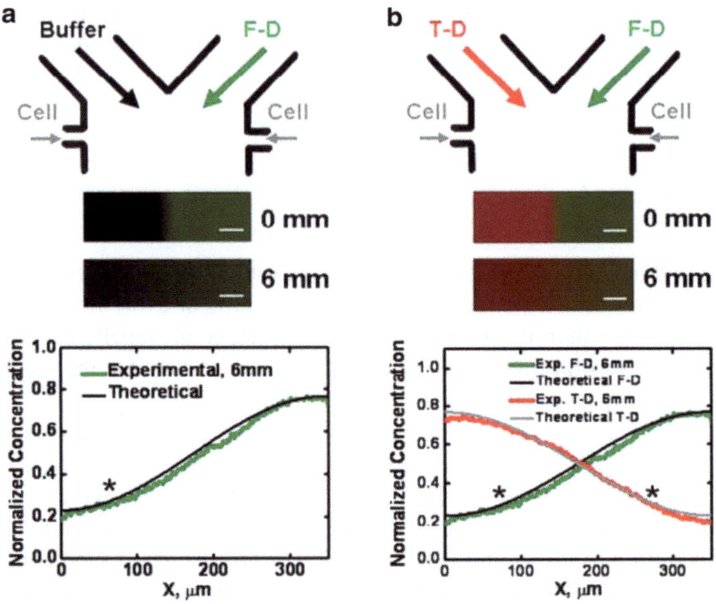

Fig. 4 Example of gradient generation in a microfluidic device. (**a**) Generation of a single gradient. Fluorescence intensities of FITC-Dextran (*F-D*) at 0 mm and 6 mm below the junction of the Y-channel are shown. The experimental and theoretical values of the gradient profile at 6 mm are in agreement. (**b**) Competing gradients. Fluorescence intensities of FITC-Dextran (*F-D*) and Texas-Red-Dextran (*T-D*) at 0 mm and 6 mm below the junction of the Y-channel are shown. Again, the experimental and theoretical values of the gradient profiles at 6 mm are in agreement [41]

trajectories of individual *E. coli*. By measuring swimming speed and bias in the swimming direction of single cells for a range of chemoattractant concentration gradients, it was possible to compute the chemotactic velocity and the sensitivity (Fig. 5) [42].

Another form of "taxis" is galvano- or electrotaxis. An applied electric field with a strength within the physiological range can induce directional cell migration of, for example, epithelial cells, endothelial cells, fibroblasts and neutrophils, and could play a role in cell positioning during wound healing. It was shown that lymphocytes respond to applied direct current (DC) electric fields. In a simple microfluidic device, human peripheral blood lymphocytes migrate towards the cathode in physiologically relevant DC electric fields (Fig. 6) [43].

3.2 *Quorum Sensing*

A more complex form of chemotaxis involves bacterial social interactions. The environmental topology of complex structures is used by *E. coli* to create travelling waves of high cell density, a prelude to quorum sensing (QS). When cells are grown to a

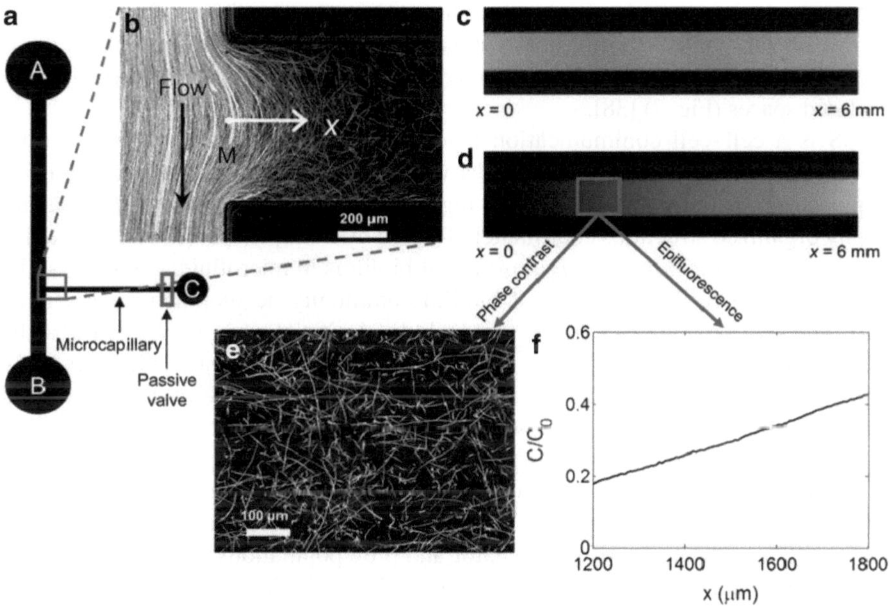

Fig. 5 Determination of the chemotactic sensitivity of *E. coli*. (**a**) Schematic of the microfluidic device. Chemoattractant and fluorescein were injected in the microcapillary via inlet *C* by means of a passive valve. (**b**) Flow in the main channel (from *A* to *B*) was used to transport *E. coli* past the entrance (*M*) of the microcapillary. (**c, d**) Epifluorescence images of the microcapillary filled with α-methylaspartate at $t = 0$ (**c**) and $t = 45$ min (**d**). (**e**) Trajectories of *E. coli*. (**f**) Concentration profile obtained from (**d**) [42]

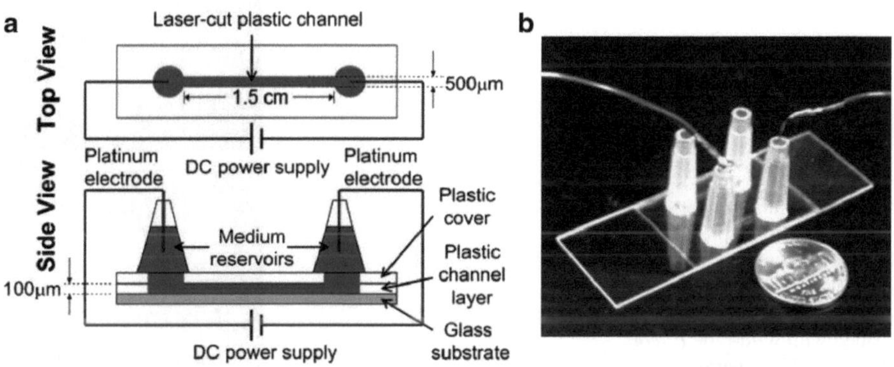

Fig. 6 (**a**) Microfluidic device for the visualisation of cell migration in an applied electric field. (**b**) Photograph of the microfluidic device. Two identical channels were configured in parallel [43]

moderate density within a confining microenvironment, these travelling waves of cell density allow the cells to find and collapse into confining topologies, which are unstable to population fluctuations above a critical threshold. This was first observed in mazes designed to mimic complex environments, then more clearly in a simpler geometry consisting of a large open area surrounding a square (250 × 250 μm) with a narrow opening of 10–30 μm. It was shown that under nutrient-deprived conditions bacteria search out each other in a collective manner and confine themselves to highly enclosed spaces (Fig. 7) [38].

QS is a cell–cell communication mechanism that is used by bacteria during biofilm formation. Biofilm formation is widely acknowledged to occur through a sequence of spatially and temporally regulated colonization events. Biofilms are highly organized structures coordinately formed by multiple species of bacteria. By developing *Pseudomonas aeruginosa* PA14 biofilms in microfluidic chambers, the interplay between the dynamics of biofilm community development, mass transport, and hydrodynamics can be shown [44]. Artificial ecosystems can also be fabricated in microfluidics. Two *E. coli* populations communicated bidirectionally through QS and regulated each other's gene expression and survival via engineered gene circuits. This synthetic ecosystem resembles a canonical predator–prey system in terms of logic and dynamics. The predator cells kill the prey by inducing expression of a killer protein in the prey, while the prey rescues the predators by eliciting expression of an antidote protein in the predator. Extinction, coexistence and oscillatory dynamics of the predator and prey populations possibly depend on the operating conditions, as experimentally validated by long-term culturing of the system in microchemostats [45].

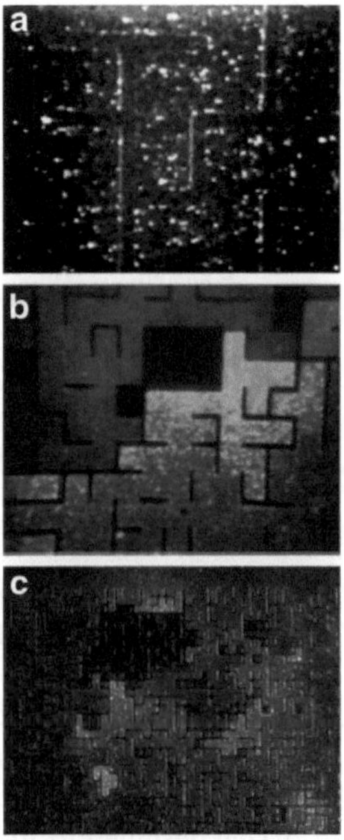

Fig. 7 Epifluoresence images of GFP expression of *E. coli* in a random maze. (**a**) Cells in the maze immediately after being loaded with wild-type *E. coli* (RP437). (**b**) Dynamic accumulation of cells at 2 h after the initial loading, showing accumulation in a "dead-end" part of the maze. (**c**) Accumulation of cells into several different confining regions. No clustering is seen in the open region [38]

3.3 Cell Sorting

Microfluidic device can be used to sort cells. Fluorescence-activated cell sorter (FACS) machines have been scaled down to chip size. Single cell manipulation at a high speed is made possible by the fast response time of an integrated piezoelectric (PZT) actuator. Cells are sorted on the detected optically expression [46]. Cells can also be separated in an electric field on the basis of electrically distinguishable phenotypes [47]. Other cell sorters look at the fitness of the cell. For example, semen can be fractionated by their ability to swim through interfaces between adjacent laminar streams into separate streamlines, which enables isolation of motile, morphologically normal spermatozoa from semen samples [48]. This resulted in a microscale sperm sorter in which high-quality sperm can be isolated and used for in vitro fertilization [49].

Microfluidic devices are slowly emerging as new platforms with extensive applications in cancer biology, clinical oncology, and high-throughput drug screening routines. These devices are particularly attractive as diagnostic tools, allowing rapid analysis of only small amounts of patient-derived cells [50]. At certain stages, cancer cells start migrating through the body. As such there is a need to isolate and enumerate circulating tumour cells (CTCs) in peripheral blood as it has clinical relevance in combating cancer [51]. CTCs are associated with the progression of the disease [52] and, as such, an important indicator of the effectiveness of treatment methods [53]. Studies have revealed that the shear modulus, stiffness, size and deformability of cancer cells [54] are distinctively different from blood cells [55]. Cancer cells within blood samples can also be isolated as blood cells are more deformable than cancer cells. Flowing blood through a PDMS pillar array with 5-μm gaps ensures that the more deformable white blood cells, with the same dimension as the cancer cells, can readily pass through the system. Cancer cells will be removed from the blood sample, ensuring a high purity of the trapped cancer cells. In this manner, fast label-free cancer screening can be performed [56].

Blood cells can also be fractionated and isolated from blood plasma without dilution by using a continuous-flow deterministic array of pillars. By using a narrow focused hydrodynamic jet, the blood cells can be fractionated as a function of their diameter. By varying the inter-pillar distances, a wide range of cell sizes can be continuously separated using this technique [57].

4 3D Cell Cultures

4.1 Importance of 3D Cell Culture

In vivo, cells have interactions with one another and their surrounding ECM. The ECM acts as a support network containing proteins such as collagen, elastin and laminin which gives the tissues their mechanical properties and promotes communication between cells embedded in the matrix. Receptors on the surface of the cells, e.g., integrins, anchor to the ECM and attract biochemical cues from the immediate surroundings [2, 58]. This complex mechanical and biochemical interplay is not reproduced when cells are cultured in 2D. 2D cell cultures are generally used for the maintenance of cells and in biological studies and impose unnatural geometric and mechanical constraints on cells. 3D cell cultures re-establish cell–cell and cell–ECM interactions and can mimic real tissue better than conventional 2D cultures [59]. For instance, a 3D culture environment has been found to promote epithelial polarity and differentiation for breast epithelial cells, a phenomenon not observed when the cells are cultured in 2D [60]. Primary hepatocytes have been found to exhibit higher levels of liver-specific functions, such as albumin synthesis, when cultured in 3D microenvironments, whereas they rapidly de-differentiate and lose their functions when cultured in 2D [61]. Cancer cells switch from a mesenchymal (spindle-shaped cells) to an amoeboid (ellipsoid-shaped cells) motility pattern when cultured in 3D, an occurrence never observed in 2D [62].

Gene expression profiles of cells grown in 2D and 3D can also be different. For example, vascular smooth muscle cells (SMCs) grown in 2D and 3D differed for 99 genes. SMCs cultured in 3D showed a higher expression level of matrix proteins (collagen I and fibrinogen). This indicated that SMCs were more active in ECM production in a 3D matrix than in 2D [63]. Similarly, microarray analysis of melanoma cells cultured as 3D spheroids reported an up-regulation of 106 genes and a down-regulation of 73 genes as compared to culture as a 2D monolayer. In particular, a number of genes encoding for chemokines and pro-angiogenic factors associated with melanoma progression and metastasis were reported to be upregulated in 3D spheroids [64]. Generally, cells cultured in 3D maintain a gene expression profile that is closer to tissues in vivo [58], showing that 3D cell cultures better mimic the in vivo environment compared to conventional 2D cultures.

4.2 Controlling the Soluble Microenvironment in Microfluidics

Cell cultures in 3D have a more in vivo like behaviour than standard 2D cultures. Adding a soluble microenvironment to the cell culture increases the functionality of the cells. Bioactive factors in the microenvironment such as cytokines and growth factors, must be properly regulated as they are involved in autocrine and paracrine signalling pathways, which control cell functions such as differentiation (e.g., sonic hedgehog [65, 66], bone morphogenetic proteins [67, 68]), proliferation (e.g., basic fibroblast growth factor, bFGF [69]) and apoptosis (e.g., transforming growth factor, TGF-β1 [70]). However, to temporally and spatially control the soluble factors in culture medium is difficult under conventional static cell culture conditions due to the large liquid volumes, the constant loss of activity of growth factors and the accumulation of metabolic waste [71].

Microfluidics, by contrast, allows for temporal and spatial control over the soluble microenvironment because of the nature of laminar flow [30, 31, 72]. Other parameters also affect the soluble factor signalling, such as cell density, exogenous growth factors, medium change frequency or even culture platforms [73]. Therefore, all these parameters have to be taken into consideration when designing a microfluidic device for cell culture and the control of the spatial and temporal environment.

Due to the high surface-to-volume ratio, the phenomena of molecular diffusion and heat transport in microfluidics resemble those found in vivo. Therefore, soluble factors in microfluidic devices can be presented in a more physiologically relevant context [32, 74], providing a more in vivo like cell culture. In key areas of drug discovery, such as preclinical testing of drugs in living cells, microfluidic devices can improve the screening results [18]. Their biomimetic nature enables more convincing data on drug testing while their microscale size allows for easy scaling-up and, hence, high-throughput drug screening possibilities [75–77]. The application of biomimicry is not yet found in conventional cell culture systems.

4.3 3D Cell Culture Platforms in Fluidics

Due to the nature of microfluidic devices, they have advantages over conventional cell culture platforms in terms of enabling precise cell-based assays. The laminar flow and the high surface area-to-volume ratio created inside microfluidic devices have made it possible to create a more controllable microenvironment that closely mimics in vivo situations [32]. Therefore, a number of microbioreactors or microfluidic devices have been developed with the aim of establishing biomimetic microenvironments for 3D cell cultures. By hydrodynamic focusing, cells suspended in hydrogels were stabilized in a 3D geometry in microfluidic devices for different biological applications. For instance, Wong et al. have employed this approach to construct a tuneable 3D microenvironment for cells in order to study intercellular communications [78], while Vickerman et al. have studied the directional migration of human adult dermal microvascular cells in response to changes in their microenvironment [79]. Microscale bioreactors are also found useful in tissue engineering applications. A perfused multiwell plate with electronic controls was developed for the 3D culturing of primary rat hepatocytes. The cells remained viable and functional after 7 days [80]. Microfluidics can contribute significantly to the construction of large tissues. Microfluidic structures ensure a steady blood supply, thereby circumventing the well-known problem of providing larger tissue structures with a continuous flow of oxygen and nutrition and with adequate removal of waste products [19]. Researchers have engineered a "synthetic capillary" using microfluidic technology. Microfluidic devices can be completely covered with endothelial cells to generate capillary-like structures that mimic the microvascular network morphologies [81, 82].

4.4 Drug Screening

Microfluidic devices are gaining attention as reliable platforms for drug testing. Due to their compact size and capability to be multiplexed and scaled-up for high-throughput screening (HTS), assays can be performed in a continuous manner, allowing for the analysis and isolation of compounds on a single chip. Since microfluidic channels are closed systems, reactions can take place in a much faster fashion without depletion of reagents. For instance, an artificial liver sinusoid within a microfluidic device was engineered for long-term culturing of both rat and human hepatocytes. For model validation, diclofenac was then used as a drug for hepatotoxicity testing. The IC_{50} (the concentration of drug which inhibits 50% of the cell functions) approached data given in the literature [82]. Gradient generators can be easily incorporated into microfluidic devices to target cells with different drug concentrations. Our group has reported the effort towards developing the HepaTox chip, which allows 3D culture and toxicity testing of hepatocytes [83]. It consists of eight cell culture compartments and a gradient generator for delivery of different drug concentrations (Fig. 8). The HepaTox chip can readily be used to determine the IC_{50} of a given drug. Due to multiplex capability, microfluidic devices can be assembled into a modular system to mimic physiological situations.

The systemic effect of a drug can be studied by culturing different cell types at the same time, followed by administration of a drug and evaluation of the pharmacokinetics. Shuler et al. developed an integrated microscale cell culture analogue (μCCA) specifically for pharmacokinetic studies (Fig. 9) [84–86]. The design is based on a simplified mathematical representation of the body as a system with several interconnected compartments with specific flow parameters. The μCCA contained liver, lung and fat cells to represent organs involved in drug metabolism. Cells were cultured in a fluidically linked fashion with culture medium being circulated as a blood surrogate; the systemic effect of a drug on the cells was then evaluated. The development of the μCCA laid the foundation for a realistic in vitro pharmacokinetic model and provided an integrated biomimetic system for culturing multiple cell types. Based on these principles, a multi-channel 3D microfluidic cell culture system with compartmentalized microenvironments was developed for potential application in human drug screening [87]. Multiple cell types were cultured on the same chip, but limited cross-talk between cell culture compartments was found, reflecting the in vivo situation where different organs have their own particular microenvironments (Fig. 10). The emergence of more sophisticated microfluidic devices has made it possible to support or even replace animal models.

4.5 Tumour Spheroids

Within 3D cell culture techniques, tumour spheroids are a more classic approach to obtain and maintain representations of human tumour models. They have, for

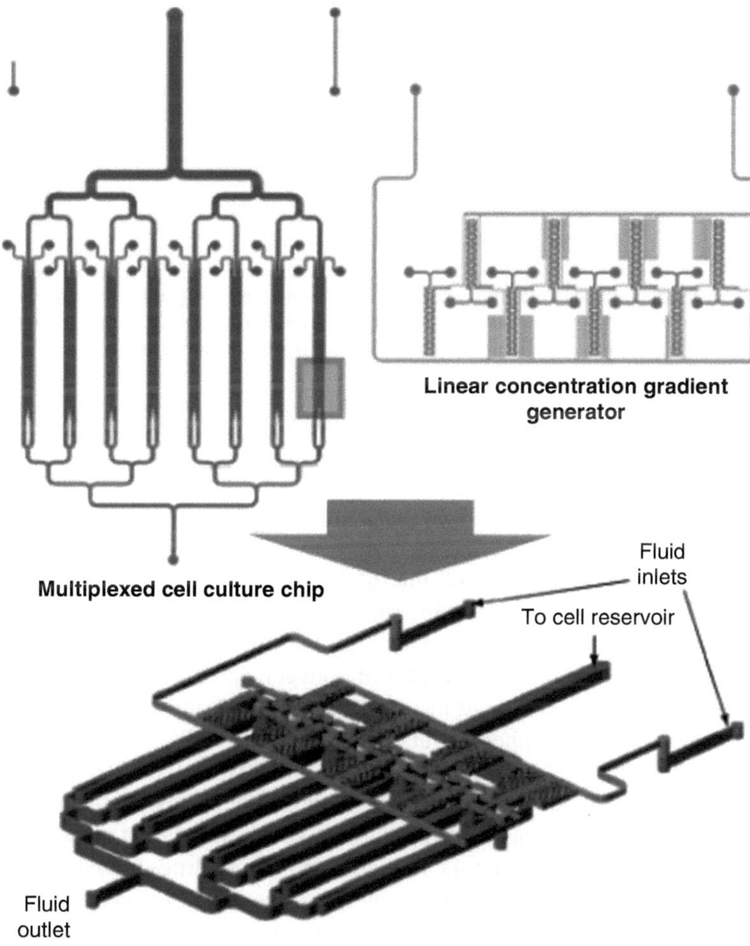

Fig. 8 The 3D HepaTox Chip was developed for the simultaneous administration of multiple drug concentrations. It consists of an eight-channel chip with 3D cell culture and a plug-in concentration gradient generator for drugs [83]

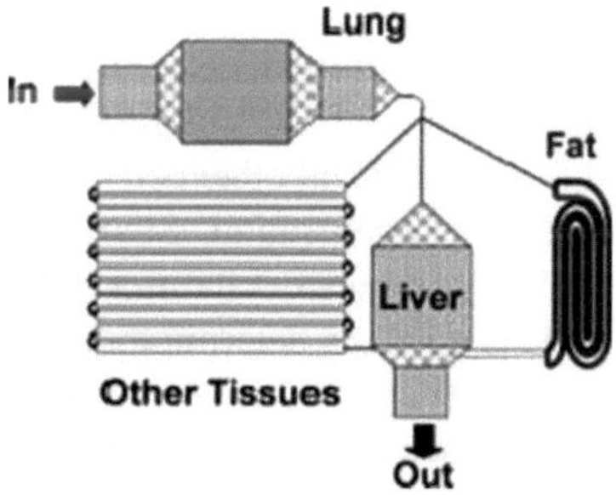

Fig. 9 Four-chamber microscale cell culture analogue . The microfabricated device comprises two depths, 20 μm (shown in *gray*) for the lung and liver chambers and 100 μm (shown in *white*) for other tissues and fat, and includes all channels. The dimensions of the chambers are in the order of millimetres [84]

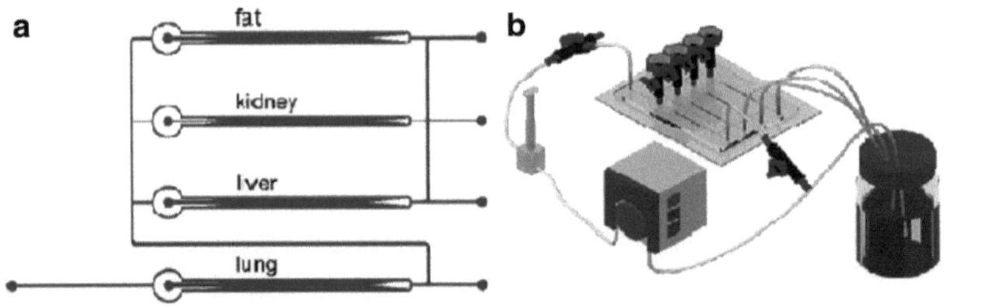

Fig. 10 (**a**) Multichannel 3D microfluidic cell culture system consisting of four chambers with 3D cell constructs, representing four organs potentially involved in drug toxicity studies. (**b**) Cells are cultured in a closed-loop with medium being circulated as a blood surrogate [87]

example, enhanced our understanding of radio- and chemoresistance of the tumours [88], and of their microenvironmental regulation [89]. Only recently has the production of tumour spheroids been explored utilising microtechnology. This gives a greater control over the size of the spheroid compared to more conventional methods. The majority of studies on spheroid formation have focused on the use of hepatocytes. By using a combination of microfabrication and microcontact printing, Fukuda et al. developed a spheroid microarray culture system with 700 wells/cm^2 [90]. The bottoms of the 300 μm diameter wells were modified by proteins, resulting in two different regions: a 100 μm centre of collagen and a non-adhesive region of polyethylene glycol (PEG) around the rest of the wells. Primary hepatocytes spontaneously formed uniform spheroids at the centre of the wells. Using different well diameters yielded differently sized spheroids, from 180 to 260 μm [91]. In this case, the wells were entirely coated with PEG. With the aid of perfusion, the microfluidic system can control the formation of spheroids in such a way that it recapitulates the in vivo state of vascularised tumours. Also, the delivery of drugs in low volumes can be achieved in this manner. Systems like that described above can progress towards production of uniform tumour models that can revolutionise anti-tumour screening techniques, and even personalise medicine [92].

5 Stem Cells

Stem cells differentiate to any cell type in embryonic and post-natal development. The term “stem cell” was coined as far back as 1868 to describe a special cell that could differentiate to all other cell types [93]. Understanding stem cell differentiation is a key in tissue regeneration. Stem cells are increasingly implanted in defective organs, replacing defunct cells in a bid to regenerate the organ [94]. Mechanical, chemical and spatial cues determine the stem cell fate and can be readily controlled in microfluidic devices, as has been shown in the previous sections (Fig. 11). As such, stem cells can be used as in vitro models for drug screening and to understand their development as a function of the microenvironment. Embryonic stem cells (ESCs) are isolated from

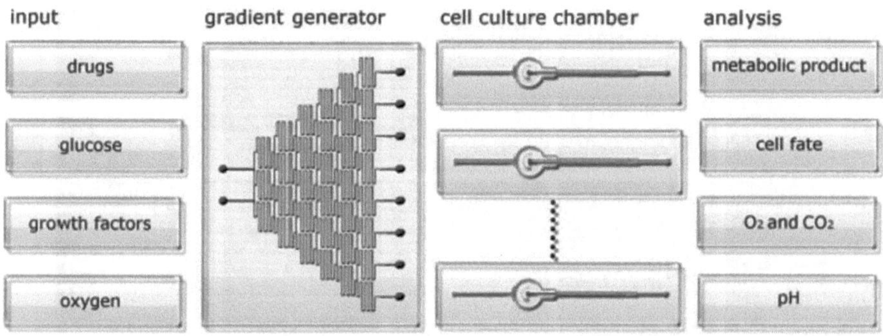

Fig. 11 Overview of the possible uses for a multiplexed microfluidic cell culture system. *Input* can be used for control of the stem cell nice and drug testing. The *gradient generator* creates a concentration profile of the input factors. The *cell culture chamber* is a perfusion chamber. *Analysis* involves monitoring the output of the device, which can include imaging with appropriate biomarkers, analysis using inline sensors or standard laboratory equipment [74]

the inner cell mass of blastocyst-stage embryos. They were first derived from mouse embryos [95–97], followed by the derivation of primate [98, 99] and human [100, 101] ESC lines. Recently, an alternative method for deriving pluripotent cells by retroviral transduction of a combination of embryonic genes into somatic cells (iPS cells) was reported by Yamanaka's group [102, 103], followed by several other groups [104–106].

5.1 Stem Cell Niche

It has been shown that stem cell self-renewal and differentiation rates are influenced by their microenvironment, termed stem cell niche [107, 108]. The stem cell niche consists of the following components: (1) extracellular matrix [109]; (2) biochemical factors including cell–cell and cell–matrix interactions, localized soluble stimuli and gradients of soluble factors like growth factors and glucose concentration [110, 111]; (3) physico-chemical factors including temperature, pH, and oxygen availability [110]; and (4) mechanical stimuli, including compression tension and shear [112]. Changing any of these factors will lead to other differentiation pathways and phenotypes.

5.2 Protein Patterns

Surface topology and spatial constraints determine the stem cell fate as well. Although protein patterns are not in the realm of microfluidics yet, it is worthwhile to contemplate that the stem cell's fate is influenced by its constrictions. By patterning fibronectin on a surface in a defined shape it is possible to influence cell differentiation. MSC were placed on two shapes, a star and a flower shape. The star shape resulted in more

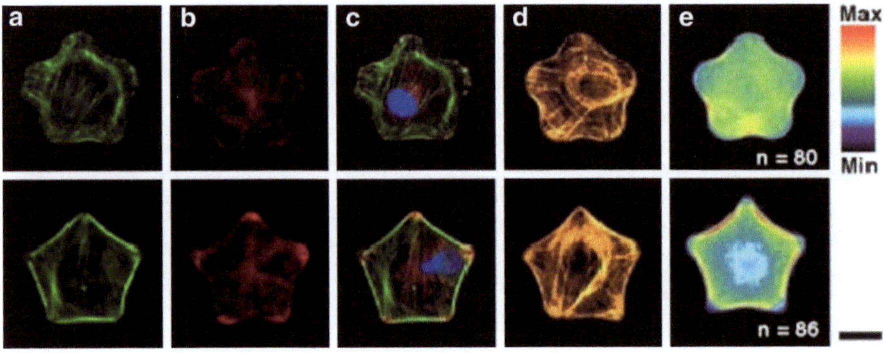

Fig. 12 (**a–c**) Immunofluorescent images of cells in flower and star shapes stained for F-actin (*green*), vinculin (*red*) and nuclei (*blue*). (**d**) Immunofluorescent images of cells in flower and star shapes stained for myosin IIa. (**e**) Fluorescent heatmaps of cells stained for myosin IIa as a quantitative measure of contractility. *Scale bar*: 20 μm [113]

Fig. 13 "Christmas tree" gradient generator. (**a**) Top view of the microfluidic device consisting of gradient-generating and observation portions. (**b**) 3D representation of the observation portion in which cells are exposed to gradients of chemoattractant [34]

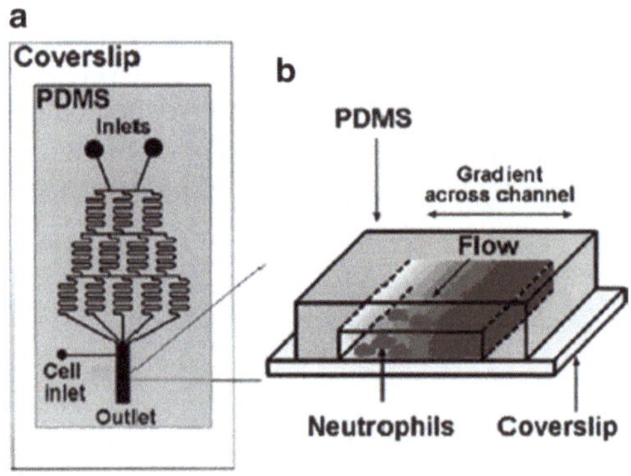

stress fibers along the edges and the MSC differentiated preferentially in osteoblasts, while the MSCs on the flower shape differentiated into adipocytes (Fig. 12) [113].

5.3 Soluble Factors

As has been mentioned in the section on 3D cell cultures (Sect. 4), soluble factors play an important part in cell functions, and in the case of stem cells, on the cell fate. Microfluidics can help to study the responses of stem cells towards the soluble factors in a precise controlled extracellular microenvironment.

A gradient generator can be used to optimise the proliferation and differentiation of stem cells in culture as a means to establish the concentration of soluble factors needed for optimal culture. After a week of culturing human neural stem cells (hNSCs) in a constant flow of a growth factor mixture containing epidermal growth factor (EGF), fibroblast growth factor 2 (FGF2) and platelet-derived growth factor (PDGF). The proliferation rate increased 6.8-fold, while the differentiation of hNSCs into astrocytes decreased (from 40% to 15%) over increasing concentration of the growth factor [114] by using a "Christmas tree" gradient network (Fig. 13) [34]. Similarly, it was shown that neural progenitors derived from human endometrial cancer (hEC) cells could be differentiated into neurons and generated a complex neural network. The neural progenitors were exposed for 8 days to stable concentration gradients of various signalling molecules (sonic hedgehog, FGF8, and bone morphogenetic protein 4). The average numbers of both neuronal cell body clusters and neurite bundles were directly proportional to sonic hedgehog concentrations in the gradient chip [27]. Not only can growth factors be presented in such a fashion, but nutrients and oxygen as well, which were presented to murine ESCs [115].

5.4 *Shear Stress*

Apart from chemical cues, shear stress also plays an important part in the proliferation and differentiation process of stem cells [116]. Continuous microfluidic perfusion systems not only offer control over the cell–media interaction but the mechanical forces applied during flow as well. As such, the flow rate has to be

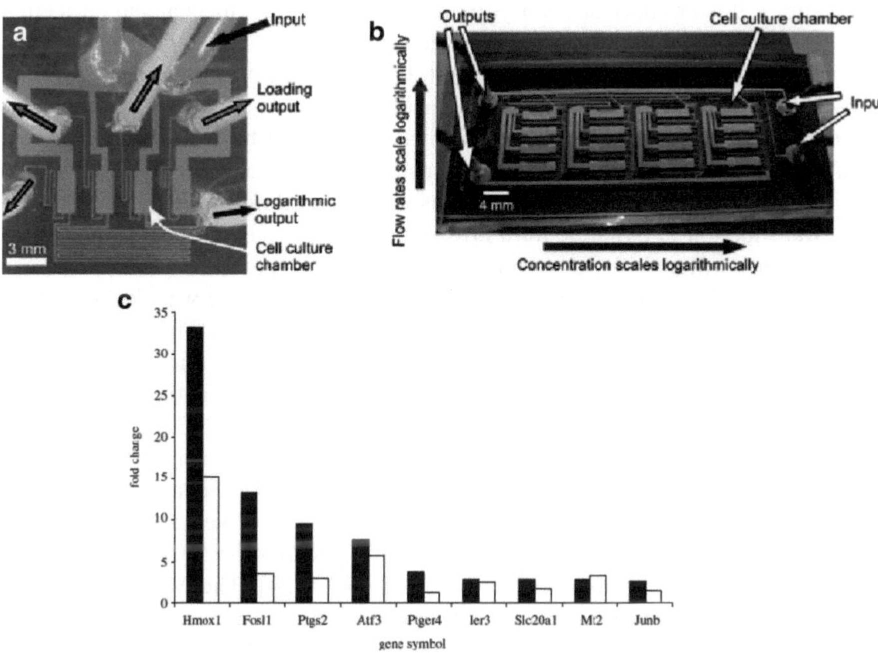

Fig. 14 Microfluidic array for a logarithmical perfusion device. (**a**) Photo of the four-channel device filled with fluorescein. The chip has four loading outputs (*gray arrows*) and two input tubes (one for loading cells and one for perfusion; *black arrows*). (**b**) Photo of multilayer 4 × 4 device filled with fluorescein. (**c**) Gene expression during dynamic fluid flow versus no-flow fold-change values at 30 min (*filled bars*) and 1 h (*open bars*) post-flow [21]

optimised to the favourable growth conditions for stem cells. Conventional systems applying such a mechanical force require the use of large cell culture surfaces, large liquid volumes, and bulky hydrostatic and compressive loading methods [117].

By using a logarithmic range of flows over four culture chambers, it was shown that murine ESCs only proliferated at higher flow rates, as did 3T3 fibroblasts (Fig. 14) [21]. It was demonstrated that circulating endothelial progenitor cells (CEPCs) can be used for cardiovascular revascularization, generation of non-thrombogenic vascular grafts, and endothelialization of cardiovascular tissue engineered implant [118]. A microfluidic platform (EPC chip) was developed that is capable of capturing EPCs, mediated by the interaction of EPCs with immobilized antibodies under variable shear stress. This device could provide a platform for isolating progenitor cells for tissue engineering applications and cell-based revascularization strategies, as well as validating these cell populations as biomarkers in cardiovascular disease [119].

Oscillatory fluid flow is a potent signal capable of regulating stem-cell differentiation. Applying cyclic mechanical loads to the skeleton will result in increased bone formation. These loads lead to dynamic pressure gradients and oscillatory flow of the interstitial fluid inside the bone, which, in turn, exposes the cells resident in the bony matrix to oscillatory fluid shear stress [120]. Loading induced oscillatory fluid flow has the potential to upregulate multiple transcription factors involved in distinct lineage pathways. For the purpose of developing regenerative medicine therapeutics, it is not only necessary to determine if specific mechanical cues elicit progenitor differentiation, but also the mechanism by which a cell transduces the physical signal into a cell fate commitment. It has been demonstrated that exposure to oscillatory fluid flow, a predicted mechanical signal in mature bone, regulates progenitor cell differentiation via the activation of the RhoA–ROCKII pathway. Additionally, it has been shown that tension within the actin cytoskeleton is required for the activation of mechanosensors or signalling mechanisms involved in the transduction of dynamic fluid flow into cell fate decisions [121].

By using a microfluidic device it can also be shown that the sheer stress can increase the proliferation of osteoblasts. A microfluidic device was designed with 150-μm repetitive channels and 300 μm chambers in PDMS. The use of PDMS ensures a good exchange with the cell culture [122, 123]. The microdevices make dynamic cell culture conditions inside a microstructured geometry possible. The proliferation and differentiation of osteoblastic cells (MC3T3-E1) were then measured inside the microdevices under static and dynamic conditions, and the results were compared with conventional cultures on a PDMS-coated Petri dish. It was found that the cell viability was around 85% up to 1–2 weeks for shear stress value under 5 mPa, while the activity of alkaline phosphatase (ALP), the enzyme marker for osteoblasts, was also increased 7.5-fold [124]. The MC3T3-E1 could also be successfully cultured in a glass microchip. The advantage of a glass chip is that the screening of ALP activity can be performed with modified osteoblasts expressing green fluorescent protein (GFP). It was found that in the case of differentiation medium containing bone morphogenetic protein 2, the ALP activity in the culture supernatant was enhanced 10-fold in the microchannel compared with static conditions in a 48-well dish [125].

6 High-Throughput Multi-array Chip

Stem cell culture will in future become the source of cells for preliminary drug screening, reducing the need for costly animal experiments [126, 127]. To address the problem of scalability, there is a need to perform tests on stem cells in a highly parallel fashion. Microfluidic devices offer this possibility, while limiting the amount of reagents and exercising precise control over the external cues. High-density microfluidic arrays with 24 × 24 cell chambers have been used for cell

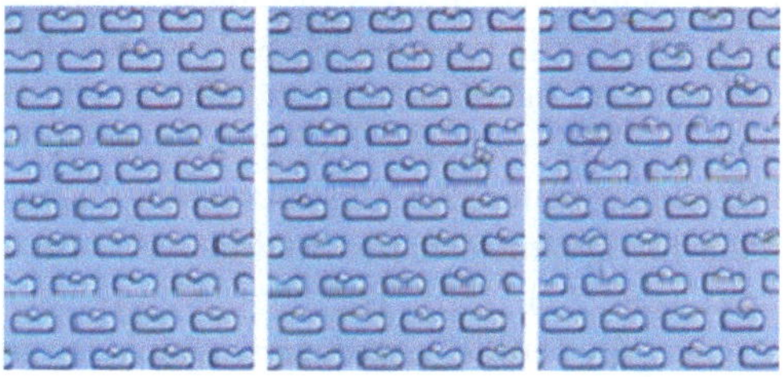

Fig. 15 Arrayed single cell culture. Images of cells cultured within the microfluidic arrays are shown. Cells were cultured under continues perfusion with an average velocity of 25 μm s^{-1} for over 24 h [130]

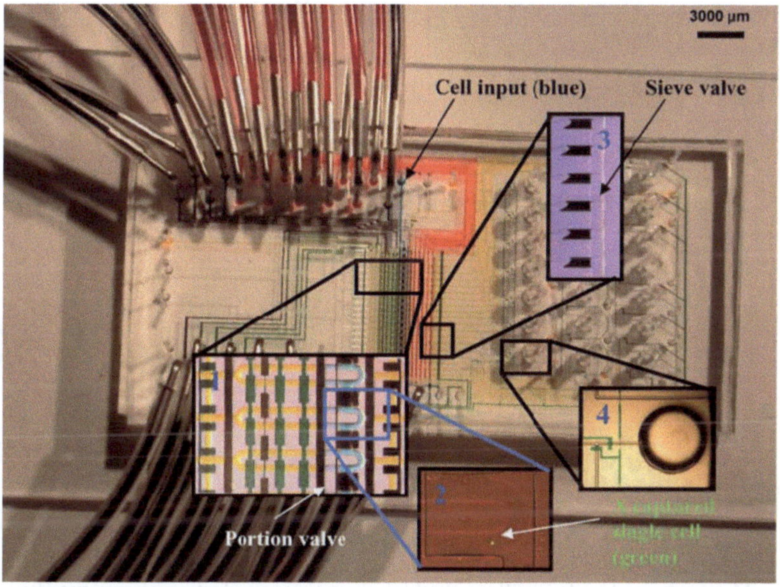

Fig. 16 Single-cell mRNA extraction microfluidic device, filled with food dye for illustration. All flow channels are filled with *yellow* food dye, multiplexer control channels are filled with *red* dye, collection and waste channels are in *blue*. *Insets 1–4* show enlargements of four important areas of the chip. After loading the cell suspension from the cell input inlet, single-cells are captured in the cell lysis module (*inset 1*) within the flow channels (*blue*). The pump valves are *green*. The separation valve is *black*. The lysis buffer is *yellow* [133]

cytotoxicity studies [128], while a 16 × 16 array of 10 nL chambers has been developed for gene expression studies using fluorescent reporter genes [129]. Single cells can also be studied in high-throughput fashion. Approximately 100 cells growing on trapping sites (Fig. 15) have been observed for over 24 h [130].

These high-throughput methods have also found their way into stem cell studies. Quake's work has demonstrated the possibility of using complex devices incorporating pumps and valves on a chip to control the distribution of reagents [131]. They have shown that human MSCs can proliferate and differentiate and that the cell's motility can be affected in such a device [132]; later they developed a microfluidic device to study the gene expression profiling in a single hESC. They measured the absolute numbers of three mRNAs by translating them to cDNA on the same device (Fig. 16) [133]. This method showed that hESC colonies are highly heterogeneous cell populations regulated by different gene networks.

As with tumour spheroid production, multiarray chips were also used to produce hESC aggregates. An anisotropically etched mould in silicon was used as a template for pyramid-shaped PDMS wells for high-throughput production of aggregates (Fig. 17) [135]. These chips were used to generate spatially and temporally synchronised hESC aggregates, which can be used for further fundamental studies in early human development processes. Another method used PEG-coated wells

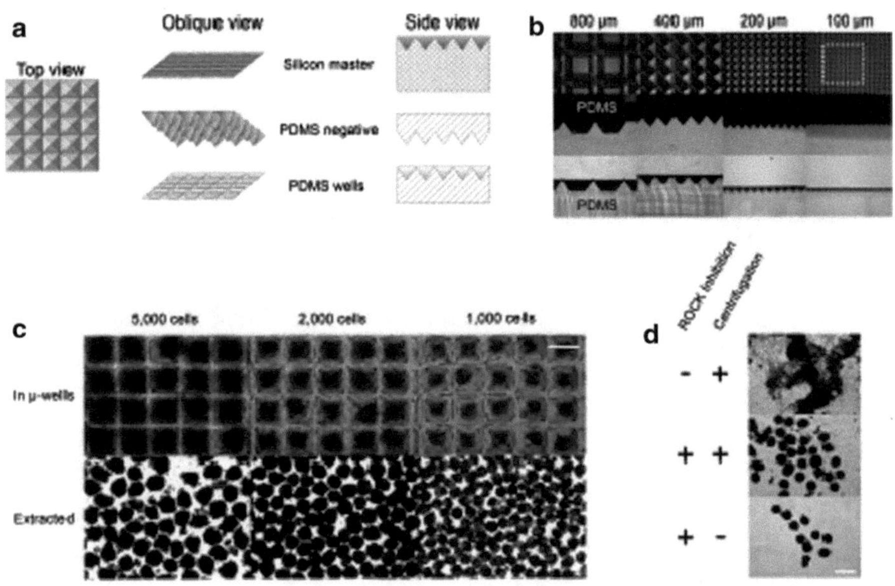

Fig. 17 Micropatterned surfaces allow ultrahigh-throughput production of size-specified aggregates. Surfaces patterned were generated in (**a**) poly(dimethylsiloxane) (PDMS) with wells of (**b**) 100, 200, 400 and 800 μM^2. (**c**) Sections of PDMS textured with 400 μM wells were inserted into wells in a 24-well plate. A single-cell suspension of hESCs was dispensed into the microwells. After 24 h, the well contents were imaged, extracted and re-imaged. (**d**) Aggregates were formed only in the presence of 10 μM of the ROCK inhibitor Y-27632, in the presence or absence of centrifugation. *Scale bars*: 400 μM [134]

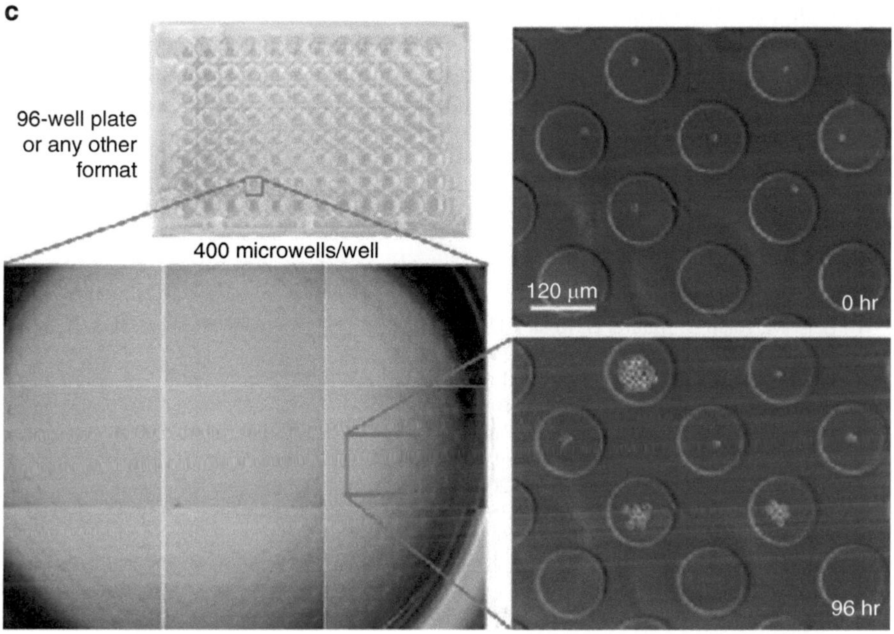

Fig. 18 Hydrogel microwell arrays can be placed on the bottom of any standard well plate to culture single haematopoietic stem cells and track their behaviour [136]

with a diameter of 150 μm [134]. The harvested aggregates had a viability of more than 95%. Other microwells of 100–130 μm in diameter have been fabricated by hydrogel casting to a PDMS mould printed with PEG-modified Protein A. This technique transfers the PEG into the bottom of the hydrogel microwells (Fig. 18). It was shown that self-renewal of haematopoietic stem cells (HSCs) and long term blood reconstitution can be induced by exposing the cells to either soluble or tethered proteins [136].

7 Conclusion

Microfluidic devices are well suited to support 3D cell cultures, while controlling their microenvironment. This allows for precise control over the viability, growth and differentiation of cells or stem cells. In the right microenvironment, cells will be functional and as such can be used as platforms for drug toxicity studies. The use of human cell lines, or better still, hESCs to mimic in vivo complex structures in vitro will enhance drug screening and reduce the need for animal testing. Microfluidics has the advantage over other systems in that they can be readily used to simulate human (disease) models, using precious human cells. This will have an advantage for drug-related studies because animal models do not accurately capture human physiology.

For example, if Michael Shuler's work on animal-on-a-chip [84] had been implemented earlier, the fatal side effects of the drug Naphthalene could have been prevented because it would have not passed the preliminary studies. This example shows the advantages of such systems and motivates us to continue with the study of cells in microfluidics.

References

1. Leist M, Raab B, Maurer S, Rösick U, Brigelius FR (1996) Free Radic Biol Med 21:297
2. Abbott A (2003) Nature 424:870
3. Hui EE, Bhatia SN (2007) Proc Natl Acad Sci USA 104:5722
4. Khetani SR, Bhatia SN (2008) Nat Biotech 26:120
5. Du Y, Chia SM, Han R, Chang S, Tang H, Yu H (2006) Biomaterials 27:5669
6. Fan YY, Chapkin R, Ramos KS (1995) In Vitro Cell Dev Biol Anim 31:492
7. Sorrell JM, Baber MA, Caplan AI (2007) Cells Tissues Organs 186:157
8. Whitesides GM (2006) Nature 442:368
9. Park JY, Takayama S, Lee SH (2010) Integr Biol 2:229
10. Bauer M, Su G, Beebe DJ, Friedl A (2010) Integr Biol 2:371
11. Gómez-Sjöberg R, Leyrat AA, Pirone DM, Chen CS, Quake SR (2007) Anal Chem 79:8557
12. Meyvantsson I, Beebe DJ (2008) Annu Rev Anal Chem 1:423
13. Meyvantsson I, Warrick JW, Hayes S, Skoien A, Beebe DJ (2008) Lab Chip 8:717
14. Tai Hyun P, Michael LS (2003) Biotechnol Prog 19:243
15. Whitesides GM, Ostuni E, Takayama S, Jiang X, Ingber DE (2001) Annu Rev Biomed Eng 3:335
16. Folch A, Toner M (2000) Annu Rev Biomed Eng 2:227
17. Toh YC, Zhang C, Zhang J, Khong YM, Chang S, Samper VD, van Noort D, Hutmacher DW, Yu H (2007) Lab Chip 7:302
18. Dittrich PS, Manz A (2006) Nat Rev Drug Discov 5:210
19. Andersson H, van de Berg A (2004) Lab Chip 4:98
20. Chen CS, Jiang X, Whitesides GM (2005) MRS Bull 30:194
21. Kim L, Vahey MD, Lee HY, Voldman J (2006) Lab Chip 6:394
22. McCue S, Noria S, Langille BL (2004) Trends Cardiovasc Med 14:143
23. Tilles AW, Baskaran H, Roy P, Yarmush ML, Toner M (2001) Biotechnol Bioeng 73:379
24. Chotard-Ghodsnia R, Haddad O, Leyrat A, Drochon A, Verdier C, Duperray A (2007) J Biomech 40:335
25. Cartmell SH, Porter BD, García AJ, Guldberg RE (2003) Tissue Eng 9:1197
26. Ong SM, Zhang C, Toh YT, Kim SH, Foo HL, Tan CH, van Noort D, Park S, Yu H (2008) Biomaterials 29:3237
27. Park JY, Yoo SJ, Hwang CM, Lee SH (2009) Lab Chip 9:2194
28. Park JH, Kim SK, Woo DH, Lee EJ, Kim JH, Lee SH (2009) Stem Cells 27:2646
29. Takayama S, Ostuni E, LeDuc P, Naruse K, Ingber DE, Whitesides GM (2001) Nature 411:1016
30. Takayama S, McDonald JC, Ostuni E, Liang MN, Kenis PJ, Ismagilov RF, Whitesides GM (1999) Proc Natl Acad Sci USA 96:5545
31. Kane RS, Takayama S, Ostuni E, Ingber DE, Whitesides GM (1999) Biomaterials 20:2363
32. Khademhosseini A, Langer R, Borenstein J, Vacanti JP (2006) Proc Natl Acad Sci USA 103:2480
33. Paul JH, Philip JL, Sabounchi P, Lin R, Lee LP (2005) Biotechnol Bioeng 89:1
34. Jeon LN, Baskaran H, Dertinger SKW, Whitesides GM, van de Water L, Toner M (2002) Nat Biotech 20:826

35. Sodunke TR, Bouchard MJ, Noh HM (2008) Biomed Microdevices 10:393
36. Taylor AM, Blurton-Jones M, Rhee SW, Cribbs DH, Cotman CW, Jeon LN (2005) Nat Meth 2:599
37. Chung S, Sudo S, Mack PJ, Wan C-R, Vickerman V, Kamm RD (2009) Lab Chip 9:269
38. Park S, Wolanin PM, Yuzbashyan EA, Lin H, Darnton NC, Stock JB, Silberzan P, Austin RH (2003) Proc Natl Acad Sci USA 100:13910
39. Nam SW, van Noort D, Yang Y, Kim SH, Park S (2007) Biochip J 1:111
40. Nam SW, van Noort D, Yang Y, Park S (2007) Lab Chip 7:528
41. Lin F, Butcher EC (2006) Lab Chip 6:1462
42. Ahmed T, Stocker R (2008) Biophys J 95:4481
43. Lin F, Baldessari F, Gyenge CC, Sato T, Chambers RD, Santiago JG, Butcher EC (2008) J Immunol 181:2465
44. Janakiraman V, Englert D, Jayaraman A, Baskaran H (2009) Ann Biomed Eng 37:1206
45. Balagaddé FK, Song H, Ozaki J, Collins CH, Barnet M, Arnold FH, Quake SR, You L (2008) Mol Syst Biol 4:187
46. Cho SH, Chen CH, Tsai FS, Godin JM, Lo YH (2010) Lab Chip 10:1567
47. Vahey MD, Voldman J (2008) Anal Chem 80:3135
48. Schuster TG, Cho B, Keller LM, Takayama S, Smith GD (2003) Reprod Biomed Online 7:75
49. Chung Y, Zhu X, Gu W, Smith GD, Takayama S (2006) Methods Mol Biol 321:227
50. Wlodkowic D, Cooper JM (2010) Curr Opin Chem Biol 14:1
51. Reuben JM, Krishnamurthy S, Woodward W, Cristofanilli M (2008) Expert Opin Med Diagn 2:339
52. Wiedswang G, Naume B (2007) Nat Clin Pract Oncol 4:154
53. Nole F, Munzone E, Zorzino L, Minchella I, Salvatici M, Botteri E, Medici M, Verri E, Adamoli L, Rotmensz N, Goldhirsch A, Sandri MT (2007) Ann Oncol 19:891
54. Weiss L (1990) Adv Cancer Res 54:159
55. Mohamed H, McCurdy LD, Szarowski DH, Duva S, Turner JN, Caggana M (2004) IEEE Trans Nanobioscience 3:251
56. Tan SJ, Yobas L, Yew G, Lee H, Ong CN, Lim CT (2009) Biomed Microdevices 11:883
57. Davis JA, Inglis DW, Morton KJ, Lawrence DA, Huang LR, Chou SY, Sturm JC, Austin RH (2006) Proc Natl Acad Sci USA 103:14779
58. Birgersdotter A, Sandberg R, Ernberg I (2005) Semin Cancer Biol 15:405
59. Pampaloni F, Reynaud EG, Stelzer EH (2007) Nat Rev Mol Cell Biol 8:839
60. Cukierman E, Pankov R, Stevens DR, Yamada KM (2001) Science 294:1708
61. Chia SM, Leong KW, Li J, Xu X, Zeng K, Er ON, Gao S, Yu H (2000) Tissue Eng 6:481
62. Wolf K, Mazo I, Leung H, Engelke K, von Andrian UH, Deryugina EI, Strongin AY, Bröcker EB, Friedl P (2003) J Cell Biol 160:267
63. Li S, Lao J, Li YS, Zhao Y, Chu J, Chen KD, Tsuo TC, Peck K, Chien S (2003) FASEB J 17:97
64. Ghosh S, Spagnoli GC, Martin I, Ploegert S, Demougin P, Heberer M, Reschner A (2005) J Cell Physiol 204:522
65. Blank U, Karlsson G, Karlsson S (2008) Blood 111:492
66. Briscoe J (2009) EMBO J 28:457
67. Alexandra S, Stuart TF, Baron MH (2004) J Cell Biochem 93:224
68. Muruganandan S, Roman A, Sinal CJ (2009) Cell Mol Life Sci 66:236
69. Kardami E, Detillieux K, Ma X, Jiang Z, Santiago J, Jimenez SK, Carttini PA (2007) Heart Fail Rev 12:267
70. Gressner AM, Lahme B, Mannherz HG, Polzar B (1997) J Hepatol 26:1079
71. Vicki IC, Philippe T, Sanga S, Sheel J, Gage FH, Bhatia SN (2004) Biotechnol Bioeng 88:399
72. Takayama S, Ostuni E, Qian X, McDonald JC, Jiang X, LeDuc P, Wu MH, Ingber DE, Whitesides GM (2001) Adv Mater 13:570
73. Walker GM, Zeringue HC, Beebe DJ (2004) Lab Chip 4:91
74. van Noort D, Ong SM, Zhang C, Zhang S, Arooz T, Yu H (2009) Biotechnol Prog 25:52

75. Acharya AP, Clare-Salzler NJ, Keselowsky BG (2009) Biomaterials 30:4168
76. Nguyen DN, Kim P, Martínez-Sobrido L, Beitzel B, García-Sastre A, Langer R, Anderson DG (2009) Biotechnol Bioeng 103:664
77. Olaharski AJ, Uppal H, Cooper M, Platz S, Zabka TS, Kolaja KL (2009) Toxicol Lett 188:98
78. Wong AP, Perez-Castillejos R, Love JC, Whitesides GM (2008) Biomaterials 29:1853
79. Vickerman V, Blundo J, Chung S, Kamm RD (2008) Lab Chip 9:146
80. Domansky K, Inman W, Serdy J, Dash A, Lim MHM, Griffith LG (2010) Lab Chip 10:51
81. Shin M, Matsuda K, Ishii O, Terai H, Mofrad MK, Borenstein J, Detmar M, Vacanti JP (2004) Biomed Microdevices 6:269
82. Lee PJ, Hung PJ, Lee LP (2007) Biotechnol Bioeng 97:1340
83. Toh YC, Lim TC, Tai D, Xiao G, van Noort D, Yu H (2009) Lab Chip 9:2026
84. Viravaidya K, Shuler ML (2004) Biotechnol Prog 20:590
85. Viravaidya K, Sin A, Shuler ML (2004) Biotechnol Prog 20:316
86. Sung JH, Kam C, Shuler ML (2010) Lab Chip 10:446
87. Zhang C, Zhao Z, Abdul Rahim NA, van Noort D, Yu H (2009) Lab Chip 9:3185
88. Shield K, Ackland ML, Ahmed N, Rice GE (2009) Gynecol Oncol 113:143
89. Rodriguez-Enriquez S, Gallardo-Perez JC, Aviles-Salas A, Marin-Hernandez A, Carreno-Fuentes L, Maldonado-Lagunas V, Moreno-Sanchez R (2008) J Cell Physiol 216:189
90. Fukuda J, Sakai Y, Nakazawa K (2006) Biomaterials 27:1061
91. Sakai Y, Nakazawa K (2007) Acta Biomater 3:1033
92. Hirschhaeuser F, Menne H, Dittfeld C, West J, Mueller-Klieser W, Kunz-Schughart LA (2010) J Biotechnol 148:3
93. Ramalho-Santos M, Willenbring H (2007) Cell Stem Cell 1:35
94. Imreh MP, Gertow K, Cedervall J, Unger C, Holmberg K, Szoke K, Csoregh L, Fried G, Dilber S, Blennow E, Ahrlund-Richter L (2006) J Cell Biochem 99:508
95. Martin GR (1981) Proc Natl Acad Sci USA 78:7634
96. Evans MJ, Kaufman MH (1981) Nature 292:154
97. Doetschman TC, Eistetter H, Katz M, Schmidt W, Kemler R (1985) J Embryol Exp Morphol 87:27
98. Thomson JA, Kalishman J, Golos TG, Durning M, Harris CP, Becker RA, Hearn JP (1995) Proc Natl Acad Sci USA 92:7844
99. Thomson JA, Kalishman J, Golos TG, Durning M, Harris CP, Hearn JP (1996) Biol Reprod 55:254
100. Reubinoff BE, Pera MF, Fong CY, Trounson A, Bongso A (2000) Nat Biotechnol 18:399
101. Thomson JA, Itskovitz-Eldor J, Shapiro SS, Waknitz MA, Swiergiel JJ, Marshall VS, Jones JM (1998) Science 282:1145
102. Takahashi K, Yamanaka S (2006) Cell 126:663
103. Takahashi K, Tanabe K, Ohnuki M, Narita M, Ichisaka T, Tomoda K, Yamanaka S (2007) Cell 131:861
104. Yu J, Vodyanik MA, Smuga-Otto K, Antosiewicz-Bourget J, Frane JL, Tian S, Nie J, Jonsdottir GA, Ruotti V, Stewart R, Slukvin II, Thomson JA (2007) Science 318:1917
105. Wernig M, Meissner A, Foreman R, Brambrink T, Ku M, Hochedlinger K, Bernstein BE, Jaenisch R (2007) Nature 448:318
106. Park IH, Zhao R, West JA, Yabuuchi A, Huo H, Ince TA, Lerou PH, Lensch MW, Daley GQ (2008) Nature 451:141
107. Viswanathan S, Benatar T, Mileikovsky M, Lauffenburger DA, Nagy A, Zandstra PW (2003) Biotechnol Bioeng 84:505
108. Prudhomme W, Daley GQ, Zandstra PW, Lauffenburger DA (2004) Proc Natl Acad Sci USA 101:2900
109. Flaim CJ, Teng D, Chien S, Bhatia SN (2008) Stem Cells Dev 17:23
110. Moeller HC, Mian MK, Shrivastava S, Chung BG, Khademhosseini A (2008) Biomaterials 29:752
111. Underhill GH, Bhatia SN (2007) Curr Opin Chem Biol 11:357

112. Sim WY, Park SW, Park SH, Min BH, Park SR, Yang SS (2007) Lab Chip 7:1775
113. Kilian KA, Bugarija B, Lahn BT, Mrksich M (2010) Proc Natl Acad Sci USA 107:4872
114. Chung BG, Flanagan LA, Rhee SW, Schwartz PH, Lee AP, Monuki ES, Jeon NL (2005) Lab Chip 5:401
115. Cochran DM, Fukumura D, Ancukiewicz M, Carmeliet P, Jain RK (2006) Ann Biomed Eng 34:1247
116. Stolberg S, McCloskey KE (2009) Biotechnol Prog 25:10
117. Angele P, Yoo JU, Smith C, Mansour J, Jepsen KJ, Nerlich M, Johnstone B (2003) J Orthop Res 21:451
118. Kaushal S, Amiel GE, Guleserian KJ, Shapira OM, Perry T, Sutherland FW, Rabkin E, Moran AM, Schoen FJ, Atala A, Soker S, Bischoff J, Mayer JE (2001) Nat Med 7:1035
119. Plouffe BD, Kniazeva T, Mayer JE Jr, Murthy SK, Sales VL (2009) FASEB J 23:3309
120. Donahue TL, Haut TR, Yellowley CE, Donahue HJ, Jacobs CR (2003) J Biomech 36:1363
121. Arnsdorf EJ, Tummala P, Kwon RY, Jacobs CR (2009) J Cell Sci 122:546
122. Charati SG, Stern SA (1998) Macromolecules 31:5529
123. Szita N, Zanzotto A, Boccazzi P, Sinskey A, Schmidt M, Jensen K (2002) Monitoring of cell growth, oxygen and pH in microfermentors. In: Baba Y, Shoji S, Berg A (eds) Proceedings 6th International Conference on Micro Total Analysis Systems (μTAS'02), Nara, Japan, 3-7 November 2002. Kluwer, The Netherlands, pp 7–9
124. Leclerc E, David B, Griscom L, Lepioufle B, Fujii T, Layrolle P, Legallaisa C (2006) Biomaterials 27:586
125. Jang K, Sato K, Igawa K, Chung UI, Kitamori T (2008) Anal Bioanal Chem 390:825
126. Johnston J (2005) Hastings Cent Rep 35:16
127. Fok S, Domachuk P, Rosengarten G, Krause N, Braet F, Eggleton BJ, Soon LL (2008) Biophys J 95:1523
128. Wang Z, Kim MC, Marquez M, Thorsen T (2007) Lab Chip 7:740
129. King KR, Wang S, Irimia D, Jayaraman A, Toner M, Yarmush ML (2007) Lab Chip 7:77
130. Di Carlo D, Wu LY, Lee LP (2006) Lab Chip 6:1445
131. Hong JW, Quake SR (2003) Nat Biotechnol 21:1179
132. Gomez-Sjoberg R, Leyrat AA, Pirone DM, Chen CS, Quake SR (2007) Anal Chem 79:8557
133. Zhong JF, Chen Y, Marcus JS, Scherer A, Quake SR, Taylor CR, Weiner LP (2008) Lab Chip 8:68
134. Karp JM, Yeh J, Eng G, Fukuda J, Blumling J, Suh KY, Cheng J, Mahdavi A, Borenstein J, Langer R, Khademhosseini A (2007) Lab Chip 7:786
135. Ungrin MD, Joshi C, Nica A, Bauwens C, Zandstra PW (2008) PLoS ONE 3:e1565
136. Lutolf MP, Doyonnas R, Havenstrite K, Koleckar K, Blau HM (2009) Integr Biol 1:59

Top Curr Chem (2011) 304: 323–338
DOI: 10.1007/128_2011_145

Published online: 21 April 2011

Microfluidic Platform for the Study of *Caenorhabditis elegans*

Weiwei Shi, Hui Wen, Bingcheng Lin, and Jianhua Qin

Abstract *Caenorhabditis elegans* is a well-established model organism for the study of various biological areas. However, the procedures generally used for this worm are limited by the low throughput, low automation, and imprecise delivery of external stimuli. Microfluidics technology is emerging as an attractive and enabling platform to overcome these problems. In this review, we summarize the current microfluidic approaches for the investigation of behavior and neurobiology in *C. elegans*, and discuss the trends of future development.

Keywords *Caenorhabditis elegans* · Microfluidics · Neurobiology · Neuron degeneration · Neuron regeneration

Contents

1 Introduction 324
2 *C. elegans* Manipulations on Microfluidics Platforms 325
2.1 *C. elegans* Manipulation in Microchannel Format 325
2.2 *C. elegans* Manipulations in Microdroplet Format 328
3 Applications in the Investigations of *C. elegans* Behavior and Neurobiology 330
3.1 Behavior Studies 330
3.2 Neurobiology Studies 332
4 Conclusion and Perspective 336
References 337

W. Shi, H. Wen, B. Lin, and J. Qin (✉)
Dalian Institute of Chemistry and Physics, Chinese Academy of Sciences, 457, Zhongshan Road, Dalian 116023, China
e-mail: jhqin@dicp.ac.cn

Abbreviations

6-OHDA	Neurotoxin 6-hydroxydopamine
C. elegans	*Caenorhabditis elegans*
MPP+	1-Methyl-4-phenylpyridinium ion
MPTP	1-Methyl-4-phenyl-1,2,3,6-tetrahydropyridine
PD	Parkinson's disease
PDMS	Polydimethylsiloxane
RNAi	RNA interference

1 Introduction

Caenorhabditis elegans has become a well-established model organism since the pioneer work introduced by Brenner in 1963 [1]. Since its establishment as a model organism, *C. elegans* has been an invaluable tool for biological research. An immense spectrum of questions can be addressed using this small nematode, making it one of the most versatile and exciting model organisms. *C. elegans* has many unique features, including small size (1 mm long, 50 μm wide), hermaphroditic, cell constant (959 in the adult hermaphrodite), transparent body, short life cycle (about 3.5 days at 20°C), and ease of cultivation on agar or in liquid. Moreover, *C. elegans* is the first multicellular organism that has a completely sequenced genome, and its genome is surprisingly similar to that of humans (40% homologous). Despite having a simple nervous system (comprised of only 302 neurons), *C. elegans* exhibits complex neuronally controlled behavioral modalities (e.g., associative learning and the response to noxious stimuli) [2–6]. All of these characteristics have made *C. elegans* an attractive model organism in genetics, neurobiology and some other areas. At present, it is widely used in the study of cell apoptosis, RNA interference (RNAi), aging, and drug discovery [7–9].

Although *C. elegans* has several unique advantages for the study of animal biology, automated methods that have been applied to this model organism are still few. Conventionally, the operations of *C. elegans* is usually performed on a Petri dish or multiwell plate, which is handled manually and is time-consuming. The worms are often cultured on agar or in liquid in large numbers, and the various behaviors of worms in response to external stimuli have to be analyzed after multiple procedures that include worm culture, stimulation, washing, sorting, and detection steps. During this process, the procedures generally used are often limited by low throughput, low automation, and the imprecise delivery of external stimuli. For fluorescence imaging analysis, the individual worm has to be handled by picking with a platinum wire and immobilizing with glue or anesthesia, and this method potentially suffers from low efficiency and potential damage to the worm after gluing. Therefore, it is desirable to explore a reliable, automated, and easily integrated technique for *C. elegans* high-throughput assay.

Advances in the application of microfluidics technology to biological assays help to automate otherwise time-consuming experiments. In recent years, microfluidics technology is emerging as an attractive and enabling platform for the study of *C. elegans*, including their behavior and neurobiology [10–17]. The unique properties of this technology lie in several aspects: first, the dimension of microfluidic channel is a perfect match with the size of the tiny worm. Second, the ability to manipulate small amounts of liquid makes it suitable for the manipulation of a single worm or the precise delivery of external stimuli. Third, the use of transparent materials such as polydimethylsiloxane (PDMS) allows for optical imaging. Last, the capability to realize large scale integration makes it possible to handle a large population of worms in parallel or in series for high throughput assay or screening in drug discovery.

In this review, we aim to illustrate the current microfluidic approaches for the investigations of behavior and neurobiology in *C. elegans*. The two kinds of microfluidics-based platforms, channel and droplet formats for worm manipulations, will be covered and the related applications offered by this technology will be demonstrated.

2 *C. elegans* Manipulations on Microfluidics Platforms

2.1 C. elegans *Manipulation in Microchannel Format*

Worm manipulation in channel microfluidics format is the basis for diverse types of applications in *C. elegans* research. The unique properties with this format include: (1) The commonly used material, PDMS, is gas permeable, transparent, nonfluorescent and biocompatible, and is suitable for worm handling and culture [18]. (2) Complex structures for different applications are easily fabricated by soft lithography [19, 20]. (3) The delivery of accurate stimuli (chemical, temperature, and mechanical) to individual worm is more controllable. (4) Multiple functional units can be assembled and diverse manipulations of worms can be processed automatically and sequentially. At present, several worm manipulation methods offered by channel-microfluidics have been proposed such as worm immobilization [21–28], cultivation (gas/liquid) [29–31], controllable stimulation [32–37], screening [28, 38], locomotion sensing [39, 40], etc.

2.1.1 *C. elegans* Immobilization

Immobilization is required for the stable imaging analysis in *C. elegans* neuron studies. Channel microfluidics provides a flexible, gentle, and reversible mechanical manner for worm immobilization instead of the conventional "glue" method [41, 42], thus avoiding the side-effects of glue or anesthesia. Currently, there are mainly two mechanical immobilization methods, which we called "tapered channel" and "microvalve control" methods. The former mainly makes use of channel geometry to immobilize individual worms. A parallel microfluidic channel array with a decreasing width (usually narrowed to ~10 μm) and a constant height (usually 30–50 μm) was designed to trap individual worms in the narrower regions of the channels. Hulme et al. [21] (Fig. 1a) fabricated a microfluidic device consisted of an array of 128 wedge-shaped microchannels to immobilize worms and a branching network of distribution channels to deliver the worms to the array. With this device, the researchers realized rapid immobilization of a large number of worms (more than 100 worms within 15 min) without mechanical damage. Also, Allen et al. [22] used a similar microfluidic chip to facilitate long-term fluorescence imaging of the immobilized *C. elegans*. The latter method mainly makes use of a microvalve to realize flexible control of the individual worms, which potentially enable the parallel or reversible immobilization of multiple worms automatically. In this method, the microvalve often contain a deformable PDMS membrane and a pressure control channel, and the deflection of the PDMS membrane is utilized to restrict the worm's movement when a proper pressure (gas/fluidic) is applied to the control channel (Guo et al. [23] and Ma et al. [26] (Fig. 1b). There are also some other *C. elegans* immobilization methods including cooling (~4°C) [28] (Fig. 2b), and increasing the CO_2 concentration [25] of the *C. elegans* environment on microfluidic chip, by which *C. elegans* lose their locomotion activities and become immobilized. In comparison with the mechanical methods, cooling and CO_2 methods result in complete immobilization of the worms. Also, some assistant approaches such as suction channel [24, 27] (Fig. 1c) and temperature-sensitive gel [31] have been developed for immobilizing *C. elegans* on microfluidics.

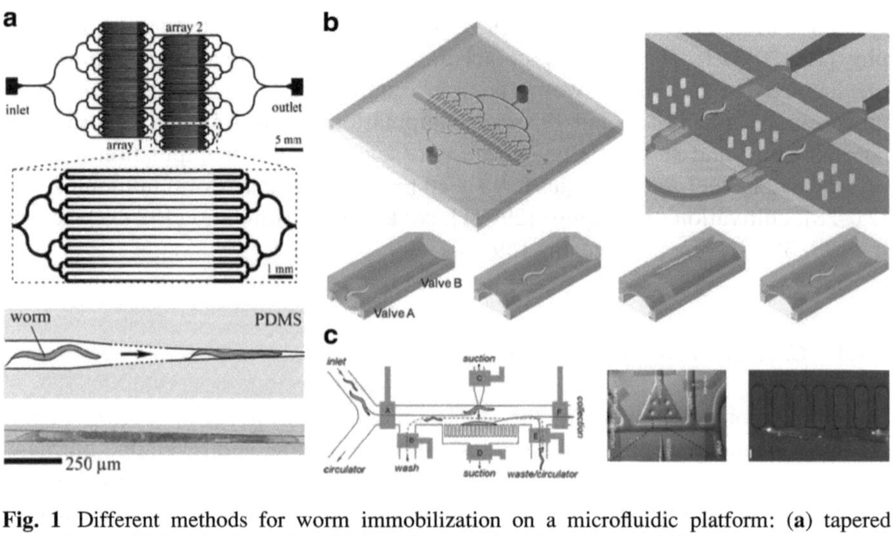

Fig. 1 Different methods for worm immobilization on a microfluidic platform: (**a**) tapered channel array [21]; (**b**) microvalves [26]; and (**c**) suction channel array [27]

2.1.2 Long-Term Cultivation

Long-term cultivation and maintenance is crucial for the investigation of many aspects of *C. elegans*, such as aging. Channel microfluidics provides a potential platform for the long term study of worms and this is mainly due to the capability to enable the exchange of nutrition and excreta of worms in microchannels, and the gas permeability offered by the PDMS material for necessary oxygen supply to the worms. Kim et al. [29] and Hulme et al. [30] realized long-term culture of *C. elegans* on microfluidic chips. The former researchers established a microfluidic cultivation system in which centrifugal force was utilized to drive liquid, and the processes of feeding and waste removal were accomplished automatically. The population of *C. elegans* cultivated in the chip was monitored over a period of 14 days. The latter used a constant flow of *Escherichia coli* suspension to deliver food and remove the waste of the worms continuously. The worms were cultured in the chambers individually, and the body size and locomotion were characterized for 20 days.

2.1.3 Microenvironment Construction

C. elegans can sense and respond to even tiny changes (mechanical or chemical) in its environment acing as a sensor. By using a well-designed microfluidic chip, the mechanical microenvironment of *C. elegans* can be constructed and their movement pattern or scope be restricted. By using the properties of channel microfluidics, various precise chemical stimuli could be delivered to *C. elegans* in any time, type and concentration. Lockery et al. [32] and Park et al. [33] used a PDMS/agar micropillar array to mimic the soil structure where *C. elegans* lived in the wild, and studied different patterns of the locomotion of worms. Gray et al. [36] (Fig. 2a) and Zimmer et al. [37] investigated the behavior of *C. elegans* in response to different concentrations of O_2 on different PDMS/agar hybrid chips. Qin et al. [43] established several microfluidic mazes and investigated the exploration and learning behaviors of wild-type *C. elegans* and the dopamine-poor mutant, cat-2. Zhang et al. [44] use a four-choice maze to study the olfactory preferences of *C. elegans*. Chronis et al. [34] and Chalasani et al. [35] realized the accurate and switchable deliver of buffer or odor stimuli to the worm's nose.

2.1.4 Sorting and Screening

C. elegans is an ideal model organism for the study of pathology and pharmacology, the evaluation of its characteristics in a high-throughput manner is significant for drug screening. Chung et al. [28] (Fig. 2b) fabricated a multivalve-integrated microfluidic device that could identify the phenotype features of the tested *C. elegans* and manipulate the worm positioning automatically with customized software. Within this device, high-throughput worm screening based on

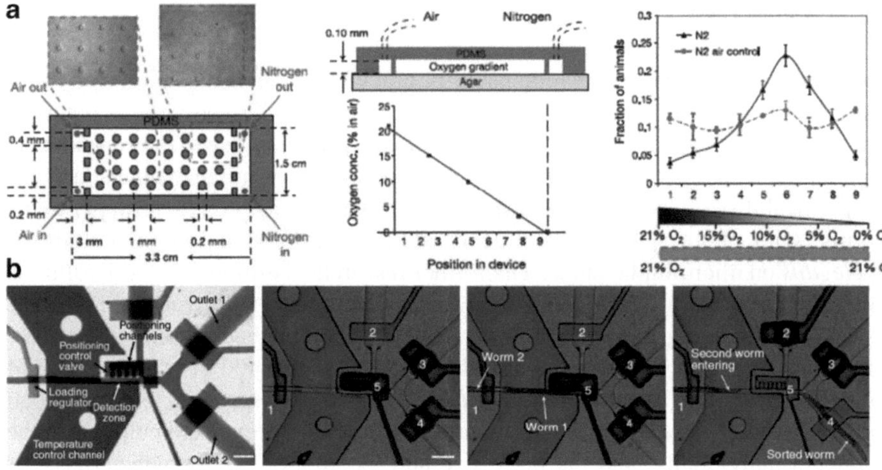

Fig. 2 (**a**) PDMS/agar hybrid chip enables the gradient generation of O_2 concentrations for investigating the behavior of *C. elegans* in response to different concentrations of O_2 [36]. (**b**) Multivalve-integrated microfluidic device that can manipulate worm position and screen mutant worms automatically [28]

morphological and intensity features was achieved at a rate of several hundred worms per hour. On this basis, Crane et al. [38] changed the automatic classification of worms into human decision-making and made the screening more flexible for unidentified mutants.

2.1.5 Locomotion Sensing

C. elegans has demonstrated free mobility behavior in liquid and on agar, and the mobility properties can be sensed by the micropillar-based mechanical sensor in channel microfluidics. Park et al. [39] fabricated micrometer-scale piezoresistive cantilevers as force–displacement sensors, and measured the mechanical properties of *C. elegans*. For the same purpose, Doll et al. [40] used SU-8 force sensing pillars to characterize the scale of interaction forces that worms generated during their locomotion.

2.2 C. elegans *Manipulations in Microdroplet Format*

Droplet microfluidics is emerging as an attractive platform for biology research [45, 46], and it has also been explored in the study of *C. elegans* in recent years. By using this method, the generated array microdroplets can serve as microreactors and have unique benefits for the flexible and high-throughput operations of individual

C. elegans. Firstly, the droplet size can be flexibly controlled ranging from nanoliters to microliters, which suits worm assay during all the developmental stages (from embryo, egg, and larvae to adult). Secondly, the generation, transportation, and manipulation of droplets are precisely controlled and can achieve dispersal and delivery of individual worms. Thirdly, the high-throughput and high-content characters of droplet microfluidics are more suitable for the analysis of large-scale worms or large-scale experimental conditions at single animal resolution. Clausell-Tormos et al. [47] encapsulated worm eggs into aqueous droplets in a gas-permeable PTFE tubing, and observed different stages including egg hatching, worm growing and egg laying within the droplet. Shi et al. [48] (Fig. 3) integrated a T-junction droplet generator and a droplet mechanical trapping array in a microfluidic chip; realized automatic and continuous operations including droplet generation, individual worm encapsulation, and droplet trapping sequentially; and investigated the mobility defects of *C. elegans* L1 larva induced by neurotoxin MPP+ on this platform. In their recent study [49], a novel floatage-based droplet trapping method was developed, and the microfluidic platform was improved by coupling with a droplet floatage trapping array and a tapered channel immobilization array. With this improved microfluidic chip, the studies of movement and fluorescence features of individual worms could be investigated. In their experiment, the mobility defects, dopaminergic (DAergic) neuron degeneration, and increase in oxidative stress of adult *C. elegans* under the treatment of neurotoxin 6-OHDA were observed.

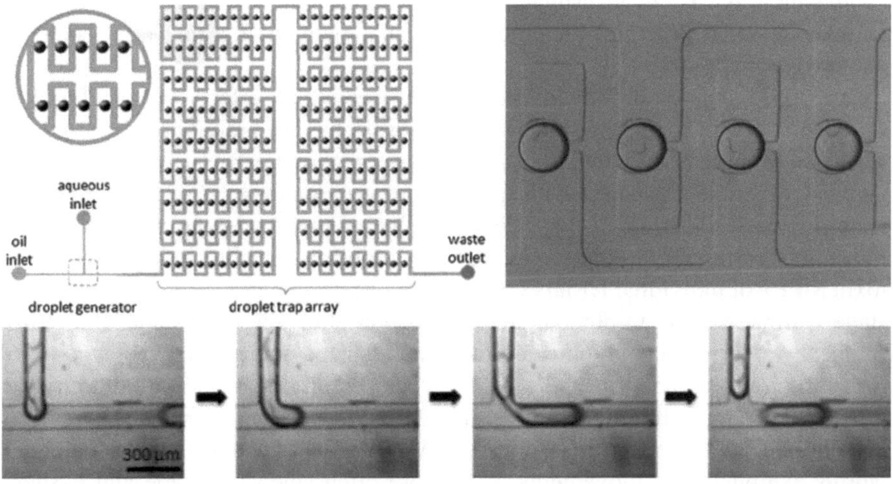

Fig. 3 Droplet-based microfluidic device integrated with a T-junction droplet generator and a droplet mechanical trapping array. With this device, individual *C. elegans* can be encapsulated into microdroplets and, after the droplets are trapped, the mobility behavior of the worms can be investigated at single-animal resolution [48]

3 Applications in the Investigations of *C. elegans* Behavior and Neurobiology

Based on the proposed microfluidics platforms and incorporated functional elements, diverse types of biological applications, including studies on behaviors and neurobiology, have been demonstrated by using *C. elegans* as a model organism. Because the worms can sense diverse kinds of stimuli, such as touching, temperature, chemicals and ions, there is increasing interest in the characterization of the corresponding behavioral phenotypes including locomotion, chemotaxis, thermotaxis, feeding, egg-laying, and mating etc. These studies are extremely useful for the screening of mutants, exploring the relationship between gene and neural circuit functions and behaviors, and screening therapeutic drugs for neurodegenerative diseases.

3.1 Behavior Studies

3.1.1 Mobility Behavior

Determining the mobility patterns is important for characterizing various *C. elegans* behaviors, including chemotaxis, electrotaxis, learning etc. On agar, wild-type *C. elegans* crawl on the smooth surface and produce sinusoidal waves along the body; in liquid, swimming of wild-type worms is usually characterized by a C-shaped posture [50]. By using microfluidics, the format for the normal movement of *C. elegans* was easy to investigate, and the various complex movement behaviors in relevant microstructures (microchannels and micropillar array, etc.) could be explored as well.

In 2005, Lange et al. [51] investigated the behavioral effects of spaceflight on *C. elegans* in a microfluidic shadow imaging system. The activity of *C. elegans* in liquid environment as a function of ambient temperature was measured, and an increase in stroke frequency with temperature was found. In 2008, Shi et al. [48] (Fig. 4a) used an array to disperse and deliver individual worms, restrict their movement within the droplets, and study the mobility behavior induced by neurotoxin MPP+ of individual L1 larva that were encapsulated in microdroplets. The L1 stage worms showed an obvious mobility defect after treatment of MPP+: stroke frequency decreased and omega- and tetanic-shape increased. Ma et al. [26] (Fig. 4b) trapped individual *C. elegans* young adults in a microfluidic channel array by switching the microvalve at the entrance, and characterized the mobility behavior of adult worms in response to MPP+. It was found that the untreated worms exhibited free movements, with more often sinewave-shape and C-shape movement states in the imaging channels, whereas the worms after MPP+ treatment were subjected to mobility defects, such as slow, titanic, and coiled movements.

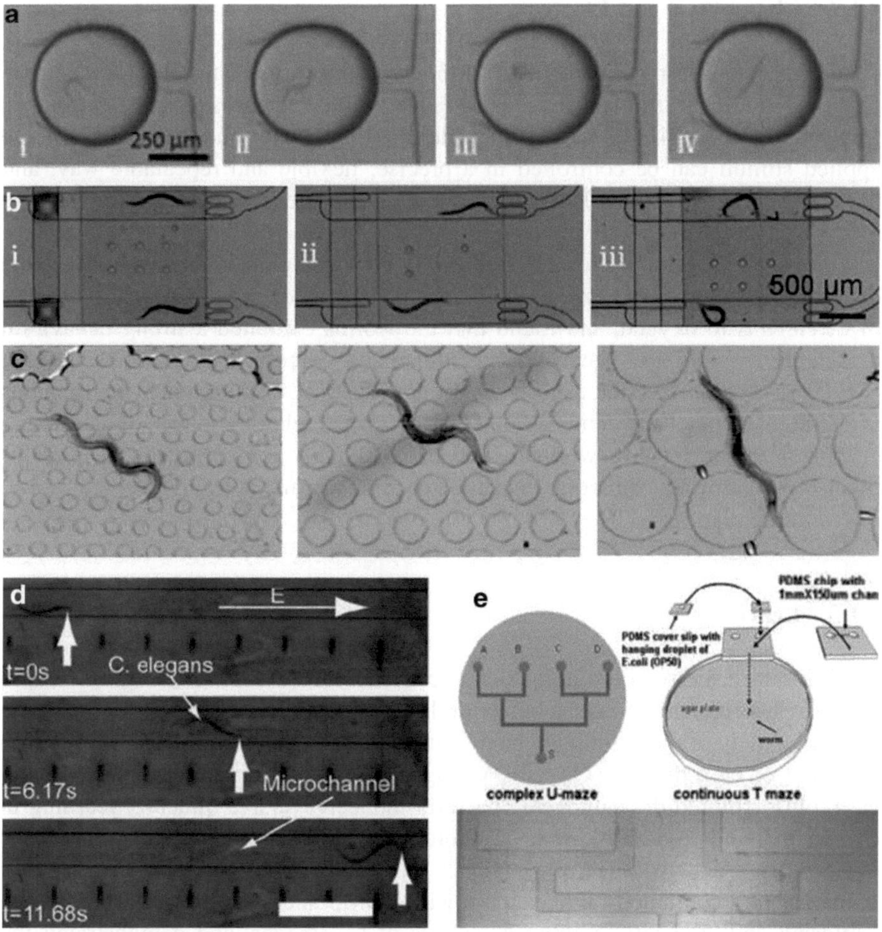

Fig. 4 Characterization of *C. elegans* behaviors on different microfluidic platforms. (**a**) Droplet-based microfluidic chip for encapsulating individual worms and studying their mobility defects induced by neurotoxin [48]. (**b**) Microvalve-based microfluidic device for investigating the mobility behavior of individual adult worms [26]. (**c**) Micropillar structures in microfluidic chip to mimic the natural soil environment and investigate the mobility behavior of the worms in this "artificial soil" [32]. (**d**) The worm's movement in response to alternating electric fields in a straight channel [52]. (**e**) Complex U-maze and continuous T-maze chip for investigating the exploring and associated learning behaviors of *C. elegans* [43]

Lockery et al. [32] (Fig. 4c) and Park et al. [33] constructed micropillar structures in microfluidic chips to mimic the natural soil environment where *C. elegans* live in the wild, and investigated the mobility behavioral of the worms in this "artificial soil". These works showed the unique advantages of manipulating individual *C. elegans* and the capability for parallel analysis of microfluidic platform.

3.1.2 Chemotaxis and Electrotaxis

C. elegans can display behavioral responses to external stimuli, for example chemical gradients and electric fields, which are called chemotaxis and electrotaxis, respectively [53, 54]. By using microfluidics, the range and conditions of the applied stimuli can be controlled in a precise, flexible and repeatable way, and the corresponding responses of large-scale *C. elegans* can be detected in real-time [36, 52, 55, 56].

Gray et al. [36] established a hybridized PDMS/agar chip, and generated a stable concentration gradient of oxygen in the chamber to study the behavioral responses of *C. elegans* to oxygen, and found that the worms exhibited a strong behavioral preference for 5–12% oxygen, avoiding higher and lower oxygen levels. Rezai et al. [52] (Fig. 4d) described the worm's movement response to alternating electric fields in a straight channel and demonstrated that 1 Hz and higher frequency of alternating current field can localize worms in the channel. This method could be used for precisely controlling, directing, and transporting worms in *C. elegans* studies on microfluidic platforms.

3.1.3 Learning

Behavioral plasticity is the unique characteristics of *C. elegans* due to its simple nervous system and complex behavioral modalities. By using the microfluidic platform, study of the learning behavior of *C. elegans* has been carried out through a variety of chip designs. On a PDMS/agar hybridized microfluidic chip, Zhang et al. [44] explored the pathogenic bacteria-induced aversive olfactory learning in *C. elegans*, and studied the role of serotonin in the ADF chemosensory neurons involved in *C. elegans* learning. Qin et al. [43] (Fig. 4e) presented a complex U-maze and a continuous T-maze chip with PDMS and agar, and investigated the exploring and associate learning behaviors of *C. elegans*, and demonstrated the effects of dopamine on learning by evaluating the behavior of the cat-2 mutant in the continuous T-maze assay. This work introduced microfluidic mazes as a valuable new tool that could potentially encourage the research of *C. elegans* behavior.

3.2 Neurobiology Studies

Despite owning a simple nervous system comprised of only 302 neurons, the neurotransmitters, synaptic proteins and ion channels of *C. elegans* are highly conserved with human beings. Moreover, owing to the transparent body of *C. elegans*, the characteristics of neurons relying on the neuron fluorescence imaging can be investigated conveniently. Combined with the unique features, the microfluidic platform has been widely studied in the area of neurobiology on

C. elegans, including neuron activity, neuron ablation and regeneration, and neuron degeneration [14, 17]. Furthermore, human neurodegenerative diseases (like Parkinson's disease and Alzheimer's disease) models of *C. elegans* have been established successfully through methods such as transgenic, mutant screening and drug inducing [57–60]. Some related researches on pathogenesis and drug screening have also been carried out.

3.2.1 Neuron Activity

Neuron activity is a fundamental area of study in neuroscience. By using the newly emerged fluorescent calcium indicators, calcium transients in individual neurons in living *C. elegans* can be tracked, and the activity of *C. elegans* nervous system can be monitored accordingly [61]. Combined with simultaneous behavior analysis, the behavior mechanisms of *C. elegans* can be studied at the neuronal level. Currently, various microfluidic devices have been developed for the study of *C. elegans* neuronal–behavior analysis [34, 35, 37, 62]. With these devices, the precise control of the delivered stimuli and environment of *C. elegans* can be achieved, and the neural functional imaging and corresponding behavior information can be investigated simultaneously.

Chronis et al. [34] realized the neural functional imaging of individual *C. elegans* in two microfluidic chips. In the first microfluidic chip (Fig. 5a), a single worm was trapped in a channel that was slightly wider than the worm body, the calcium transients in the AVA neurons were imaged, and the relationship between the activity of AVE neurons and the direction of the traveling body wave of *C. elegans* were investigated. In the second microfluidic chip, a single worm was immobilized in a tapered channel with its nose exposed to a flowing stream, and the calcium changes in ASH neurons were monitored. By switching the flow between stimulus and control solutions, the relationship between ASH neuron activity and the olfactory behavior of *C. elegans* were studied. Also, Chalasani et al. [35] used a similar design to characterize the role of *C. elegans* olfactory neurons (AWC) and interneurons (AIB and AIY) in information processing of food- and odor-evoked behaviors. Zimmer et al. [37] (Fig. 5b) developed a specialized imaging microfluidic device to deliver O_2 stimuli to adult worms that were immobilized in a two-layer PDMS chip and studied the calcium transients in individual URX and BAG sensory neurons responding to O_2 upshift and downshift. The results indicated that URX and BAG were activated by increases and decreases in O_2 levels, respectively.

3.2.2 Neuron Degeneration

Due to the unique properties of its nervous system, *C. elegans* has been well studied for establishing models of human neurodegenerative diseases (like Parkinson's disease, PD) and revealing the pathological mechanisms of these diseases [57–60]. Neurotoxins 6-hydroxydopamine (6-OHDA), 1-methyl-4-phenyl-1,2,3,6-tetrahydropyridine

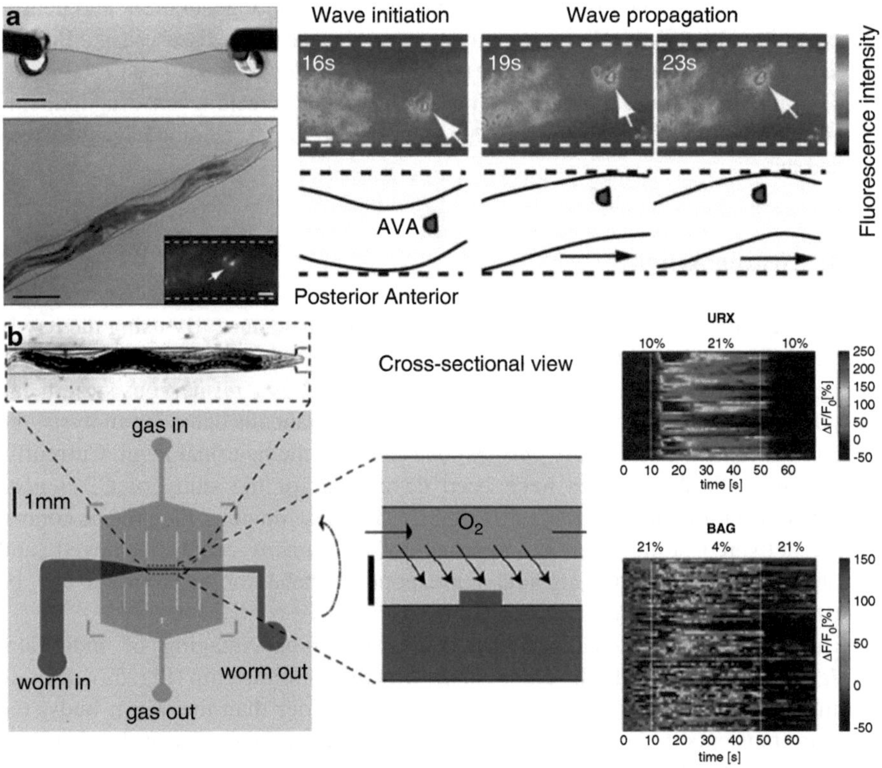

Fig. 5 Monitoring the activity of nervous system and simultaneous behavior of *C. elegans*. (**a**) Characterization of the relationship between AVE neuron activity and direction of the traveling body wave of *C. elegans* in a microfluidic channel [34]. (**b**) Calcium transients in individual URX and BAG sensory neurons of *C. elegans* respond to O_2 upshift and downshift [37]

(MPTP) or 1-methyl-4-phenylpyridinium ion (MPP+, the active metabolite of MPTP) have been used to inducing DAergic neuron degeneration in *C. elegans* and to simulate many symptoms of PD, like slow movement, rigidity, and tremor. Combining the established *C. elegans* PD model with microfluidics technology, it is possible to find the pathogenesis of PD and discover potential preventive and therapeutic drugs.

Ma and others [26] (Fig. 6a) investigated the mobility defects and DAergic neurons degeneration in individual mutant UA57 worms induced by MPP+ on a program-controlled microvalve-based microchannel array. The possible mechanism of these symptoms were discussed, and it was found that the mobility defects of *C. elegans* are closely related to DAergic neurons loss caused by MPP +. This was the first time that the mobility behavior and DAergic neurons features of individual *C. elegans* had been monitored simultaneously in real-time. Then, on an integrated droplet-based microfluidic device, Shi et al. [49] (Fig. 6b) characterized three types of response, including mobility defects, DAergic neurons degeneration, and oxidative stress increase, in UA57 and CL2166 worm strains induced by 6-OHDA.

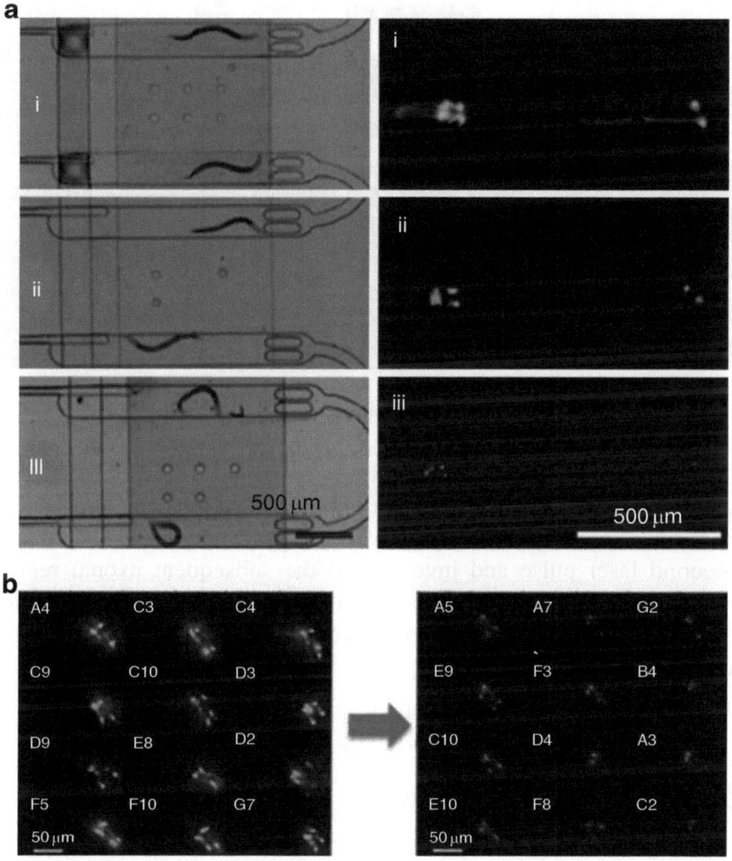

Fig. 6 Neurotoxins (**a**) MPP+ and (**b**) 6-OHDA induced degeneration of DAergic neurons in mutant UA57 worms [26, 48]

The results indicated that the increased oxidative stress might be involved in the process of neuron degeneration, leading to mobility defects.

3.2.3 Neuron Ablation and Regeneration

In neuroscience, understanding of the biological functions and regeneration mechanisms of specific neurons plays a vital role in developing new therapies for human neurological disease. Aiming at this problem, several researches were carried out by the newly emerged femtosecond laser surgery for severing axons (axotomy) in *C. elegans* [63]. This process often requires complete immobilization of the worms and precise control of the environment. In such a case, microfluidics is an ideal platform for neuron ablation operations and investigation of the subsequent axonal regeneration activity.

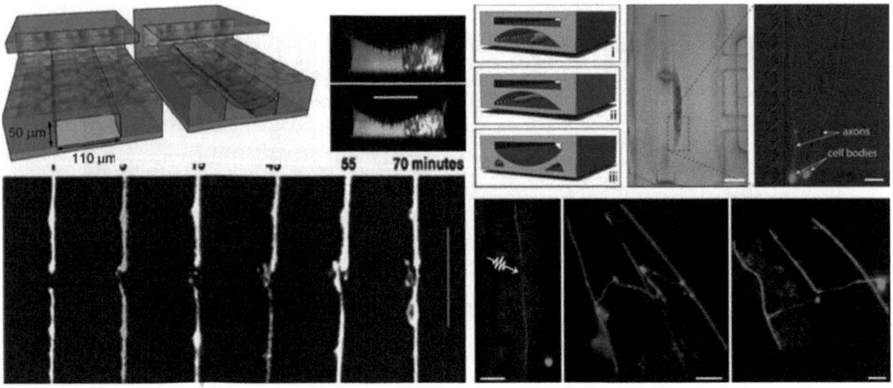

Fig. 7 *C. elegans* neuron ablation and regeneration on microfluidics [23, 24]

Guo et al. [23] (Fig. 7a) and Zeng et al. [24] (Fig. 7b) fabricated microvalve-based and tapered-channel-based microfluidic devices for providing the precision immobilization of *C. elegans* required by the surgery; they then severed axons with a femtosecond laser pulse and investigated the subsequent axonal regeneration. Chung and others [64] used the cooperation of a cooler and microvalves to restrict the movement of *C. elegans*, and exploited software to realize automatic image identification and cell ablation.

4 Conclusion and Perspective

The unique properties of microfluidics technology have proved it to be as an important platform for the study of the morphology, development, behavior, and neurobiology of *C. elegans*. Although great progress has been made in this area, the investigations are still in the early stages. Some crucial issues should be considered in terms of technically challenging and practical application. The design and performance of microfluidic platforms have to be improved to meet the requirement of biological research, and will ensure successful applications. The new microfluidic approaches have to be explored to interface with the existing system for a versatile, automated, and high-throughput assay. Also, the new applications have to be identified by the collaboration of scientists from bioengineering, MEMS, micro/nano and biomedical sciences. We believe the utility of this technology will offer invaluable biological information in future.

Acknowledgements This paper was supported by the National Nature Science Foundation of China (No. 20635030 and 90713014), Key Project of Chinese National Programs for Fundamental Research and Development (973 program, No. 2007CB714505 and 2007CB714507), Knowledge Innovation Program of the Chinese Academy of Sciences (KJCX2-YW-H18), and Instrument Research and Development Program of the Chinese Academy of Sciences (YZ200908).

References

1. Brenner S (1974) Genetics 77:71
2. Sulston JE, Horvitz HR (1977) Dev Biol 56:110
3. Sulston JE, Schierenberg E, White JG, Thomson JN (1983) Dev Biol 100:64
4. White JG, Southgate E, Thomson JN, Brenner S (1986) Philos Trans R Soc Lond B Biol Sci 314:1
5. Bargmann CI (1993) Annu Rev Neurosci 16:47
6. Bargmann CI (1998) Science 282:2028
7. Fraser AG, Kamath RS, Zipperlen P, Martinez-Campos M, Sohrmann M, Ahringer J (2000) Nature 408:325
8. Kim JK, Gabel HW, Kamath RS, Tewari M, Pasquinelli A, Rual JF, Kennedy S, Dybbs M, Bertin N, Kaplan JM, Vidal M, Ruvkun G (2005) Science 308:1164
9. Lin K, Hsin H, Libina N, Kenyon C (2001) Nat Genet 28:139
10. Zhang JXJ (2007) HFSP J 1:220
11. Ben-Yakar A, Bourgeois F (2009) Curr Opin Biotechnol 20:100
12. Ben-Yakar A, Chronis N, Lu H (2009) Curr Opin Neurobiol 19:561
13. Feng XJ, Du W, Luo QM, Liu BF (2009) Anal Chim Acta 650:83
14. Wang JY, Ren L, Li L, Liu WM, Zhou J, Yu WH, Tong DW, Chen SL (2009) Lab Chip 9:644
15. Chronis N (2010) Lab Chip 10:432
16. Crane MM, Chung K, Stirman J, Lu H (2010) Lab Chip 10:1509
17. Taylor AM, Jeon NL (2010) Curr Opin Neurobiol 21:1
18. McDonald JC, Duffy DC, Anderson JR, Chiu DT, Wu HK, Schueller OJA, Whitesides GM (2000) Electrophoresis 21:27
19. Xia YN, Whitesides GM (1998) Annu Rev Mater Sci 28:153
20. Whitesides GM, Ostuni E, Takayama S, Jiang XY, Ingber DE (2001) Ann Rev Biomed Eng 3:335
21. Hulme SE, Shevkoplyas SS, Apfeld J, Fontana W, Whitesides GM (2007) Lab Chip 7:1515
22. Allen PB, Sgro AE, Chao DL, Doepker BE, Edgar JS, Shen K, Chiu DT (2008) J Neurosci Methods 173:20
23. Guo SX, Bourgeois F, Chokshi T, Durr NJ, Hilliard MA, Chronis N, Ben-Yakar A (2008) Nat Methods 5:531
24. Zeng F, Rohde CB, Yanik MF (2008) Lab Chip 8:653
25. Chokshi TV, Ben-Yakar A, Chronis N (2009) Lab Chip 9:151
26. Ma H, Jiang L, Shi WW, Qin JH, Lin BC (2009) Biomicrofluidics 3:44114
27. Rohde CB, Zeng F, Gonzalez-Rubio R, Angel M, Yanik MF (2007) Proc Natl Acad Sci USA 104:13891
28. Chung KH, Crane MM, Lu H (2008) Nat Methods 5:637
29. Kim N, Dempsey CM, Zoval JV, Sze JY, Madou MJ (2007) Sens Actuators B Chem 122:511
30. Hulme SE, Shevkoplyas SS, McGuigan AP, Apfeld J, Fontana W, Whitesides GM (2009) Lab Chip 10:589
31. Krajniak J, Lu H (2010) Lab Chip 10:1862
32. Lockery SR, Lawton KJ, Doll JC, Faumont S, Coulthard SM, Thiele TR, Chronis N, McCormick KE, Goodman MB, Pruitt BL (2008) J Neurophysiol 99:3136
33. Park S, Hwang H, Nam SW, Martinez F, Austin RH, Ryu WS (2008) PLoS One 3:e2550
34. Chronis N, Zimmer M, Bargmann CI (2007) Nat Methods 4:727
35. Chalasani SH, Chronis N, Tsunozaki M, Gray JM, Ramot D, Goodman MB, Bargmann CI (2007) Nature 450:63
36. Gray JM, Karow DS, Lu H, Chang AJ, Chang JS, Ellis RE, Marletta MA, Bargmann CI (2004) Nature 430:317
37. Zimmer M, Gray JM, Pokala N, Chang AJ, Karow DS, Marletta MA, Hudson ML, Morton DB, Chronis N, Bargmann CI (2009) Neuron 61:865
38. Crane MM, Chung K, Lu H (2009) Lab Chip 9:38

39. Park SJ, Goodman MB, Pruitt BL (2007) Proc Natl Acad Sci USA 104:17376
40. Doll JC, Harjee N, Klejwa N, Kwon R, Coulthard SM, Petzold B, Goodman MB, Pruitt BL (2009) Lab Chip 9:1449
41. Goodman MB, Hall DH, Avery L, Lockery SR (1998) Neuron 20:763
42. Lewis JA, Wu CH, Berg H, Levine JH (1980) Genetics 95:905
43. Qin JH, Wheeler AR (2007) Lab Chip 7:186
44. Zhang Y, Lu H, Bargmann CI (2005) Nature 438:179
45. Song H, Chen DL, Ismagilov RF (2006) Angew Chem Int Ed 45:7336
46. Teh SY, Lin R, Hung LH, Lee AP (2008) Lab Chip 8:198
47. Clausell-Tormos J, Lieber D, Baret JC, El-Harrak A, Miller OJ, Frenz L, Blouwolff J, Humphry KJ, Koster S, Duan H, Holtze C, Weitz DA, Griffiths AD, Merten CA (2008) Chem Biol 15:427
48. Shi WW, Qin JH, Ye NN, Lin BC (2008) Lab Chip 8:1432
49. Shi W, Wen H, Lu Y, Shi Y, Lin B, Qin J (2010) Lab Chip 10:2855
50. Gray J, Lissmann HW (1964) J Exp Biol 41:135
51. Lange D, Storment CW, Conley CA, Kovacs GTA (2005) Sens Actuators B Chem 107:904
52. Rezai P, Siddiqui A, Selvaganapathy PR, Gupta BP (2010) Lab Chip 10:220
53. Ward S (1973) Proc Natl Acad Sci USA 70:817
54. Sukul NC, Croll NA (1978) J Nematol 10:314
55. Luo L, Gabel CV, Ha HI, Zhang Y, Samuel ADT (2008) J Neurophysiol 99:2617
56. Rezai P, Siddiqui A, Selvaganapathy PR, Gupta BP (2010) Appl Phys Lett 96:153702
57. Dimitriadi M, Hart AC (2010) Neurobiol Dis 40:4
58. Nass R, Miller DM, Blakely RD (2001) Parkinsonism Relat Disord 7:185
59. Braungart E, Gerlach M, Riederer P, Baumeister R, Hoener MC (2004) Neurodegener Dis 1:175
60. Schmidt E, Seifert M, Baumeister R (2007) Neurodegener Dis 4:199
61. Kerr R, Lev-Ram V, Baird G, Vincent P, Tsien RY, Schafer WR (2000) Neuron 26:583
62. Wang Y, Wang J, Du W, Feng X, Liu B-F (2011) Anal Bioanal Chem 399:3475
63. Yanik MF, Cinar H, Cinar HN, Chisholm AD, Jin YS, Ben-Yakar A (2004) Nature 432:822
64. Chung K, Lu H (2009) Lab Chip 9:2764

Index

A
Absorption spectrometry, 121
Acetyl isoamyl acetate synthesis, lipase-catalyzed, 51
Acoustic microstreaming, 56
2-Acrylamido–2-methyl–1-propanesulfonic acid (AMPSA), 160
Active sheath flow, 8
Advection, 42
AgPDMS, 101, 108
Aluminum oxide membranes (AOM), 211
Aminopropyltriethoxysilane, 137
Amperometry, 125
Antibodies, 85, 119, 133
Antiresonant reflecting optical waveguides (ARROWs), 189
Aptamers, 135
Arrayed microvalves, 17
ATP analysis, 124

B
Ba–Ti–O-type nanoparticles, 95
Beads, 138
Beta-amyloid peptide, 184
Bioluminescence, 122
Biosensors, 117
Bone marrow stem cells (BMSCs), 298
Bottom-up analysis, 261, 266
Bovine serum albumin (BSA), 281

C
Calcium and titanium precipitates (CTPs), 95
Cancer cells, 306
Capillary array electrophoresis (CAE), 207
Capillary electrophoresis (CE), 80, 121, 271
Capillary number, 48
Carbon nanotubes (CNTs), 161
Ca–Ti–O-type composites, 95
Cavitation microstreaming, 56
CD4-positive T-cells, 182
Cell–cell communication, 304
Cell cultures, 3D, 295, 305
Cell lysis, 207
Cell sorting, 139, 179, 305
Chaotic advection micromixers, 42
Charged coupled devices (CCDs), 178
Chemiluminescence, 122, 181, 195
Chemotaxis, 301
Chip gel electrophoresis protein profiling (CGE-PP), 285
Circulating endothelial progenitor cells (CEPCs), 313
Circulating tumor cells (CTCs), 305
Collagen, 305
Complementary metal–oxide–semiconductor (CMOS) chips, 195
Conductometry, 125
Confocal fluorescence detection, microchip, 178
Controlled trafficking, 70
CPDMS, 101
C-reactive protein (CRP), 124
Cytochrome c, 122, 275, 281, 289

D

Debye layer extension, 157
Deep reactive ion etching (DRIE), 211
Deionzation, membrane-induced, 157
Demultiplexer (DEMUX), 19
Depletion, 153
Detection, inside chip, 185
Diallyldimethylammonium (DADMA), 160
Dielectrically controlled sorting devices, 13
Dielectrophoresis (DEP), 14, 54, 153, 161, 209
  flow switching, 14
Dielectrophoretic disturbance, 54
Diffusion, Fick's law, 30, 31
Distributed feedback (DFB), 194
DNA amplification, 82
  analysis, 203
    monolithic integrated polycarbonate, 249
  detection/sensing, 153, 159
  diagnostics/forensics, 203
  hybridization, FCS detection, 161
  microarrays, 155, 203, 229
  separation matrices, 225
  sequencing separations, 178
Droplet-based microfluidics, 30
Droplet-based microreactors, 82
Droplets, breakup, 71
  compartmentalization, 81, 85
  content characterization, 79
  control, GERF, 105
  detection, soft conducting electrodes, 101
  fusion, 73
  generation, 70, 71
  hyphenation, 80
  incubation, 78
  logic, 108
  microfluidics, 69, 140
  micromixers, 50
  shuttle hybridization, 234
  sorting, 76
  splitting, 74
  trapping, 78
Drug screening, 308

E

Einstein–Stokes equation, 31
Elastin, 305
Electrochemical detection, 117
Electrochemical quartz crystal microbalance (EQCM), 128
Electrochemiluminescence, 122, 124
Electrokinetic instability (EKI) (disturbance), 52
Electrokinetic molecular concentration, 161
Electrolytes, 131
Electrophoresis, 80
  two-dimensional, 261, 263
Electrorheological fluid (ERF), 91
Electrorheological particles, conducting-polymer-based, 98
  inorganic–polymer hybrid, 97
Electrowetting, 27, 141
Electrowetting on dielectrics (EWOD), 54, 93, 143
Electrowetting shaking, 54
Embryonic stem cells (ESCs), 310
End-labeled free-solution electrophoresis (ELFSE), 227
Enzyme-based biosensors, 131
Enzyme electrodes, 131
Enzyme immunoassay (EIA), 134
Enzyme-linked immunosorbent assay (ELISA), 85, 124
Epidermal growth factor (EGF), 312
Ethylene glycol dimethacrylate (EGDMA), 283
Evanescent wave-based waveguide, 186
Evanescent waves, absorbance, 176

F

Fabry–Perot cavity, 193
Fetal bovine serum (FBS), 299
Fiber-optic localized plasma resonance (FO-LPR), 135
Fibroblast growth factor 2 (FGF2), 312
Flow control, 1
Flow focusing, 4, 39, 71
Flow rates, 298
Flow switching sheath flow device, 15
Fluid dynamics separation, 6
Fluorescein, 177
Fluorescence-activated cell sorter (FACS), 139, 305
Fluorescence detection, 121, 176

Fluorescence resonance energy transfer (FRET), 130
Fluorescent correlation spectroscopy (FCS), 161
Fluorophores, 177
Focusing enhanced mixers, 33
Free-solution conjugate electrophoresis (FSCE), 227
Free-space optical detection, microchip, 171, 173

G
Galvano-/electrotaxis, 302
Gamma-globulins, 133
Gene expression profiles, 306
Genetic analysis, 203
Giant electrorheological fluid (GERF), 91, 93, 95
  microvalves, 102
Glutathione, 181
Glycoproteins, 269
Gold binding peptide (GBP), 283

H
Hepatitis B, 300
High-throughput multiarray chip, 315
High-throughput screening, 82, 305, 308
Horseradish peroxidase (HRP), 181
Human serum albumin, 181
Hybrid ER particles, 97
Hydrogel, transition, 15

I
Immunoassays, 133
Immunoglobulins (IGs), 133
Immunosuppressive acidic protein, 181
Induced vortex molecular concentration, 157
Integrated microsystems, 203
Integrins, 306
Interdigitated ultramicroelectrode array (IDUA), 126
Internal reflection spectroscopy, 122
Intra-droplet content manipulation, 70
Ion current, 158, 163, 166
Ion-selective electrodes (ISEs), 126
Ion-selective membranes, 153
Ion-sensitive field-effect transistor (ISFET), 127
Isoelectric focusing, 177, 261, 264
Isotachophoresis, 54, 180, 280

L
Laboratory-on-a-chip (LOC) systems, 2, 29, 92, 171, 195, 207
Lactate dehydrogenase (LDH), 132
Lactate oxidase, 132
Laminar flow, 1, 30, 299
Laminin, 305
Laser-induced fluorescence (LIF), 176
Lectin affinity chromatography, 269
Lensless optofluidic microscopy (OFM), 196
Ligase detection reaction (LDR), 228
  continuous-flow (CFLDR), 248
Light-emitting diode (LED), 121, 176
Limiting current, 153
Liquid-core waveguides (LCWs), 186
Liquid gradient refractive index (L-GRIN) lens, 192
Localized surface plasmon resonance (LSPR), 125, 182
Logic control, 91
Logic gate, 109
Luminol–peroxide, 181
Lysis, 208

M
Magneto-hydrodynamic (MHD) disturbance, 55
Mass-sensitive sensors, 127
Membrane, ion-selective, 160
  depletion/vortex phenomena, 158
  reactor, 280
  sensors, polarization/Warburg impedance signals, 164
  synthesis, on-chip, 159
Metalloproteinase–8 (MMP–8), 283
Metal oxide semiconductor field effect transistor (MOSFET), 127
Micellar electrokinetic capillary (MEKC) electrophoresis, 273
Microarrays, readout, 247

Microarrays, readout, 247 (*cont.*)
substrate materials, 230
Microcantilvers, 129
Microcapillary electrophoresis (CE), 203, 207, 222
Microcapillary gel electrophoresis (-CGE), 273
Microcavity fluidic dye laser, 194
Microchannels, 1, 3, 70
microdroplet mixing, 50
Microchip, cross section, 175
Microdroplet-based mixers, 47
Microdroplets, 30, 69, 91
Microelectromechanical systems (MEMS), 1
Microelectroporation device, cell lysis, 209
Microenvironment, 295
Microflow systems, (bio)chemical applications, 20
high-throughput, 1
Microfluidic chip, 117
Microfluidic genetic analysis (MGA) system, 241, 242
Microfluidics, 1, 26, 69, 91, 140, 203, 261, 267, 295
Microfluidic systems, integrated, 234
Microfluidic thermal heating, 218
Microlens, 121, 189
Micromixers, active, 26, 32, 51
centrifugal, 57
elecrokinetic instability (EKI) (disturbance), 52
parallel lamination, 34
passive, 26, 32, 33
pumpless, 56
sequential lamination, 37
T-/Y-shaped, 34
Micromixing, 26
Microparticle synthesis, 83
Micro-PCR devices, 203, 213
Micropumps, 140
Microreactors, droplet-based 82
Microscale cell culture analogue (CCA), 308
Microsolid phase extraction (micro-SPE), 209
Microstreaming, acoustic, 56
Micro total analysis systems (TAS), 1, 26, 173, 207
Microvalves, 1, 102
classification, 142
GERF, 102
pressure-driven, 17
Microvortices, 158
Mixing, inside droplets, 75
microfluidic devices, 31, 57
Molecular imprinting, 136
Molecular receptors, 135
Molecular sensing, 154
Multicell sorting device, 11
Multidimensional protein identification technology (Mud-PIT), 266
Multidrug-resistant tuberculosis (MDR-TB), 250
Multimode droplet sorting device, 13
Multiphase flow, 47
Multiphase microfluidics, 47
Multiple-droplet sorting device, 13
Multiple flow control systems, 17
Multiplex bead array assay (MBAA), 138
Multiplexed valve control system, 18

N

Nanohole microscopy, 196
Naphthalene–2,3-dicarboxaldehyde (NDA), 282
Nucleic acids, analyses, 206
extraction/purification/preconcentration, 209
sensing, nanomembrane-based, 153

O

On-chip electrodes, 156
On-chip membrane synthesis, 159
Optical detection, 120, 171, 195
Optically controlled flow devices, 14
Optically induced dielectrophoretic (ODEP) force, 16
Optical waveguide, planar, 186
Optofluidic (inside-chip) detection, 171
Optofluidic laser, 193
Optofluidic microscopy (OFM), 196
Organic photodiodes (OPDs), 195
Orthogonality, 261
Overlimiting current, 157

P

Particle-sorting devices, 16
PCR, 56, 69, 153, 203, 250, 252, 264
  continuous-flow (CFPCR), 203, 238, 248
  droplet, 112
  microfluidic, 212
PCR-CE, integrated sample cleanup, 245
Peclét number, 33
Peptides, MS, 275
  separation, 121, 275
Peroxalate–peroxide, 181
Peroxyoxalate CL, 181
Photomultiplier, solid-state, 123
Photomultiplier tubes (PMTs), 121
Piezoelectricity, 127
Pinched flow fractionation devices, 7
Planar serpentine micromixer (PSM), 51
Platelet-derived growth factor (PDGF), 312
Pneumatically controlled flow device, 8
Point-of-care (POC), 91, 171, 182, 195
Polar-molecule-dominated ERFs (PM-ERFs), 94
Poly(dimethylsiloxane) (PDMS), 2, 72, 101, 121, 175, 208, 279
  quartz chip, 175
Poly(ethylene glycol) (PEG), 137
Poly(methylmethacrylate) (PMMA), 97, 187, 273
Poly(vinylferrocenium), 132
Polyallylamine synthesis, 156
Polyaniline (PANI), 97
Polypyrrole (PPY), 97
Polystyrene (PS), 97
Polystyrene sulfonate, 156
Polyvinylpyrrolidone (PVP), 132
Portable microfluidic instruments, 180
Post-translational modifications, 263
Potentiometry, 125
Pressure field disturbance, 52
Prostate-specific antigen (PSA), 128
Protein analysis, 265
  integrated microsystems, 276
Proteins, 130
  crystallization, 82
  digestion, 274
  engineering, 82
  separation, 270
Proteolytic enzymes, 274
Proteome/proteomics, 261, 263
Pumping, 140

Q

Quantum dots (QDs), 123, 176
Quorum sensing, 302

R

Raman spectroscopy, 79, 124, 184
Red blood cells (RBCs), 137
Refractive index change, 182
Reverse transcription-PCR (RT-PCR), 236
Reynolds number, 30
Ring resonator, 193
RNA, 164
  hybridization sensor, integrated smart, 167

S

Sanger sequencing, 207
Selectivity enhancement, 165
Semiconducting polymers, 98
Semiconductive solids, 93
Serpentine laminating micromixer (SLM), 48
Shear stress, 50, 100, 128, 298, 313
Sheath flow, 4
Shotgun proteomics, 266
Silicon photodiode, 195
Single cell analysis, 181, 315
  droplet-based, 69, 85
Single cell culture, array, 315
Single-cell mRNA extraction, 315
Single mismatch (SNP) discrimination, 165
Slab gel electrophoresis (SGE), 222
Slanted-groove micromixer (SGM), 37
Solid-phase extraction, 203, 205
Solid-phase microreactors, 275
Solid-phase reversible immobilization (SPRI) chip, 239
Sol–gel silica fabrication, on-chip, 156
Spiral flow, 3
Split-and-recombine (SAR) micromixers, 37, 48
Sr–Ti–O-type composites, 95
Staphylococcal enterotoxin B (SEB), 124

Stem cells, 295, 310
Streptavidin–biotin, 140
Strouhal number, 33
Styrene-acrylonitrile, 97
Surface-enhanced Raman scattering (SERS), 124, 184
Surface modification, 137
Surface plasmon resonance (SPR), 120, 182
Switches, 14, 108, 110, 189
Systematic evolution of ligands by exponential enrichment (SELEX), 135

T

TB analysis, integrated/modular microfluidic chip, 251
Thermal disturbance micromixers, 56
Thermoreversible gelation polymer (TGP), 15
Titania, 94
Top-down analysis, 261, 265
Transducers, 117
2,4, 6-Trinitrotoluene (TNT), 136
Tumor spheroids, 308
Turbulent flow, 31
Twisting flow, 3

U

Ultrasound disturbance, 55
UV/Vis absorption spectroscopy, 173

V

Valves, 140
  actuation, horizontal, 9
  control system, multiplexed, 18

W

Warburg impedance spectrum, 164
Waveguide-based cavity ring-down absorption spectroscopy, 176
Waveguides, 186
  evanescent wave-based, 186
  interference-based, 189
  liquid-core (LCWs), 186
  photonic crystal (PC), 189
Winslow effect, 93

Z

Zeolite, 94
ZnAl-based enzyme nanohybrid, 133

MIX
Papier aus verantwortungsvollen Quellen
Paper from responsible sources
FSC® C105338

If you have any concerns about our products,
you can contact us on
ProductSafety@springernature.com

In case Publisher is established outside the EU,
the EU authorized representative is:
Springer Nature Customer Service Center GmbH
Europaplatz 3, 69115 Heidelberg, Germany

Printed by Libri Plureos GmbH
in Hamburg, Germany